Uniscience Series on Fine Particle Science and Technology

Editor

**John Keith Beddow, Ph.D.**
Professor of Chemicals and Materials Engineering
Division of Materials Engineering
University of Iowa
Iowa City, Iowa

**Advanced Particulate Morphology**
J. K. Beddow and T. P. Meloy

**Separation of Particles From Air and Gases, Volumes I and II**
Akira Ogawa

**Particle Characterization in Technology**
Volume I: Applications and Microanalysis
Volume II: Morphological Analysis
J. K. Beddow, Editor

# Particle Characterization in Technology

## Volume I
## Applications and Microanalysis

Editor
**John Keith Beddow, Ph.D.**
Professor of Chemical and Materials Engineering
College of Engineering
Division of Materials Engineering
The University of Iowa
Iowa City, Iowa

CRC Series on Fine Particle Science and Technology

Editor-in-Chief
**John Keith Beddow**

**CRC Press, Inc.**
**Boca Raton, Florida**

**Library of Congress Cataloging in Publication Data**
Main entry under title:

Particle characterization in technology.

(CRC series on fine particle science and technology)
Bibliography.
Includes index.
1. Particles. 2. Bulk solids. 3. Particle size determination. I. Beddow, John K. II. Series.
TA418.78.P36 1984 620'.43 83-14363
ISBN 0-8493-5784-5 (v. 1)
ISBN 0-8493-5785-3 (v. 2)

This book represents information obtained from authentic and highly regarded sources. Reprinted material is quoted with permission, and sources are indicated. A wide variety of references are listed. Every reasonable effort has been made to give reliable data and information, but the author and the publisher cannot assume responsibility for the validity of all materials or for the consequences of their use.



Direct all inquiries to CRC Press, Inc., 2000 Corporate Blvd., N.W., Boca Raton, Florida, 33431.


Second Printing, 1985

International Standard Book Number 0-8493-5784-5 (v. 1)
International Standard Book Number 0-8493-5785-3 (v. 2)

Library of Congress Card Number 83-14363
Printed in the United States

# INTRODUCTION

The problem of particle characterization is one of the central themes of particulate technology. It has as its ultimate goal the development of the ability to predict from a knowledge of the particle characteristics, the properties and behavior of particles individually and also en masse. Our present stage of knowledge indicates that our goal should be judged as a high ideal — we are indeed a long way from our ultimate goal. However, there is clearly no cause for despair. This is because of the great progress that has been made in the last few years. Some of the more important of these are the subject of the chapters of these two volumes.

Not so very long ago we had to be satisfied with a low power look at particulate materials. There is now available a remarkable set of techniques under the collective title of microanalysis. Some of the more significant of these methods are reviewed in Volume I by Gavrilovic (SIMS), Kaufmann and Wieser (LAMMA), and Davidson (electron microscopy and spectroscopy).

The remainder of Volume I presents an important exposition of some of the most significant areas where particle characterization is applied. The technological fields include pharmaceutical materials by Hersey, bulk solids by Smith and Lohnes, and explosions by Palmer. In each of these fields the authors describe a great deal of work that has to be done to be able to state the properties of the bulk particulate material with quantitative exactness. Some information to relate these properties to particle characteristics is already available and reported in this and Volume II. The remainder will be generated in the future. The other fields of application reported out in this volume include applications of fractal analysis by Kaye, fine particle characterization methods in liquids by Lieberman, practical aspects of electrozone size analysis by Karuhn and Berg, and finally, a brief description of the well-known COULTER® Counter by Kinsman.

Morphological analysis has been with us since the early 1970s, but it made its great leap after the 1977 research workshop at the University of Iowa, which was funded by the National Science Foundation, Particulates Processing and Multiphase Flow Program under Dr. Morris Ojalvo. The reader may compare the proceedings of that workshop (*Advanced Particulate Morphology,* Beddow and Meloy, Eds., CRC Press, 1980) with this volume in order to grasp the giant steps forward that have been made in a very few years. The first section of Volume II deals with both theory and methods of morphological analysis. In particular there are reviews by Luerkens, Beddow, Vetter, Gotoh, and Meloy. PIAS instrumentation is described by Chang. For the moment, the general consensus favors the use of the Fourier series in morphological analysis. The use of the Fast Fourier Transform would speed up the rate of data acquisition. The FFT is fully discussed by Kanatani (Chapter 4). The task of data analysis is of primary importance, and Inoue has written a chapter on the potential use of a relatively new computer language called FORTH. It is believed that its adoption at this time in place of FORTRAN and the like may help reduce some of our current difficulties. To date, morphological analysis has been concerned exclusively with the profile of the particle. In addition, there is much information to be obtained from within the particle and also from within clusters and agglomerates. There is therefore an important need to extend the technique so that we may deal with textural features of particles. Leong's chapter on color as a morphic feature represents a beginning step in this direction.

Data analysis is to morphological analysis as the identification of chemical elements is to chemical analysis. Various topics are discussed in the second volume including a multivariate pattern recognition approach by Ehrlich and Full. This deals with ways in which morphic mixtures may be analyzed for their end members. The Fourier series is an infinite series. How far one should go before making the decision to truncate is dealt with by Lenth in his chapter. Bezdek and Solomon demonstrate further use of fuzzy sets in morphological analysis. (The reader is referred to Bezdek's recent book, *Pattern Recognition With Fuzzy Objective Function Algorithms,* Plenum Press, New York, 1981 for additional information.) The chapter by Hua (Chapter 10) underscores the need for introducing texture as a feature in morphological analysis. In this chapter Hua shows that it is not possible to differentiate between alumina clusters (designated O and N) merely on the basis of the information in their outer profiles. This section of the book is brought to a conclusion by the chapter dealing with data presentation in the form of histograms by Full, Ehrlich, and Kennedy.

Applications of morphological analysis include a review by Beddow, relation of the size and shape of abrasive particles to the rate of wear that they cause in the case of steel samples by Vetter, and identification of the quantitative shape characteristics associated with particles produced by different processes by Guo and Ramakrishnan. The study reported by Carmichael shows that not only can three-dimensional morphic features account for the sedimentation behavior of particles of different regular geometries, but their orientations are also accounted for. Sieving or screening are normally perceived as methods for separating particles on a basis of their size differences. Rajpal et al. reports that this technique may also be used for shape separation of particles. Chang et al. describes correlations obtained between bulk properties (including flow-time, Hausner ratio, and angle of internal friction) and the morphic features of a number of materials. Hua et al. (Chapter 18) discusses the modeling of crystallization using morphological analysis.

## THE EDITOR-IN-CHIEF

**John Keith Beddow** received his Ph.D. in Metallurgy from Cambridge University, Cambridge, England, in 1959. Currently Secretary of the Fine Particle Society, he is a member of the Faculty at the University of Iowa, where he heads a small research group in fine particle science with emphasis on morphological analysis. Dr. Beddow is an active lecturer and author. He has also been active as a Consultant in metallurgy, powder metallurgy, and powder technology for numerous corporations. He is also president of Shape Technology, Ltd. His present research activities are in particle morphological analysis. Dr. Beddow is married, with four daughters, and has resided in the U.S. since 1966.

## THE EDITOR

**John Keith Beddow** received his Ph.D. in Metallurgy from Cambridge University, Cambridge, England, in 1959. Currently Secretary of the Fine Particle Society, he is a member of the Faculty at the University of Iowa, where he heads a small research group in fine particle science with emphasis on morphological analysis. Dr. Beddow is an active lecturer and author. He has also been active as a Consultant in metallurgy, powder metallurgy, and powder technology for numerous corporations. He is also President of Shape Technology, Ltd. His present research activities are in particle morphological analysis. Dr. Beddow is married, with four daughters, and has resided in the U.S. since 1966.

## ADVISORY BOARD

## CONTRIBUTORS

**John Keith Beddow, Ph.D.**
Professor of Chemical and Materials Engineering
College of Engineering
Division of Materials Engineering
The University of Iowa
Iowa City, Iowa

**Robert H. Berg, M.S., P.E.**
President
Particle Data, Inc.
Particle Data Laboratories, Ltd.
Elmhurst, Illinois

**James C. Bezdek, Ph.D.**
Head
Department of Computer Science
University of South Carolina
Columbia, South Carolina

**Gordon Butters, Ph.D.**
BP Chemicals, Ltd.
Research and Development Department
South Wales Division
Sully, South Glamorgan, Wales

**Gregory Carmichael, Ph.D.**
Associate Professor
Chemical and Materials Engineering Program
College of Engineering
Division of Materials Engineering
The University of Iowa
Iowa City, Iowa

**Chi-Ren Chang, Ph.D.**
College of Engineering
Division of Materials Engineering
The University of Iowa
Iowa City, Iowa

**S. Collins**
Image Processing Laboratory
Information Engineering
The University of Iowa
Iowa City, Iowa

**David L. Davidson, Ph.D.**
Institute Scientist
Southwest Research Institute
San Antonio, Texas

**Robert Ehrlich, Ph.D.**
Department of Geology
University of South Carolina
Columbia, South Carolina

**William E. Full, Ph.D.**
Department of Geology and Geography
Wichita State University
Wichita, Kansas

**John Gavrilovic, Ph.D.**
Senior Research Scientist
Walter C. McCrone Associates, Inc.
Chicago, Illinois

**Keishi Gotoh, Ph.D.**
Professor
Department of Energy Engineering
Toyohashi University of Technology
Toyohashi, Japan

**Andy Guo**
Fine Particle Research Group
College of Engineering
Division of Materials Engineering
The University of Iowa
Iowa City, Iowa

**John A. Hersey, Ph.D. (deceased)**
Sigma Industrial Professor of Pharmaceuticals
Victorian College of Pharmacy
Melbourne, Australia

**Louise Hua**
Fine Particle Research Group
College of Engineering
Division of Materials Engineering
The University of Iowa
Iowa City, Iowa

**Toshio Inoue, Ph.D.**
Professor
Department of Mineral Development Engineering
Faculty of Engineering
University of Tokyo
Tokyo, Japan

**Ken-ichi Kanatani, Ph.D.**
Associate Professor
Department of Computer Science
Gunma University
Gunma, Japan

**Richard F. Karuhn, B.S.**
Director
Particle Data Laboratories, Ltd.
Elmhurst, Illinois

**Prof. Dr. Raimund Kaufmann**
Lehrstuhl für Klinische Physiologie der Universität
Düsseldorf, West Germany

**Brian H. Kaye, Ph.D.**
Professor of Physics
Director
Fineparticle Research Institute
Laurentian University
Sudbury, Ontario, Canada

**Stephen Kennedy, Ph.D.**
Postdoctoral Fellow
Department of Geology
University of South Carolina
Columbia, South Carolina

**Shepard Kinsman**
909 Eastridge Village Drive
Miami, Florida

**Mitsuyoshi Kumamoto**
Graduate Student
Department of Energy Engineering
Toyohashi University of Technology
Toyohashi, Japan

**Russell V. Lenth, Ph.D.**
Associate Professor
Department of Statistics and Actuarial Science
The University of Iowa
Iowa City, Iowa

**Poh-Leng Leong, S.M.**
Operations Research
Massachusetts Institute of Technology
Cambridge, Massachusetts

**Alvin Lieberman, M.S.**
Technical Specialist
Particle Measuring Systems, Inc.
Boulder, Colorado

**Robert A. Lohnes, Ph.D.**
Professor
Civil Engineering Department
Iowa State University
Ames, Iowa

**David W. Luerkens, Ph.D.**
Research Engineer
E. I. DuPont de Nemours & Company
Atomic Energy Division
Savannah River Laboratory
Aiken, South Carolina

**Thomas P. Meloy, Ph.D.**
Particle Analysis Center
West Virginia University
Morgantown, West Virginia

**K. N. Palmer**
Head of Fire Protection Division
Fire Research Station
Borehamwood, England

**Samir Rajpal**
Fine Particle Research Group
College of Engineering
Division of Materials Engineering
The University of Iowa
Iowa City, Iowa

**Prof. Dr. P. Ramakrishnan**
Department of Metallurgical Engineering
Indian Institute of Technology
Bombay, India

**C. Seemuth**
Audio Visual Center
The University of Iowa
Iowa City, Iowa

**David L. O. Smith, Ph.D.**
Assistant Professor
Civil Engineering Department
Iowa State University
Ames, Iowa

**Kenneth Solomon, Ph.D.**
Research Associate
U.S. Salinity Laboratory
Riverside, California

**Paul Swanson, Ph.D.**
John Deere & Company
Moline, Illinois

**Arthur F. Vetter, M.S.**
Department of Chemical and Materials Engineering
College of Engineering
Division of Materials Engineering
The University of Iowa
Iowa City, Iowa

**Prof. Dr. Paul Wieser**
Institut für Physik
Universität Hohenheim
Stuttgart, West Germany

**Ming-Jau Yin**
College of Engineering
Division of Materials Engineering
The University of Iowa
Iowa City, Iowa

ACKNOWLEDGMENTS

As Editor of these two volumes, I would like to extend my sincere thanks to the many authors who have contributed so splendidly to the final outcome. Particulate technology is such a diverse field and its activists are so very busy. This is the main reason why there are so few books and why most of us have to go into the original literature if we wish to obtain information. I hope that the contributing authors to this book have by their outstanding efforts given the reader invaluable new information and ideas and that in the future, these will contribute to substantial programs and applications. If this is so, then they will have earned our gratitude.

# DEDICATION

This book is dedicated to John Arthur Hersey who was Sigma Industrial Professor of Pharmaceutics at the Victorian College of Pharmacy in Melbourne, Australia. He died unexpectedly in Melbourne on February 14, 1982. He was well known in North America, Europe, and the Pacific Region. He will best be remembered to particle technologists for his brilliant work on ordered mixing. His chapter in Volume I is entitled, "Characterization of Pharmaceutical Particulate Materials".

# TABLE OF CONTENTS

Volume I

Volume II

THEORY AND METHODS



DATA ANALYSIS

APPLICATIONS

# *Part I*
# *Microanalysis*

Chapter 1

# SURFACE ANALYSES OF SMALL INDIVIDUAL PARTICLES BY SECONDARY ION MASS SPECTROSCOPY

**John Gavrilovic**

## TABLE OF CONTENTS

# I. INTRODUCTION

In the early days, when the chemical and physical properties of granular materials were required they were determined based upon an analysis of the bulk material itself. Thus, in the development of a new product and/or in the identification of a contaminant or failure mechanism, the physical and chemical properties of the material were an averaged or integrated result. As the state-of-the-art in chemical analyses advanced and as the development and the design of new materials and products became more sophisticated, a bulk analysis would no longer suffice to correctly characterize a material. The determination of the properties of granular material advanced from the analysis of large bulk samples to the analysis of a number of smaller samples taken from the bulk. Unfortunately, this advancement in analysis techniques provided data which were, again, integrated over a number of individual particles and still did not provide an accurate characterization of the material. Such analyses often revealed misleading information as the surface layers (absorption layers) of a particle are very seldom consistent from particle to particle. Thus, to learn more about the surface layers or thin surface films on a material, the analysts had to advance their techniques to the analyses of individual small particles and eventually to the surface analyses of individual small particles.

Because each individual particle within a material is unique, the bulk analysis as well as the surface analysis of an individual particle will yield a significant increase in knowledge of its properties. This information is essential to the researcher in the development of his new product or in the solving of a contamination problem. In analyzing individual particles, the analyst not only can determine the composition of the individual particle on a layer by layer basis, but he can identify the possible interactions of the particles based upon the inhomogeneities within their surface layers. Such a depth distribution allows the analyst to individually observe the changes within the surface structure of small particles.

It is obvious that over the years the analytical techniques available for characterizing a material have progressed from the chemical analyses of a bulk material to the analyses of individual small particles. In perspective, the sample size has diminished from a large macrosample of material, to a few representative particles from the material, to an individual small particle. That is, from standard wet chemistry-size samples, to microanalyses, and finally to the bulk and surface analyses of single particles down to one femtogram or less in size.

# II. SECONDARY ION MASS SPECTROSCOPY INSTRUMENTATION

## A. Instrumentation

Surface analyses of bulk powders have been possible for some time by the same techniques applied to large solid samples. In many cases, however, only very small samples of small particles are available, and frequently only a single particle is available. The surface analytical methods available for bulks are then inadequate for small particles due to the very small area available for analysis. Analytical techniques for surface analysis of small single particles have been developed for practical application only in the past 15 to 20 years. Among the most advanced microanalytical techniques at the forefront of technology today are Auger electron spectroscopy and secondary ion mass spectroscopy (SIMS). This chapter will discuss the surface analysis of small individual particles through the use of SIMS.

Two different designs of SIMS instruments, the ion microscope and the ion microprobe, are available for the bulk and surface analyses of single small particles.

The first of these designs, **the ion microscope,** is the direct imaging analyzer developed by Castaing and Slodzian.[1] Within the ion microscope, a relatively large 3- to 300-μm ion beam strikes the sample and a magnified secondary ion image is produced through ion transfer optics. Any portion of this image can be selectively isolated, examined, and analyzed in detail for chemical composition, isotopic ratios, etc. The spatial resolution of an ion microscope is better than 1 μm while the mass resolution depends upon the mass spectrometer design. New-generation instruments are capable of mass resolutions of 8000 to 10,000, a tenfold increase over earlier instruments.

The second design, **the ion microprobe,** a system designed by Liebl,[2] is based upon an electrostatic ion optics system that focuses a primary ion source into a 2-μm ion beam. This primary ion beam may then be rastered over either a large or small area to generate secondary ions from the sample area of interest. The secondary ions are then analyzed using a double-focusing mass spectrometer system.

Each of these instrument designs obviously has advantages and disadvantages in comparison to the other. Ideally, when performing an analysis on a small particle one would like to use a magnified secondary ion image for the location of the small particle. However, in performing surface analyses the quality of the light optics is not only important but essential, since use of the ion optics system would remove the surface layer of interest before the analysis could be performed. Unfortunately, the light optics within the ion microscope are poor, thus the ion microprobe has the advantage.

### B. Bulk and Surface Analyses

Through the advancement and refinement of SIMS techniques it is now possible to analyze both the surface and the bulk properties of individual small particles, with good spatial resolution and high sensitivity. Using the SIMS technique, secondary ions are generated during the interaction of an energetic ion beam (or, more recently, a focused laser beam)[3] with a solid surface. Within the bombarding beam, high-energy ions such as $^{16}O^-$, $^{32}O^+$, $^{40}Ar^+$, and $^{133}Cs^+$ are focused on a small area and rapidly sputter away the surface layers. The interaction of the bombarding beam with the solid is extremely complex and is discussed in a later section. During such an interaction only a small fraction of the atoms sputtered away, typically about 10%, will yield positive or negative ions. These positive or negative ions, when selectively extracted and passed through a mass spectrometer, will provide compositional information for the various atomic layers sputtered away. Specifically, the secondary ions generated will contain compositional information about the top layer and all successive layers etched away. The thickness of the layer will depend upon the etching rate of the primary beam.

By drastically reducing the primary ion beam intensity which, in turn, reduces the primary beam etching rate, analyses can be performed on the upper 5 to 30 atomic layers of a sample. The depth resolution in SIMS, in general, is limited by the escape depth of the secondary ions during sputtering. This depth is a few atomic layers thick and represents the lower limit for surface analysis. By controlling the sputtering process, successive atomic layers can be examined and the analysis then becomes a depth profile analysis of the solid.

### C. Limitations

In mass spectrometry the background count rates are normally of the order of one count per second or less while, in comparison, the signal for an associated mass peak may reach over a million counts per second. Due to such high signal/noise ratios, SIMS is potentially an extremely sensitive analytical method with sensitivities, in some cases, reaching to the parts per billion range. Such extremes, however, cannot be achieved under the restricted conditions encountered in the surface analyses of small particles. In these instances, the sensitivity is severely hampered by the small sample size.

Another, almost as severe limitation has its origin in the methods of preparing particles for surface analyses. Particles exposed to high-energy ion bombardment tend to dissipate some of the absorbed energy in various forms (by becoming electrically charged, by heating,

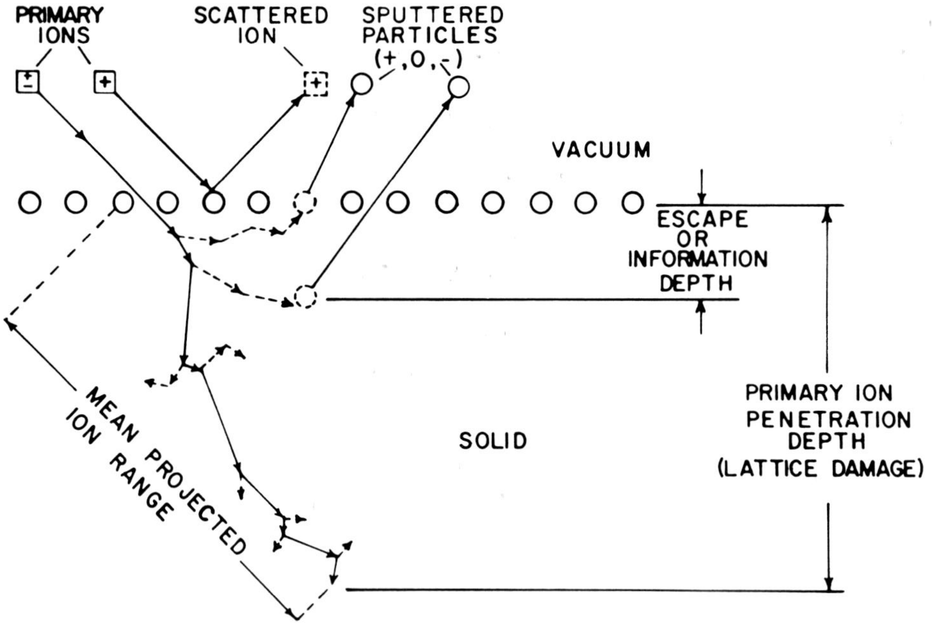


FIGURE 1. Schematic representation of the various processes resulting from the collision of a 1-keV ion beam with a solid surface. (From Honig, R. E., *Thin Solid Films*, 31, 89, 1976. With permission.)

etc.) which can render the particle unstable on its substrate. The common methods used in mounting very small particles for other microanalysis techniques can, therefore, not be used for particles when surface analysis is desired. Instead, skill and ingenuity are necessary to overcome the mounting problems.

In spite of numerous limitations, SIMS is a very viable analytical technique for determining the surface chemical composition of particles down to the submicrometer size range. In some cases, such as the detection of minute quantities of specific elements on small particles, SIMS remains a very useful and by far the most sensitive surface analytical method available.

## III. INTERACTION OF SMALL PARTICLES WITH THE ION BEAM

### A. Interaction Phenomena

Primary ion beams acquire their energy from a high accelerating potential that may vary from 1 to 20 kV. Such a beam sputters atoms from the surface of a solid or small particle giving rise to several processes as shown in the schematic of Honig[4] (see Figure 1). Most of the sputtering occurs on the surface of the particle but some ions may be implanted 30 to 80 Å below the surface. Neutral atoms, positive and negative ions, and electrons are all created in the small plasma cloud above and around the small particles being bombarded. Some of the various effects that occur during the interaction of the ions with a solid are given by Benninghoven[5] in Table 1. The table summarizes the emission processes and surface zone changes.

The reactions that occur in the plasma cloud around the solid particle are complex and not entirely understood.[6-8] Although some of the ions generated on the sample surface are neutralized by collision with electrons many other ions tend to combine with each other or with the ions from the primary beam, especially when reactive primary ions are used. Energies within the primary beam impact area are greater than 1000 eV while the energies of the chemical bond of a compound on a solid surface are well below that level. In addition, the plasma temperature is estimated to be 10,000 to 12,000 K. Therefore the breakdown of

**Table 1**
**EFFECTS OF ION IMPACT ON A SOLID SURFACE: EMISSION PROCESSES AND CHANGES IN THE SURFACE ZONE**

| Emission processes | Changes in the surface zone of the target |
|---|---|
| Atomic and molecular particles | Loss of surface particles |
| Neutrals | Sputtering |
| Positive ions | Recoil implantation |
| Negative ions | Implantation |
| Excited particles | Primary ions |
| Electrons | Surface atoms (recoil) |
| Surface processes (e.g. Auger deexcitation) | Lattice destruction |
| Bulk processes (e.g., ionization) | Imperfections |
| Photons | Amorphization |
| Gas phase processes | Chemical effects |
| Surface processes | Breaking of bonds |
| Bulk processes | Bond formation |

From Benninghoven, A., *Chemistry and Physics of Solid Surfaces,* Vanselow, R. and Tong, S. Y., Eds., CRC Press, Cleveland, 1976. With permission.

chemical compounds on the sample surface and within the plasma cloud is almost complete. Many extremely short-lived molecular fragments (such as $CH^+$, $CH^+{}_2$, $CH^+{}_3$, etc.) may be detectable due to the fast propagation time of the secondary ions through the mass spectrometer. When organic compounds are analyzed, their molecules are broken down to relatively small fragments with most of them having mass to charge (m/e) ratios below 50.

As the sputtering process proceeds, the sample is gradually eroded away with the erosion rate of the small particle being dependent upon numerous factors:

- The incident energy of the primary ions
- The type of primary ions
- The polarity of the primary ions
- The density of the ions in the beam (ions/$\mu m^2$)
- The shape and size of the particle and surface morphology
- The chemical composition of the sample

Further, the rate of sputtering is not necessarily indicative of the rate of ionization of the sample. For example, when using the SIMS technique with an oxygen or argon primary ion beam gold is one of the least detectable elements due to its low secondary ion yield. In comparison to chromium the sputtering rate of gold is quite high, about 10 times that of chromium, but under identical operating conditions the secondary ion yield of chromium is $1.5 \times 10^6$ counts per second vs. only 50 counts per second for gold. An example of the variability in sputtering rates for various materials is given in Table 2 from Laznowsky.[9]

### B. Primary Beam Selection

Selection of the primary ion beam is of fundamental importance in the analyses to be performed. Due to the various effects of ions impacting a solid surface (described in Table 1), specifically ion implantation and the chemical reaction of the surface layer with the primary beam, the analyses are performed on somewhat unstable chemical compounds formed during bombardment by the particle and the ion beam. In general, it is best to use a reactive gas (such as $^{32}O^+$, $^{16}O^-$ or $^{133}Cs^+$) as the primary beam, as the secondary ion yield is enhanced considerably. Oxygen, an electronegative element, enhances the production

**Table 2**
**SPUTTERING RATES OF VARIOUS TARGET MATERIALS UNDER BOMBARDMENT BY AN ARGON PRIMARY ION BEAM**[a]

| Target material | A/sec | A/min | μm/hr |
|---|---|---|---|
| Alumina | 2.2 | 130 | 0.8 |
| Aluminum | 7.4 | 4 | 2.7 |
| Chromium | 3.3 | 200 | 1.2 |
| Gold | 27.0 | 1600 | 9.7 |
| Iron | 5.4 | 320 | 1.9 |
| Molybdenum | 6.6 | 400 | 2.4 |
| Niobium | 5.0 | 300 | 1.8 |
| Silicon | 6.0 | 360 | 2.2 |
| $SiO_2$ | 7.0 | 420 | 2.5 |
| Silver | 34.0 | 2000 | 12.0 |
| Titanium | 3.3 | 200 | 1.2 |
| Zirconium | 5.3 | 320 | 1.9 |

[a] Operating conditions: ion energy, 1.0 keV; current density, 1.0 nA/cm$^2$.

From Laznovsky, W., *Res. Dev.*, August 1975. With permission.

of secondary ions from electropositive species whereas cesium, an electropositive element, enhances negative ions.[10] Positive primary ions are used for electrically conductive particles while for large, nonconductive particles positive ions create a positive charge that interferes with collection of the secondary ions. Two methods are employed to overcome this positive charge problem. First, an electron beam may be directed at the particles to neutralize the positive charge and thus enable the use of positive primaries or second, a negative primary ion beam such as $^{16}O^-$ can be used to efficiently compensate for the charge buildup on the particle surface.

The ionization efficiency, referred to in the previous section, is dependent not only on the association of elements present within the material being analyzed,[11-12] but to a certain degree on their chemical bonding and on their recombination potential. Recombination can occur with any or all of the elements composing the sample as well as the combination of the elements with the primary ions. In some instances diatomic ions are formed; in other instances, the oxygen in the primary ion beam will combine with cations and anions present in the sample and form various oxides. When a primary ion beam such as oxygen interacts with a simple chemical system such as calcium fluoride ($CaF_2$), one observes in addition to fluorine, the ions of HF, HFO, CaF, CaFO, CaO, CaOH, CaH, Ca, HF, $Ca_2O$, etc. The presence of hydrogen comes from the water vapor present as moisture in the vacuum system and from the water absorbed on the surface of small particles. Fortunately, most of these ions are present in quantities several orders of magnitude smaller than calcium $^{40}Ca^+$ or fluorine $^{19}F^-$.

As the selection of the primary ions has a fairly significant effect on the bulk analysis of small particles, ions of reactive gases such as oxygen and cesium are used for such analysis. However, when one analyzes partially oxidized samples, argon and possible nitrogen ions are the choices for the primary beam. In analyzing the surface of very small particles ($\ll 10$ μm) the first few monolayers are almost always oxidized, thus the choice of the primary

**Table 3**
**NUMBER OF COUNTS FROM THE TOTAL VOLUME OF SMALL PARTICLES OF VARIOUS DIAMETERS**

| Particle size (μm) | No. of counts |
|---|---|
| 100 | $2 \times 10^{13}$ |
| 10 | $2 \times 10^{10}$ |
| 1.0 | $2 \times 10^{7}$ |
| 0.1 | $2 \times 10^{4}$ |
| 0.01 | 20 |

ion beam is less critical. Oxygen for electropositive, and cesium for electronegative elements will, however, ensure maximum sensitivity.

## IV. SURFACE ANALYSIS OF SMALL PARTICLES

### A. General Considerations

Surface analyses of solids by secondary ion mass spectrometry are usually performed on samples that, for practical purposes, are infinite. That is, the ratio of the sample area to area of the primary ion beam is very large. Due to the large sample area only a portion of the surface is etched away during the analysis and sufficient material remains for analyses. In the case of Static SIMS, the density of the beam and thus the sputtering rate are so low that during the entire analysis only a fraction ($10^{-2}$ to $10^{-3}$) of a monolayer is removed and analyzed. This effectively increases the sample survivability and renders small samples almost infinitely large.

Although SIMS analysis of small individual particles is restricted due to the small sample size and thus the total number of atoms available, bulk analyses can routinely be performed. Assuming a 1% secondary ion yield and a 10% collection transmission for the mass spectrometer system, the number of counts that a small particle will yield at 100% consumption under an ion beam has been calculated for various-sized particles (see Table 3). It can be seen from Table 3 that even a submicrometer-size particle can provide a sufficient count rate for the bulk analysis of a number of elements.

For surface analysis it is desirable to utilize a primary ion beam with a cross-sectional area larger than the projected area of the small particle being analyzed:

1. To reduce the number of ions striking a small particle
2. To compensate for small drifts in the position of the ion beam
3. To benefit from a fraction of the condensed particles

In performing the surface analysis of small particles it is necessary, however, to maintain a reasonable sputtering rate in order to obtain a sufficiently detectable secondary ion signal. Although the above two requirements are obviously conflicting, adequate counts for a full mass spectrum can usually be obtained by removing from 10 to 30 Å of a large solid sample. The amount of sample etched away will, of course, vary with the sensitivity desired and the elements in question. A statistically valid number of ions of any element must be generated and collected. Since this number is dependent upon ionization efficiency those elements with high secondary ion (SI) yields will require removal of smaller amounts of material. Thus, for surface analyses one can easily see that the amount of material removed from the

**Table 4**
**NUMBER OF COUNTS FROM A MONOMOLECULAR SURFACE FILM ON A SMALL PARTICLE**

| Particle size (μm) | No. counts from a thin film |
|---|---|
| 100 | $3.14 \times 10^{10}$ |
| 10 | $3.14 \times 10^{7}$ |
| 1 | $3.14 \times 10^{4}$ |
| 0.1 | 31 |

*Note:* Counts based on 1% ion yield and 10% transmission in an ion microprobe analyzer.

surface is very critical. Table 4 gives an example of the number of counts that can be expected during the surface analysis of a single atomic layer on various-sized small particles. The data presented within the table assume an element with a good SI yield. When the surface film or layer consists of elements such as carbon or nitrogen, a particle must be several orders of magnitude larger to provide a sufficient number of SI to discriminate above background.

For surface analysis it is desirable to utilize ion beams with diameters larger than the particle diameter in order to minimize the effect of beam drift and to benefit from the condensation phenomena. Assuming an ion beam with a two-dimensional Gaussian distribution and a diameter of $3\sigma$ where $\sigma$ is the standard deviation, the relative percentages of primary ions striking a particle as a function of particle/beam diameter can be calculated. A diagram showing percentage of the total primary ions within the cross section of the beam which strike a projected area of a particle diameter is given in Figure 2.

Another phenomenon that occurs when an ion beam impinges the surface of a material is that of condensation. This phenomenon is partially caused by the collector electrode which in effect is biased to attract ions of one polarity. Ions of opposite polarity will be repelled by the collector electrode, either back onto the sample or onto the substrate surface. They will then tend to form a thin film of condensed negative or positive ions depending on the polarity of the collector electrode. The film, which can be up to a few hundred angstroms thick will consist of neutralized sample ions, recombination products of the sample and the primary ions, and neutrals from the sample. The rate of formation of the film is dependent upon a complicated function of the chemical composition of the sample, the polarity and magnitude of the collector bias, and the primary ion species.

Normally, when an analysis is performed for positive ions using a large primary beam, one would expect the loss of negative ions. However, in the case of microprobe type (small beam) instruments this is not necessarily true as a significant fraction of the ions condense on the substrate around the particle or on the surface of the sample being analyzed (Figure 3). In the surface analysis of small particles, the condensation phenomenon is often very important and gives an advantage in the analysis of trace elements by providing additional material. During the bulk analysis of a 5-μm $ThO_2$ particle with an argon beam for 240 sec, the condensation effect resulted in a surface layer that was subsequently analyzed under the following conditions: beam diameter, 5 μm; beam current, 1 nA; and counting time, 10 sec. In another case, where a small particle was mounted on a gold substrate, the gold surface was analyzed on a point-by-point basis and revealed that the condensation of ions

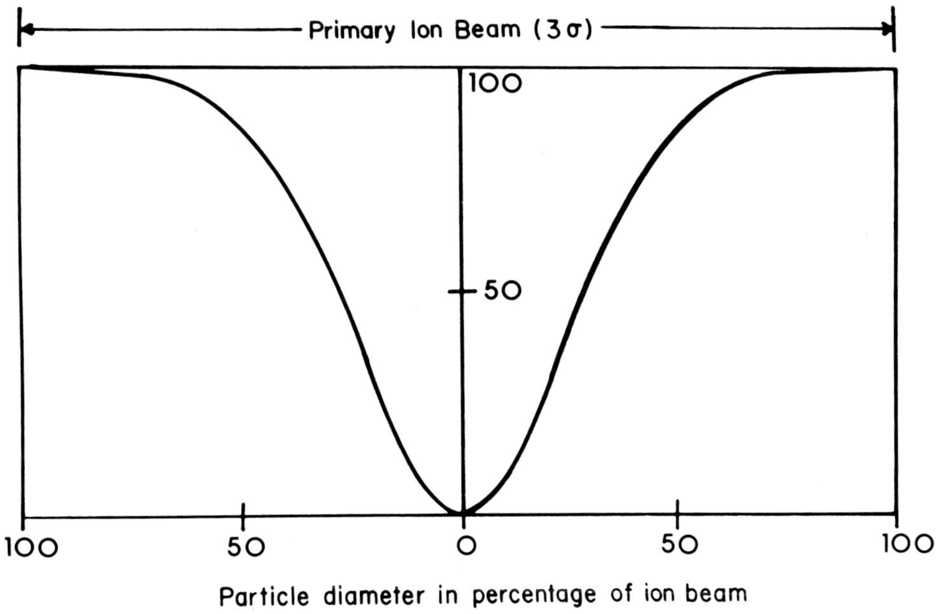


FIGURE 2. Microparticle size vs. percentage of primary ion beam.

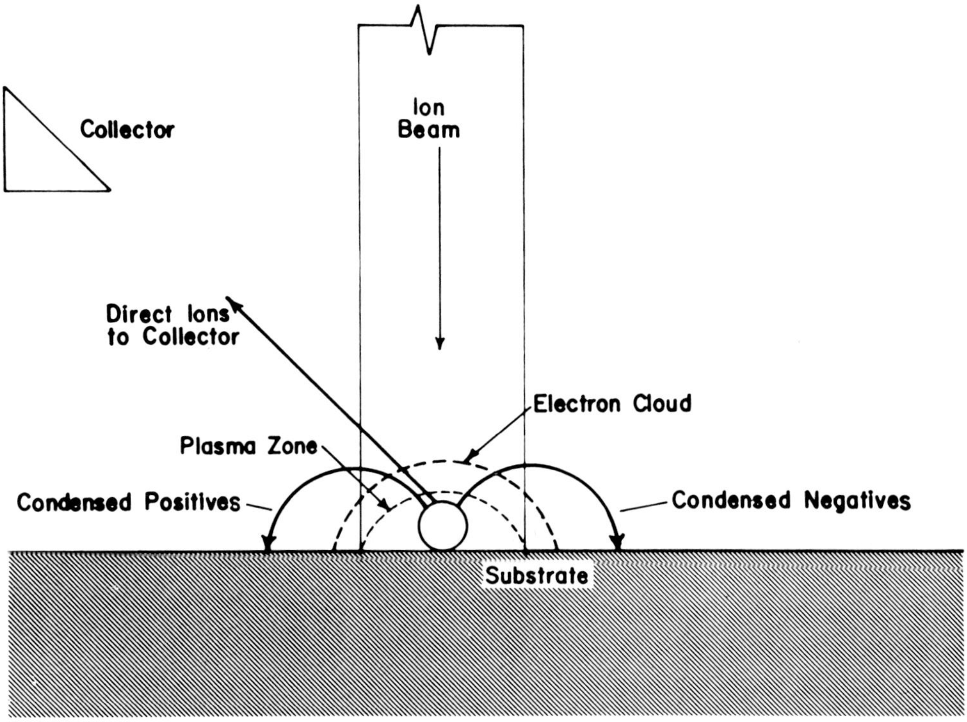


FIGURE 3. Diagram showing condensation effect created during the analyses of small particles.

was significant even at a 100-μm distance from the small particle (see Figures 4 and 5). The condensation phenomenon has an additional implication for SIMS analysis of small particles. That is, when analyzing for minor and trace elements individual particles should not be mounted closer than the condensation region.

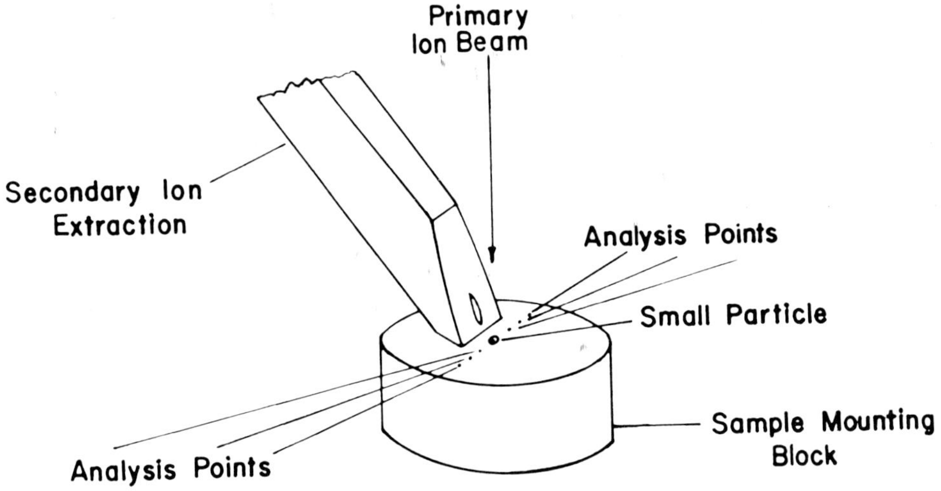


FIGURE 4. Diagram showing points of analysis in studying the condensation effect.

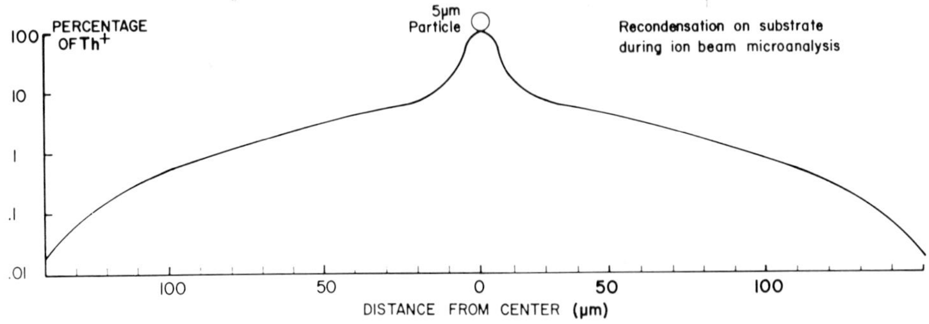


FIGURE 5. Diagram showing condensation phenomena during the analysis of a 5-μm $ThO_2$ particle.

## B. Quantitative Analyses

One of the most serious problems in establishing semiquantitative or quantitative assessments of the chemical results obtained by SIMS is the extreme variation of the secondary ion yield of various elements. This problem is indicated in the diagrams by Storms et al.[13] which show the secondary ion yields from pure elements for primary ion beams of oxygen ($^{32}O^+$) and cesium ($^{133}Cs^+$) (see Figures 6 and 7). Secondary ion yields vary five to six orders of magnitude and present a real challenge when attempting to quantify the resultant data using a theoretical argument.[14,15] Recently, Deline[16] and others expressed a simple relationship between positive and negative ion yields and the sputtering rate:

$$\frac{n^{\pm}}{n^{0}} \propto \left(\frac{1}{S}\right)^{x\pm} \propto [P]^{x\pm} \tag{1}$$

where P is the surface or near-surface concentration of the primary ion, S is the linear sputtering rate in Å/sec, n ± is the positive or negative ion yield, n° is the number of sputtered atoms, and x ± is a matrix affect exponent for positive or negative ions.

The advantage of Deline's formula[16] is that if a constant concentration of the implanted ions near the surface could be obtained there would be no matrix effect in SIMS. Unfortunately, the near-surface concentration is inversely proportional to the substrate sputtering yield and depends upon (1) the partial pressure of the reactive gas from the residual gas in

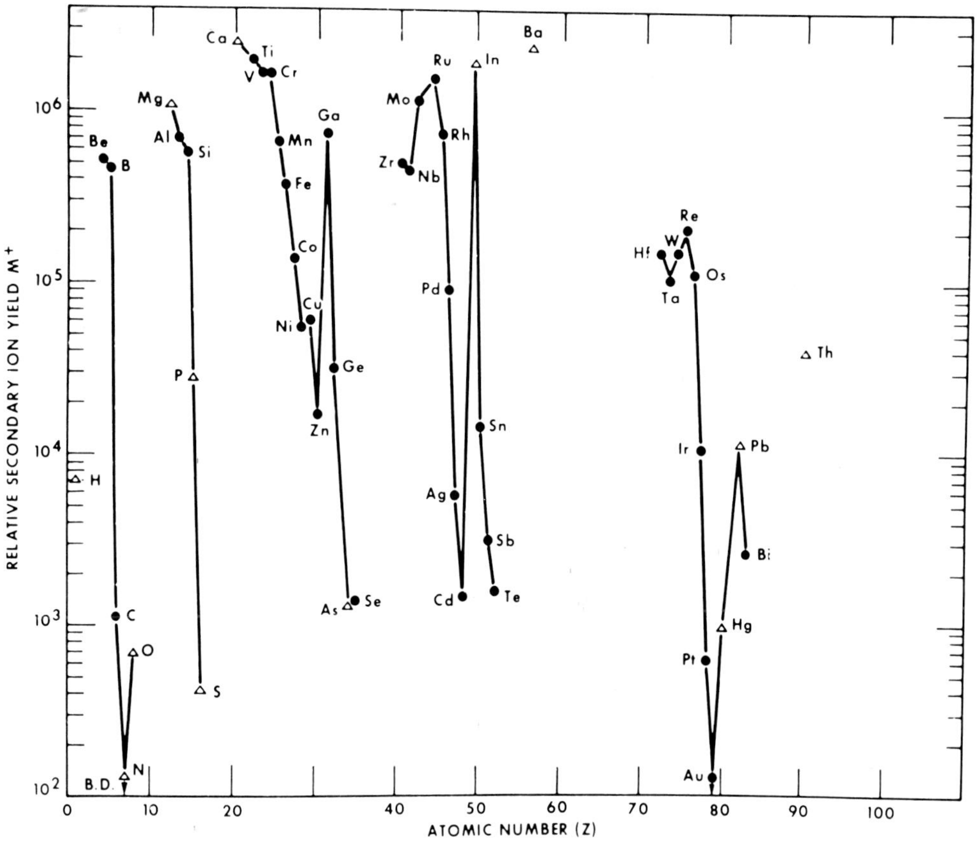


FIGURE 6. Relative secondary positive ion yield ($M^+$) due to $O^-$ bombardment vs. atomic number of secondary ion: (●) pure element; (△) compound. (From Storms, H. A., *Anal. Chem.*, 49, 13, 1977. With permission.)

the sample chamber, (2) the substrate sticking coefficient, and (3) the recoil implantation efficiencies which vary from matrix to matrix. Therefore, one of the most reliable methods for converting SIMS intensities to atomic percentages is to use the sensitivity factors[17] long employed in mass spectrometry:

$$R.S._M = \frac{M_c/R_c}{\%M/\%R} \tag{2}$$

where R.S. is relative sensitivity factor of an element M, $M_c$ is the count rate for element M in a standard, $R_c$ is the count rate of reference element R in the standard, and %M and %R are atomic percents of elements M and R in the standard. An example of the wide range of sensitivity factors seen in SIMS analyses is given in Table 5.

## C. Particle Mounting

The mounting of small particles for secondary ion mass spectrometry follows well-established procedures described in *The Particle Atlas* by McCrone and Delly.[18] The choice of substrate and the application of the sample mounting medium are more critical for SIMS than for other ultramicroanalytical techniques.

For bulk analyses the mounting medium is usually a thin collodion film, which is removed from the sample or substrate surface during the first few seconds of analyses by sputter etching or ion cleaning of the small particles. In the surface analyses of small particles, it

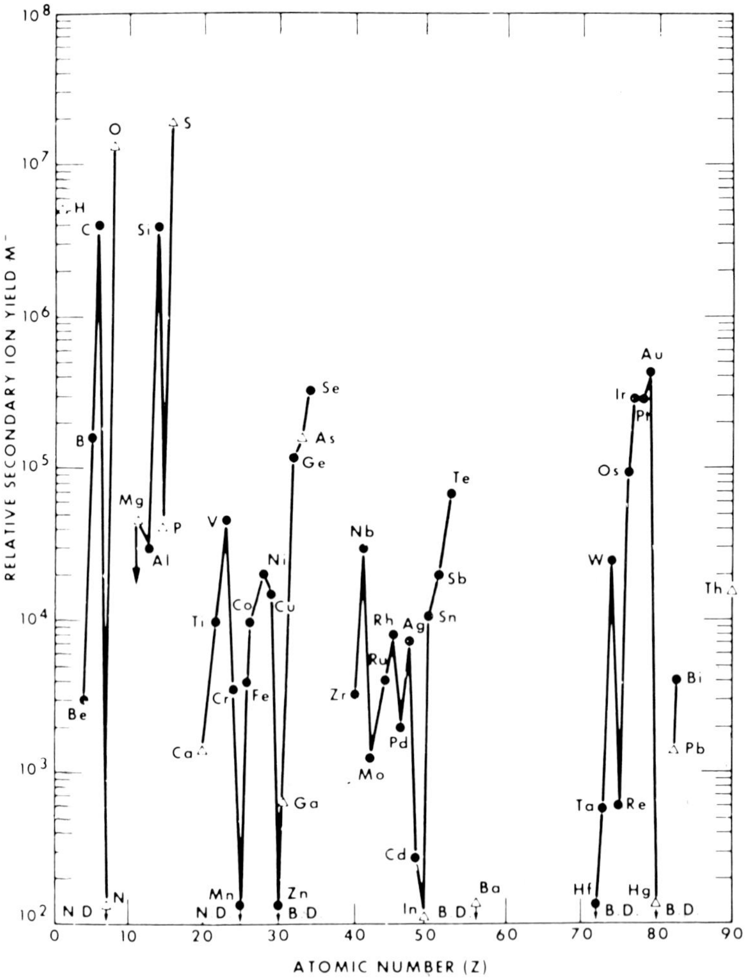


FIGURE 7. Relative secondary negative ion yield ($M^-$) from $Cs^+$ bombardment vs. atomic number of secondary ion: relative intensity is given in units counts/sec/$10^{-9}$ A; (●) pure element; (△) compound; B.D. = barely detectable; N.D. = not detected. (From Storms, H. A., *Anal. Chem.*, 49, 13, 1977. With permission.)

is the first few monolayers that are of concern and therefore must be left uncontaminated and intact by the mounting technique. For surface analyses, particles larger than 10 μm are usually placed on an almost dry spot of collodion or another suitable substrate. With this technique, the very minute spot of collodion does not creep and/or coat the upper part of the particle. Thus the surface analysis, performed with a negative ion oxygen beam several times larger than the projected area of the particle, is collected with little interference from the mounting medium. Very small particles ($\leqslant$10 μm) will usually stay on a substrate for a few seconds provided that the current density of the primary ion beam is sufficiently low, less than $3 \times 10^{-4}$ A/cm$^2$. During this time, while several surface layers are sputtered away and the secondary ions analyzed in a mass spectrometer, the small particles are usually "welded" to the substraıe (as explained later).

In the surface analysis of small particles, the areas analyzed are always much larger than the particle for reasons mentioned earlier. As the substrate is an integral part of the area analyzed, it is obvious that the substrate must be carefully selected. The choice of substrate is dictated by a number of factors:

**Table 5**
**SENSITIVITY FACTORS OF VARIOUS ELEMENTS RELATIVE TO SILICON**

| Element | Sensitivity factor |
|---|---|
| Si[a] | 1.0 |
| Na | 0.2 |
| Ca | 0.1 |
| Ti | 0.2 |
| Fe | 0.5 |
| Cu | 1.4 |
| Zn | 5.5 |
| Bi | 5.5 |
| Pb | 2.0 |
| O | 418.0 |

[a] Reference element.

- Its purity
- Its low vapor pressure
- Its surface smoothness and flatness
- Its very low secondary ion yield
- Its limited number of isotopes (preferably one)

The two substrates satisfying most of these conditions are high-purity gold and nuclear-purity vitreous carbon. In general, a gold substrate is the choice for the analysis of the lighter elements while carbon is the choice for the heavier elements. Gold, available in the form of high-purity beads (99.9999%), is pressed into a thin plate 300 to 600 μm thick using two quartz discs. As the gold plates have a high reflectance, small particles mounted on their surface can easily be observed under incident illumination. Due to the relatively high etching rate of gold, shown in Table 2, small particles are virtually "welded" to the substrate within several seconds (see Figure 8).[18] Occasionally, gold ions will combine with other elements such as K or Ca and form larger fragments that have high SI yields.

$$^{197}Au^+ + {}^{39}K^+ \rightarrow {}^{236}AuK^+ \tag{3}$$

$$^{197}Au + Ca^{40} \rightarrow {}^{237}AuCa^+ \tag{4}$$

Creation of these large molecular ions present mass spectral interferences in the high-mass ranges. A carbon substrate, on the other hand, does not yield many interference peaks in the high-mass ranges, as most elements with extremely high SI yields have low masses (Na, K, Ca) and their recombination fragments with substrate atoms are also low masses. The hard and brittle nature of vitreous carbon presents minor problems for the microscopist to easily mark the position of small particles mounted on its polished surface.

## D. Examples of Surface Analyses

It is evident when reviewing the literature that the surface analyses of very small individual particles is a relatively new advancement in the state of the art. In fact, the literature is sparse, if not void, in the referencing of examples of the surface analyses of small individual

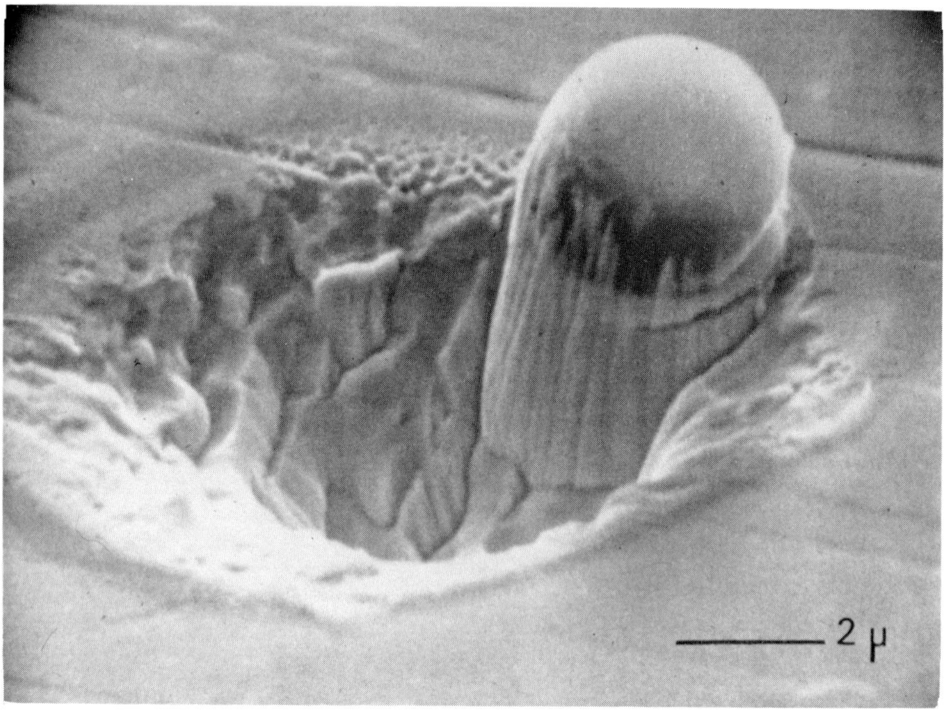


FIGURE 8. Photomicrograph of a fly ash microsphere after ion bombardment.

particles. From the hundreds of analyses performed, two examples are summarized. The first example concerns the analyses of coatings applied to sulfide minerals while the second example refers to the analyses of air pollutant deposits.

Sulfide minerals are usually separated from the bulk of a silicate gauge by their affinity and association with minute air bubbles created during the process called flotation separation. To increase the nonwetting properties of mineral particles, certain agents such as potassium or sodium ethylxanthate are added in small quantities (20 ppm) during their processing. Such reagents form incomplete molecular layers on the mineral surfaces and increase their nonwetting properties. After such a treatment, several small particles of treated pyrite, iron sulfide mineral, and chalcopyrite, copper iron sulfide, were isolated and mounted on a gold substrate and analyzed in an ion microprobe for the surface concentration of two isotopes, $^{23}Na^+$ and $^{56}Fe^+$. While the small particles were bombarded with a stationary ion beam, the computer-controlled mass spectrometer recorded the intensities of sodium and iron. The data collected on these treated particles were then compared to the data observed during the analysis of untreated particles. A significant increase in the $^{23}Na^+$ count rate was observed on the treated sample vs. the untreated particle. In reverse fashion, the counts of $^{56}Fe^+$ were significantly lower on the treated surface. These respective data have been summarized in Table 6. This study of the sodium ethylxanthate coatings on small particles of pyrite and chalcopyrite has shown the utility of SIMS in analyzing the surfaces of small mineral particles.

The second example concerns gas-solid interactions and the respective adsorbed layers on effluent. In a number of processes, especially gas-solid reactions, the surface of a small particle is the most significantly affected part. Dust particles, for example, have a high affinity for pollutants to collect on their surfaces and they attract compounds such as lead halides from automotive exhaust. A number of other toxic compounds having relatively high vapor pressures also tend to deposit predominantly on the surfaces of small particles such

**Table 6**
**RELATIVE INTENSITIES OF IRON AND SODIUM ON THE SURFACE OF SMALL PYRITE AND CHALCOPYRITE PARTICLES**

| Mass | Element | Counts | % Relative | Comments |
|---|---|---|---|---|
| 23 | Na | 98,343 | 21.9 | Untreated pyrite |
| 56 | Fe | 349,284 | 78.1 | |
| 23 | Na | 198,020 | 45.9 | Treated pyrite |
| 56 | Fe | 233,152 | 54.1 | |
| 23 | Na | 97,807 | 32.4 | Untreated chalcopyrite |
| 56 | Fe | 203,614 | 67.6 | |
| 23 | Na | 362,677 | 75.6 | Treated chalcopyrite |
| 56 | Fe | 117,244 | 24.4 | |

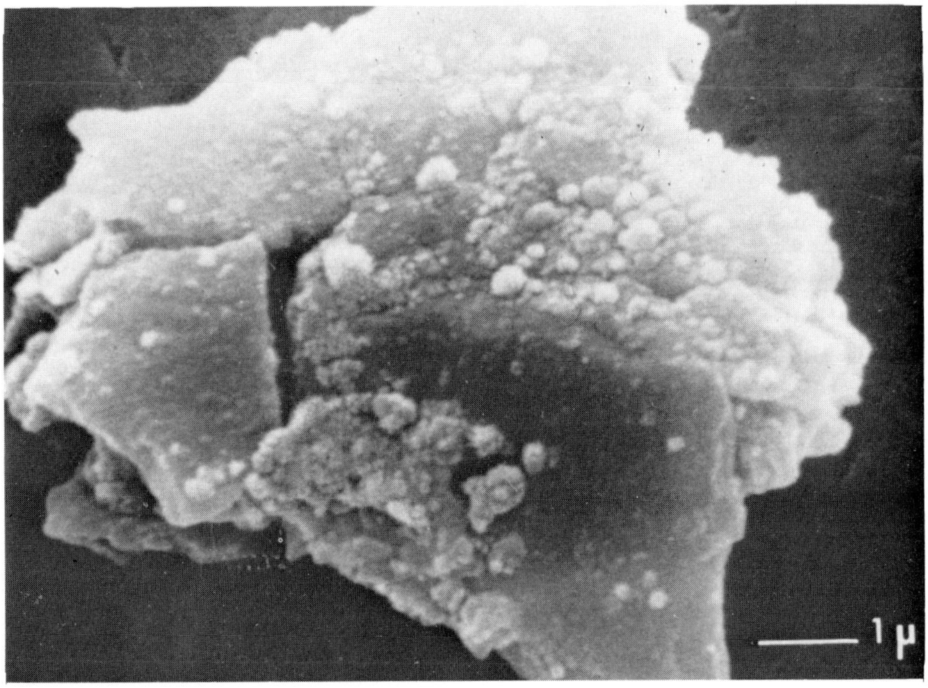


FIGURE 9. Scanning electron micrograph of an airborne contaminant consisting of an organic matrix with many small particles occluded to its surface.

as flyash and particulate effluents. Air particulates from urban areas have almost invariably a lead-enriched surface layer from the leaded fuels of auto exhausts. As the adsorbed layers are relatively thin the mass spectra of lead isotopes from such particles are normally detectable only during the first few seconds of the sputter etching process. As the primary beam continues to etch through the sample the lead-enriched layer disappears and one can observe significant changes in surface structure of the small particles. Figures 9 and 10 illustrate the effects of the ion beam on small particles. From the scanning micrographs one can see that the submicroscopic occlusions have been removed and the rough edges are smoothed.

The utility of the SIMS technique in identifying thin surface films or layers has been demonstrated on numerous individual small particles of varying composition.

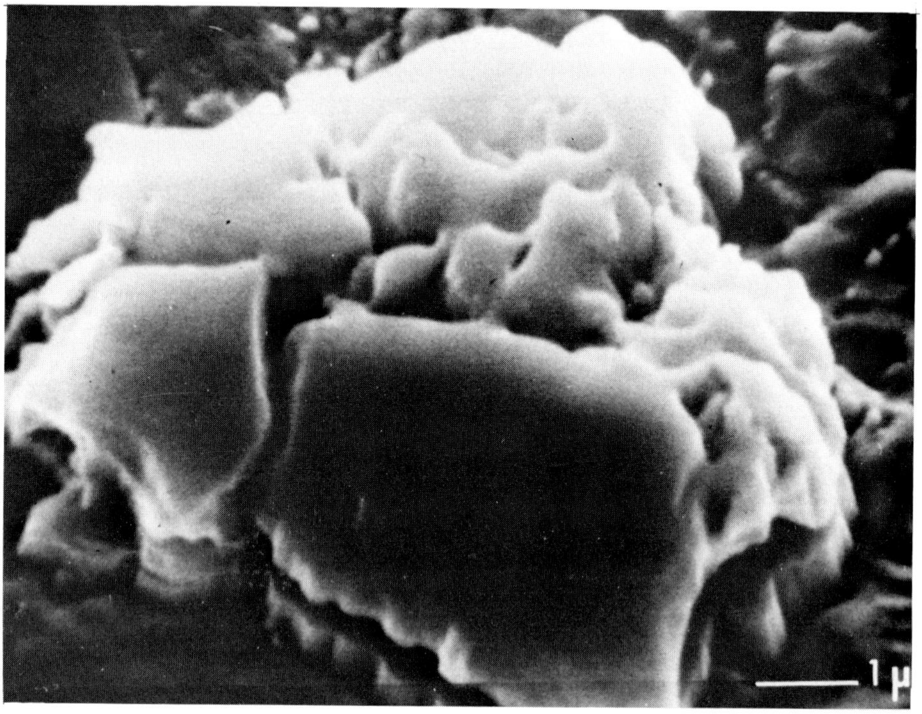


FIGURE 10. Scanning electron micrograph of the particle of Figure 9 after 20 sec of ion bombardment.

## V. FUTURE DEVELOPMENTS

Due to the destructive nature of the primary ion beam, the application of SIMS in the surface analysis of very small individual particles is very limited. Any future development in the improvement of collection efficiency and the transmission of mass spectrometers will have beneficial effects on elemental detection. The most significant analytical improvement for the surface analysis of single particles may come from the development and application of ultrafast collection systems, such as utilizing time-of-flight mass spectrometers with on-line computer processing of the results. These systems may not have the mass resolving power of present instrumentation, but the advantages of such developments will more than compensate for this by revealing a real time display of mass spectra. It is anticipated that the displays will already be computer-corrected for spectral interferences and the enormous differences in secondary ion yields. The availability of real-time spectral displays will enable the analyst to observe the surface composition of small particles as well as the subsequent compositional changes resulting from the removal of successive atomic layers.

While waiting for such significant developments, SIMS has already demonstrated its important role in the compositional studies of the surfaces of individual small particles.

## ACKNOWLEDGMENT

The effort of D. A. Brooks for his review and suggestions in the preparation of this chapter is greatly appreciated.

## REFERENCES


1. **Castaing, R., Jouffrey, B., and Slodzian, G.,** Sur les Possibilities d'Analyse Locale d'un Echantillon par Utilisation de son Emission Ionique Secondaire, *Acad. Sci.*, 122, 1010, 1960.
2. **Liebl, H.,** Ion microprobe mass analyzer, *J. Appl. Phys.*, 38, 5277, 1967.
3. **Hillenkamp, F. et al.,** A high sensitivity laser microprobe mass analyzer, *Appl. Phys.*, 8, 341, 1975.
4. **Honig, R. E.,** Surface and thin film analysis of semiconductor materials, in *Thin Solid Films,* Elsevier Sequoia S.A., Lausanne, Switzerland, 1976, 90.
5. **Benninghoven, A.,** Surface analysis by means of ion beams, in *Chemistry and Physics of Solid Surfaces,* Vol. 1, Vanselow, R. and Tong, S. Y., Eds., CRC Press, Cleveland, 1976, 207.
6. **Harrison, D. E.,** Fundamentals of ion surface interaction, Japan-U.S. joint seminar on quantitative techniques, in *Secondary Ion Mass Spectrometry,* Honolulu, October 1975.
7. **Wehner, G. K.,** SEM, *Vol. 1, Johari, O. and Corvice, I., Eds., Illinois Institute of Technology Research Institute, Chicago, 1975.*
8. **Kishinewskii, M. E.,** Mechanism for secondary ion emission, *Sov. Phys. Tech. Phys.*, 20, 799, 1975.
9. **Laznowsky, W.,** Advances in low-energy ion beam technology, 47 *Res. Dev.*, August 1975.
10. **Katz, W.,** Applied surface analysis, in *Proc. 16th Annu. Conf. Microbeam Anal. Soc.*, Geiss, R. H., Ed., San Francisco Press, San Francisco, 1981, 287.
11. **Shimizu, N.,** Energy filtering and quantitative SIMS analysis of silicates for major and trace elements, in *SIMS II,* Benninghoven, A. et al., Eds., Springer-Verlag, New York, 1979, 64.
12. **Steele, I. and Hutcheon, I.,** Ion probe analysis of natural olivine, in *Proc. 14th Annu. Conf. Microbeam Anal. Soc.*, Newbery, D., Ed., San Francisco Press, San Francisco, 1979, 338.
13. **Storms, H. A., Brown, K. F., and Stein, T. D.,** Evaluation of a cesium positive source for SIMS, *Anal. Chem.*, 49, 2023, 1977.
14. **Andersen, C. A.,** Quantitative theory of sputtered ion mass analysis, in Proc. 6th Natl. Conf. Electron Probe Analy., Paper 8, Pittsburgh, Pa., July 1971.
15. **Andersen, C. A. and Hinthorne, J. R.,** Thermodynamic approach to the quantitative interpretation of sputtered ion mass spectra, *Anal. Chem.*, 45, 1421, 1973.
16. **Deline, V. R., Katz, W., and Evans, C. A.,** Mechanism of the SIMS matrix effect, *Appl Phys. Lett.*, 33(9), 832, 1978.
17. **McHugh, J. R.,** in *SIMS, Methods of Surface Analysis,* Wolsky, S. P. and Czanderna, A. W., Eds., Elsevier, Amsterdam, 1975.
18. **McCrone, W. C. and Delly, J. D.,** *The Particle Atlas,* Vol. 1, 2nd ed., Ann Arbor Publications, Ann Arbor, Mich., 1972.

Chapter 2

# LASER MICROPROBE MASS ANALYSIS IN PARTICLE ANALYSIS

**Raimund Kaufmann and Paul Wieser**

## TABLE OF CONTENTS

# I. INTRODUCTION

This chapter deals with laser microprobe mass analysis (LAMMA), a new microanalytical technique based on laser ion source mass spectrometry that employs high-intensity laser pulses to vaporize and partly ionize small amounts of materials. Under certain conditions the ions thus produced are accessible for mass spectrometrical analysis.

In particle technology, aerosol research, and other related fields the chemical composition of the particulate material is of main interest. Many effects caused by the particulate material are intimately associated with the chemical composition of the particles. However, in most samples of real world particles the composition of individual particles is rather inhomogeneous. For instance, an atmospheric aerosol consists of an external mixture of particles emitted by different sources, but particles originating from the same source may differ with respect to both their main constituents and their content of trace components, depending on the mechanisms involved in particle formation and the history of the particles while in the airborne state.

At the surface of each individual particle interactions such as phase transitions, chemical reactions, and charge depositions occur between the particle and its environment. In addition, the particle surface is the origin of surface forces responsible for what is described as adhesion. Thus, certain chemical components may be found either concentrated in or deprivated from the particle surface layer. Taken together, it is obvious that knowing the average bulk composition of the particulate material under investigation does not allow one to characterize the physical and/or biophysical properties of many particle populations. What we are looking for, instead, is some sort of a "chemical histogram" determined in a number N of particles which should be large enough to be representative and statistically meaningful.

The above arguments indicate that in addition to the conventional approaches of analytical chemistry (which only yield information on the mean composition of a particle population), a single-particle approach such as provided by highly sensitive microprobe techniques is desirable.

Most of the microprobe techniques actually in use are still far from satisfying the various analytical requirements needed for particle characterization. An ideal instrument should have the following features:

- Besides appropriate spatial resolution, an instrument used in particle analysis should give as much analytical information as possible about the atomic and molecular constituents (including isotope discrimination) of a particle.
- The sensitivity should be high enough to detect trace substances, possibly down to the parts per million range.
- The analysis should be quantitative or at least semiquantitative.
- The measuring procedure and data processing should be fast enough to accumulate the analytical data of large particle collections in a reasonable period of time.
- The viewing should provide as much information as possible on the morphology of the particles under investigation down to the submicron size range.
- With regard to information depth, the ideal instrument should be able to discriminate between surface layers and the core material of the particle.

Clearly, such an instrument does not yet exist. Instead, a variety of techniques are available, most of them complying with only one or another of the requirements outlined above. However, in comparison with other microprobe techniques such as the electron microprobe, ion microprobe, and laser Raman microprobe, the recently introduced technique of laser microprobe mass analysis appears to come rather close to fulfilling the above criteria (see also Table 1). It is a rather fast and also highly sensitive method of analyzing individual

**Table 1**
**ANALYTICAL CHARACTERISTICS OF A NUMBER OF MICROANALYTICAL TECHNIQUES[50]**

| Method | EPMA | SIMS | AES | MOLE | LAMMA |
|---|---|---|---|---|---|
| Resolution | | | | | |
| Lateral, μm | 1 | 1 | 0.3 | 0.5—1 | 0.5 |
| Depth, μm | 1 | $5.10^{-4}$ | $10^{-3}$ | ? | ? |
| Elements detected | B-U | All | Li-U | Molecules | All |
| Isotopes | No | Yes | No | — | Yes |
| Molecular information | Some | Yes | Some | Yes | Yes |
| Absolute sensitivity | $10^{-14}$—$10^{-16}$ | $10^{-19}$ | $10^{-15}$ | $10^{-8}$ | $10^{-18}$—$10^{-20}$ |
| Relative sensitivity (ppm) | 100 | 1 | 1000 | $10^{4}$—$10^{5}$ | 1 |

From Adams, F., Block, P., Natusch, D. F. S., and Surkyn, P., Proc. Int. Conf. Environ. Pollution, Thessalonica, Greece, September 20 to 24, 1981. With permission.

particles between about 0.2 to 20 μm in diameter. The mass spectrum of either positive or negative ions obtained by a single laser pulse contains valuable qualitative information on the atomic and, to a certain extent, the molecular composition of individual particles. Although quantitative analysis appears basically feasible, the present insufficient understanding of the processes occurring during laser-induced ion formation prevents or at least restricts a quantitative interpretation of the mass spectra obtained.

Nevertheless, a rapidly increasing number of investigations demonstrate the usefulness of laser microprobe mass analysis in single-particle analysis.

Besides a brief description of the LAMMA instrument, this chapter attempts to review work done by means of laser microprobe mass analysis in single-particle analysis.

## II. LASER MICROPROBE MASS ANALYSIS (LAMMA)

Several reviews have been published[1-5] on the subject of laser microprobe techniques based on either emission spectroscopy or mass spectrometry. Most of the instrumentation so far described are of either poor optical resolution or moderate absolute or relative sensitivity. Only recently, extensive papers reviewing "true" laser microprobe systems such as the laser Raman microprobe[6] or the laser microprobe mass analyzer[7] have been published.

### A. Physical Principle

#### *1. General Considerations*

One of the main differences between laser light and the radiation from conventional light sources is the fact that laser radiation can be concentrated to a very high field strength (or power density) of up to $10^{14}$ W/cm$^2$ and more during extremely short periods of time (down to a picosecond). With such a pulse of high-power radiation, solid state matter can be instantaneously converted into a plasma of ionized nuclei and electrons which, in the most extreme case, may gain enough kinetic energy to undergo atomic fission. However, at lower power densities the interaction of laser light with matter can also be used to form an ion source for mass spectroscopy comparable, to a degree, to an electric spark ion source. Since laser light can easily be focused down to a very small spot size by simple optical lenses, the smallest diameter of which is limited by diffraction of, say, 0.5 to 1 μm in diameter, one can (unlike in other ion sources) easily design a "point source" of ions of high brightness which, after all, is the most important physical requirement for a microprobe instrument. Since duration of the laser pulse is very short (typically ~ 20 to 30 nsec in the LAMMA instrument), heat conduction from the point of interaction (focus) to the surrounding area

**Table 2**
**IRRADIANCE FOR DAMAGING THRESHOLD IN VARIOUS METALLIC AND NONMETALLIC MATERIALS**

Q-Switched Laser ($\tau$ = 25 nsec) at 3 Different Wavelengths

| | Irradiance ($10^6$ W/cm$^2$) | | |
|---|---|---|---|
| **Sample** | **266 nm** | **355 nm** | **532 nm** |
| Au | 9.4 | 11.7 | 16.8 |
| Ag | 7.5 | 19.7 | 7.6 |
| Al | 9.1 | 32.2 | 24.1 |
| C | 5.1 | 8.7 | 5.4 |
| Si | 3.2 | 9.3 | 8.0 |
| SiO | 211 | 2,000 | 5,830 |
| Epon 0.1 | 224 | 3,190 | 16,800 |
| Epon 10 | 264 | 2,960 | 18,100 |

From Floren, Th., Schädigungsverhalten von Festkörpern bei Bestrahlung mit Riesenimpuls-Lasern verschiedener Wellenlänge, Thesis, University of Frankfurt, Frankfurt, West Germany, 1980.

of the specimen can be disregarded. Interaction is strictly limited to the focus diameter. So, in principle, the probing part of a laser microprobe consists of an appropriate impulse laser and a high-power optical light microscope which is used at the same time to image the specimen and to focus the incident laser light onto a selected spot of interest.

*2. Interaction of the Laser with Solid State Matter*

If laser light is focused on solid-state material, the predominant energy transfer process depends strongly on irradiance (i.e., the power density in the focus) and may cover classical absorption at the low irradiance end of a large variety of nonlinear optical processes occurring at irradiances $\geq 10^6$ to $10^7$ W/cm$^2$. The total amount of energy deposited per unit volume determines the occurrence and amount of physical parameter changes within the irradiated material such as heating, phase transition processes, or ionization.

For a long time most models and theoretical approaches dealing with plasma formation by high-power laser radiation were based on local thermodynamic equilibrium (LTE) models.[8,9] Such a model has also been applied to the ion formation process in secondary ion mass spectrometry (SIMS) by Anderson and Hinthorne.[10] For the purpose of LAMMA analysis the LTE model appears to hold true, at least in that domain of irradiance in which volume evaporation is the predominant process.[11,12]

However, there is increasing evidence that laser-induced ion formation can also occur under nonequilibrium conditions. This is indicated, e.g., by the work of Fürstenau[13,14] who has investigated the intensity distributions of carbon cluster ions and laser-induced ion spectra from organic solids. Here, it is obvious that ion formation occurs under conditions which exclude a ''thermal generation'' process. Irradiances employed are usually less than $10^8$ W/cm$^2$ down to $10^3$ W/cm$^2$, which are orders of magnitude below the threshold for thermal damaging of dielectric solids (see Table 2). The mass spectra merely contain parent molecules and some larger fragments, indicating a nonequilibrium energy repartition such that intermolecular bonds are affected to a lesser extent than intramolecular ones.[15]

For the praxis of laser-induced mass spectrometry, it appears advisable to clearly distinguish between two rather different modes of ion formation corresponding to two regimes of irradiance or power density (D) at the irradiated volume: (1) nonthermal laser desorption (LD) as a "soft" surface ionization process at irradiances of $D < 10^8$ W/cm$^2$, and (2) thermal volume evaporation and ionization (laser pyrolysis) with $D > 10^8$ W/cm$^2$. There is usually no clear-cut borderline between the two basic processes since all the results obtained with laser microprobe systems indicate a strong influence of sample preparation, the geometry of the sample as well as of the laser beam, on the type of interaction process.

In particle analysis, under realistic conditions, there will usually be a predominance of volume evaporation such that in a first approach the LTE model and its implications for quantitative analysis might be applicable (see Section II.C.2.b).

*3. Mass Spectrometry*

As for the analytical part, the advantages of a mass spectrometric system are obvious. Since all the ions present in the laser-induced microplasma are essentially formed at the same time (probably during the rising part of the laser pulse lasting 5 to 10 nsec), time-of-flight (TOF) mass spectrometry appears to be the method of choice. This kind of mass spectrometer is rather simple and straightforward. All ions formed are accelerated to the same energy of, say, 3 keV. Between the energy (Ue) of an ion of mass (m) and its velocity (v) we find the simple relationship of

$$eU = m\,v^2/2 \qquad \text{or} \qquad v = \sqrt{\frac{2eU}{m}} \tag{1}$$

Taking 2 Ue as a constant we can see that the velocity (v) of an ion of mass (m)

$$v \simeq \sqrt{\frac{1}{m}} \tag{2}$$

is proportional to the square root of the inverse of its mass. This, in turn, means that the traveling time (t) of an ion of mass (m) for a given flight path (d) is given by

$$t \simeq d \cdot \sqrt{m} \tag{3}$$

Consequently, a TOF spectrometer simply consists of a device for accelerating the ions to a uniform energy, a drift tube of length (d) in which the ions are separated according to their mass (m), and an ion detector at the end of the drift tube where the different ions arrive in discrete packs, one pack after the other. The detector, usually a secondary electron multiplier (SEM), delivers an analog signal at its output, the amplitude of which corresponds to the number of ions present in each pack. The main advantage of the TOF mass spectrometer, as compared to other mass spectrometers, is that all sorts of ions are analyzed in one cycle. This means that all analytical information about the area of interest is obtained and recorded simultaneously. Since the TOF spectrometer basically provides for a high transmission (1 to 10% of the ions formed reach the detector), the system is expected to be fairly sensitive.

## B. Short Description of the LAMMA Instrument

Since extensive technical descriptions of the LAMMA instrument have been published elsewhere,[16,17] only a brief summary is given here. In the actual instrument (see Figure 1a) a frequency-quadrupled Nd: Yag laser ($\lambda$ = 264 nm, peak half-width $\simeq$ 15 nsec) is used. Energy available at this wavelength is about 100 $\mu$J. This power laser, together with a small

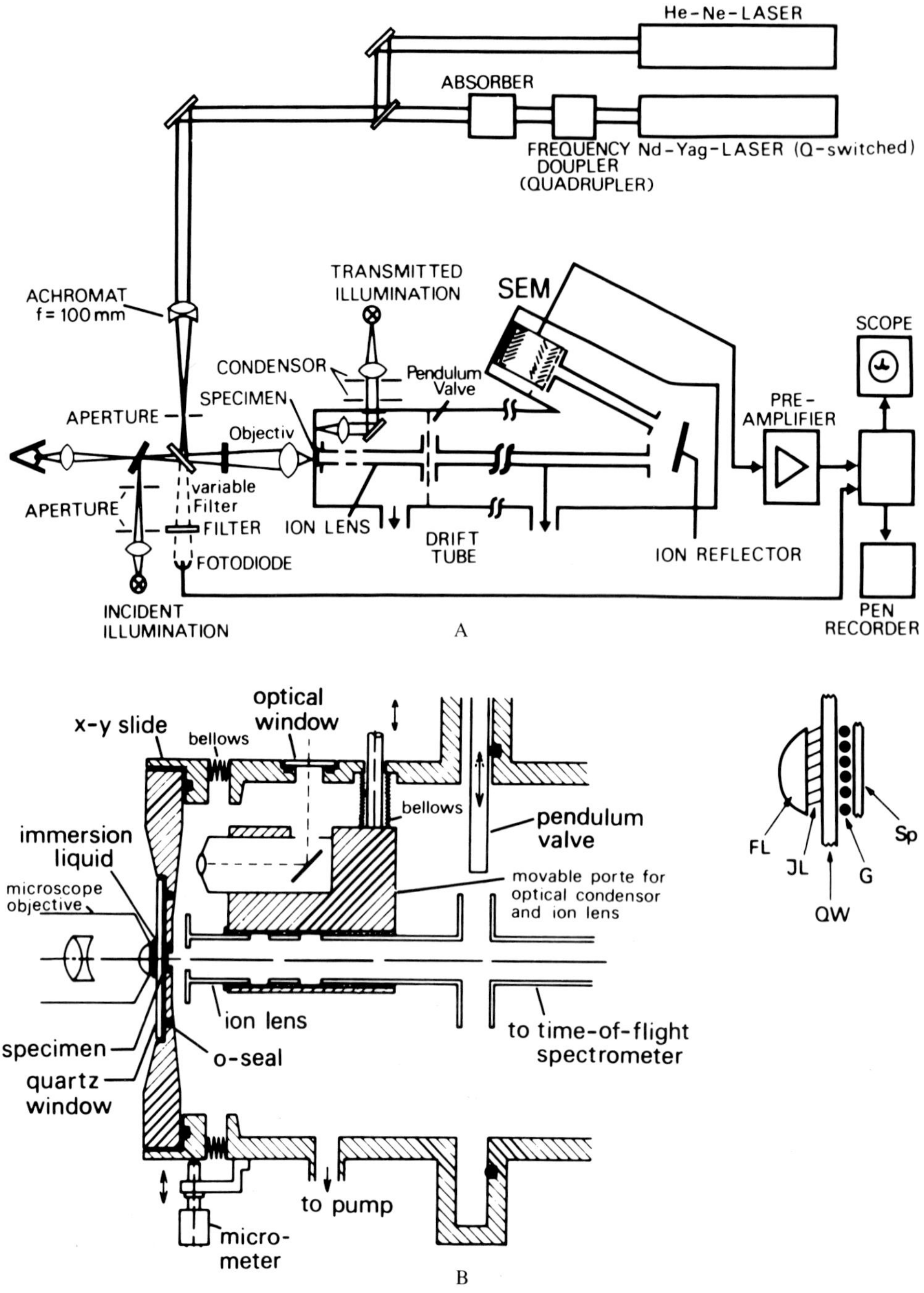


FIGURE 1. (A) Schematic representation of LAMMA 500 setup. (B) Arrangement of optical and ion optical components around the sample stage in the LAMMA 500 instrument. Inset shows the situation at the site of the sample: field lens of the microscopic objective (FL), immersion liquid (IL), quartz window (QW), supporting grid (G), and specimen (Sp).

He-Ne laser, is mounted horizontally on top of an optical platform. A system of adjustable mirrors allows a coalignment of the beams of the two lasers. After a 90° downward deflection, the laser light is fed into an incident light microscope. Unlike conventional microscopes, the optical axis of this microscope lies horizontally, aligned with the axis of the time-of-

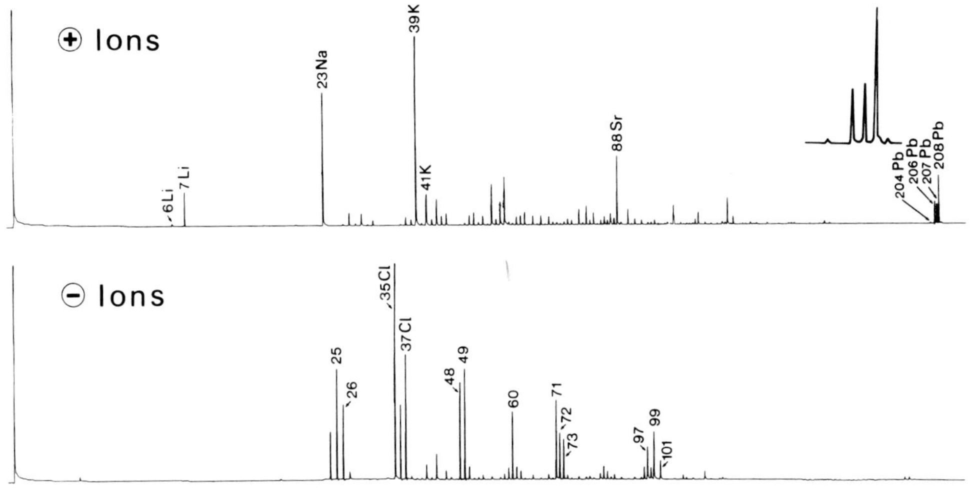


FIGURE 2. LAMMA spectra of positive and negative ions obtained from a thin (0.3 μm) section of Spurr's low-viscosity medium containing Li, Na, K, Sr, and Pb at $10^{-2}$ *M* concentration. Inset shows enlarged view of Pb isotope signals to demonstrate mass resolution attained.

flight spectrometer. Microscope objectives and other diffracting elements are made from UV-transparent materials (Zeiss Ultrafluar).

The specimen is located underneath a thin (0.15 mm) quartz cover slide which serves as a vacuum seal and optical window at the same time (see Figure 1b). This configuration permits the use of all light microscope objectives, including 100 × immersion (glycerol) objectives. Visualization of the specimen is possible by either incident or transmitted light, including UV microscopy at 260 nm. This not only improves the imaging resolution of the microscope down to about 0.15 μm, it also provides for better optical classification of inhomogeneous particle samples. During analysis, the specimen can only be seen by incident illumination because the optical condensor underneath the specimen has to be (pneumatically) exchanged for an electrostatic ion lens (see Figure 1b).

To select the area of interest the vacuum flange, on which the specimen is mounted, can be manually shifted in the x- and y-direction via two micrometer screw-drives, whereas focusing in the z-direction is achieved by a screw-drive acting on the microscope. The area of interest is indicated by focusing a red target (He-Ne-) laser into the microscopic field of view. The typical diameter of an analyzed area is about 0.5 μm with the 100 × objective and 1 μm with the 32 × objective.

The time-of-flight spectrometer consists of an accelerating and focusing electrostatic lens underneath the specimen, a field free drift tube of 1.20 m in length, an ion reflector at the end of this tube reflecting the ions to an angle of nearly 180°, and a second drift tube of about 60 cm in length. An open (Venetian blind) secondary electron multiplier is used as ion detector. The ion reflector, which had not been used in earlier versions of the LAMMA instrument, compensates for the spread of initial ion energy (0 to 50 eV). This spread originally limited mass resolution to about 400 to 500 m/Δm. In the current version, mass resolution has been improved to 800 to 1000 m/Δm without the requirement of energy filters which, of course, would decrease the overall transmission and hence, sensitivity of the instrument. Typical transient times in this spectrometer are 11 μsec for 7 Li and 60 μsec for 208 Pb (see Figure 2).

The output of the SEM can be collected either at the 12th or 17th dynode, depending on whether strong or weak signals are to be detected. The analog output signal is fed to a digital transient recorder of 2048 addresses and a dynamic range of 8 bits. Its fastest point-to-point resolution is 10 nsec. A sample interval of 50 nsec provides for about 5 to 6 data points per mass signal (mass number 50). The stored mass spectrum can either be further processed by a computer or can be recorded by the usual pen recorder (brush).

### C. Performances of the LAMMA Instrument

*1. Detection Limits For Element Analysis*

Detection limits for a number of elements are listed in Table 3. The values demonstrate that one of the features of the LAMMA instrument is the combination of both high-relative as well as absolute sensitivity. With an estimated ionization probability of $\simeq 10^{-2}$ to $10^{-1}$ for the laser pyrolysis mode and a spectrometer transmission of $\simeq 10^{-1}$, a few thousand atoms of an element present in the sampled volume will usually suffice for detection. There is, of course, an interdependence of absolute and relative sensitivity with respect to the evaporated volume. With larger volumes of evaporated material, relative detection limits improve. Therefore, the data given in Table 3, which have been determined under some sort of standard conditions (sample volume $\simeq 10^{-13}/cm^3$), may serve for orientation purposes rather than for delineating rigid or ultimate limits of the technique. Detection of trace elements in organic matrices may be a problem if their signals are obscured by those of fragment ions of nominally equal mass (for matrices of known composition background subtraction can at least partially remedy this limitation).

In contrast to most of the commonly used ionization methods in mass spectrometry, laser-induced ionization yields positive and negative ion signals of about equal intensities. In the LAMMA instrument, spectra of both ion species can be recorded for successive shots with changes in polarity of only a few voltages. The influence of the target matrix on the signal intensity of atomic ions appears to be small, at least when compared to other methods such as secondary ion mass spectrometry (SIMS).

*2. Quantitation*

In the present state of the LAMMA technique, and with the lack of a theory unifying all physical effects of ion formation, quantitative analysis with LAMMA is difficult particularly if one looks for **absolute quantitation.** As it is especially difficult or sometimes impossible to determine the sampled mass volume, the only way to absolute quantitation would be the use of internal standard reference elements. Under favorable analytical conditions (thin sections of epoxy-resin embedded material coated with known amounts of reference material[18]), such an approach can be realized. However, in particle analysis, particularly when analyzing geometrically and chemically nonhomogeneous particle populations, this approach is excluded.

For **relative quantitation** the situation is not so restricted provided that linearity between signal intensity and concentration can be assumed. That LAMMA fulfills this prerequisite has been shown, e.g., by Kaufmann et al.[19] This, in turn, opens the possibility of establishing relative sensitivity factors either empirically by means of standard reference materials of similar thermal, optical, and chemical properties as the unknown sample, or by applying, for instance, the LTE model, provided that realistic guesses can be made as to the relevant parameters such as partition functions, electron density, and plasma temperature.

***a. Relative Sensitivity Factors (Empirical Approach)***

The relative sensitivity factor (RSF) can be defined for an element or isotope x as

$$(RSF)_x = \frac{I_x \cdot C_R}{C_x \cdot I_R} \tag{4}$$

with I and C denoting the measured peak intensities and concentrations of the element x and of the reference element R, respectively.

**Table 3**
**DETECTION LIMITS (ppmw) FOR ELEMENT ANALYSIS (POSITIVE IONS)**

| | Glass matrices[a] | Epoxy resin[b] |
|---|---|---|
| Li | 3 | 0.4 |
| Be | 25 | |
| B | 9 | |
| Na | | 0.2 |
| Mg | 8 | 0.4 |
| Al | | 0.3 |
| Si | 10 | |
| K | 0.8 | 0.1 |
| Ca | | 1 |
| Ti | 1 | |
| V | | 0.5 |
| Co | | 10 |
| Cu | | 15 |
| Rb | | 0.5 |
| Sr | | 0.5 |
| Y | 4 | |
| Nb | 7 | |
| Ag | | 1 |
| Cd | | 20 |
| Jn | 7 | |
| Cs | 6 | 0.5 |
| Ba | 6 | 0.5 |
| La | 4 | |
| Ce | 5 | |
| Pr | 3 | |
| Ho | 14 | |
| Tm | 11 | |
| Lu | 6 | |
| Ta | 6 | |
| Pb | 10 | 0.4 |
| Th | 6 | |
| U | 6 | 2.0 |

[a] NBS standard reference glasses of the K 610-616 series. Limits defined by element signals exceeding three times S.D. of the background noise. No further conditions given.[11]

[b] Epoxy resin (Spurr's low-viscosity medium) reference standards containing trace amounts of organometallic complexes. Working conditions: $10^{-13}$ g evaporated material, irradiance set at five times the damaging threshold, signal intensities averaged over five analyses must exceed three times the background noise.[21]

Relative sensitivity factors applicable in LAMMA analysis of various materials have been published by Kaufmann et al.[20] for trace metals in epoxy-resin reference material, by Kaufmann et al.[21] for microspheres and fibers made from NBS reference glasses K 309 and K 491, and by Surkyn and Adams[11] for the NBS reference glass standards K 411, K 309, K

**Table 4**
**RELATIVE SENSITIVITY FACTORS (RSF) FOR TRACE ELEMENTS DISTRIBUTED IN EITHER NBS K 610 REFERENCE GLASS OR IN AN EPOXY RESIN (SPURR'S LOW-VISCOSITY MEDIUM)**

| Isotope | K 610[11] | Epoxy resin[11,21,45] | | |
|---|---|---|---|---|
| 7Li | 0.19 | 0.48 | 0.49 | 0.64 |
| 39K[a] | 1.19[a] | 3.22 | 4.14[a] | 2.70 |
| 41K | 4.16 | | 8.95 | |
| 86Sr | 1.00 | 2.00 | 1.73 | 1.99 |
| 88Sr | 0.82 | | 1.89 | |
| 208Pb | 1.00 | 1.00 | 1.00 | 1.00 |

[a] 39K data unreliable, most probably due to detector nonlinearity. See also text.

961, K 610, and K 612, and for the SRM 1633 fly ash standard. In Table 4 some RSFs measured by the above authors are summarized.

There is an obvious problem with the data published by Surkyn and Adams.[11] These authors found systematic differences between RSFs determined for isotope pairs of the same element (such as 40/44 Ca, 28/30 Si, or 54/56 Fe). They claim that this is due to a systematic inaccuracy in the measured isotopic abundance ratio, in the sense that with an increasing natural abundance ratio the measured peak intensity ratios increasingly underestimate the expected values. The authors suppose some nonlinearity in the detection system which, according to our experience, must indeed be taken into account particularly when dealing with ion currents driving the SEM beyond the linear part of its dynamic range. Careful checking of Surkyn and Adams' data revealed that they used the K 961 reference standard material with element concentration in the percent range, which, under normal working conditions of the instrument, must have driven the detector into nonlinearity. Interestingly enough, the data published by the same authors for K 610 and K 612 glasses containing most of the measured isotopes only as trace contaminants are (with the exception of 39/41 K) not affected by this problem.

***b. Relative Sensitivity Factors (LTE Approach)***

With the assumption of a local thermodynamic equilibrium (LTE) in the laser-induced microplasma the degree of ionization ($\alpha_x$) of an element x is given by

$$\alpha_x = \frac{n_x^+}{n_x^+ + n_x^0} \approx \frac{n_x^+}{n_x^0} \qquad (\text{if } n_x^+ << n_x^0) \tag{5}$$

(where $n_x^0$ and $n_x^+$ are the number densities of neutral atoms and positively charged ions, respectively), and can be calculated using the well-known Saha-Eggert equation:

$$\frac{n_x^+}{n_x^0} = \frac{2\,Q_x^+}{Q_x^0 \cdot n_e} \left(\frac{2 \cdot \pi \cdot m_e \cdot kT}{h^2}\right)^{3/2} \cdot \exp\left(\frac{W - \Delta W}{kT}\right) \tag{6}$$

where $n_e$ is the electron density, T is the plasma temperature, $Q_x^0$ and $Q_x^+$ are the partition functions of atoms and ions, respectively, $m_e$ is the electron mass, h is Planck's constant,

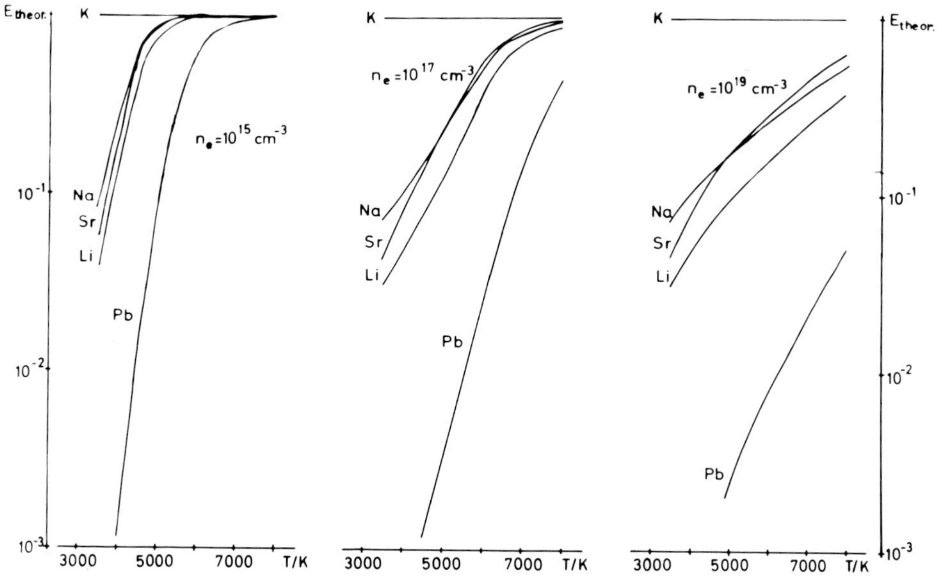


FIGURE 3. Relative sensitivity factors (RSF) normalized to potassium (K), as dependent on plasma temperature (T) and electron density ($n_e$) and predicted by the LTE model for epoxy resin standard material doped with organometallic complexes of the relevant metals.

W is the ionization energy, and ΔW is the coulomb depression. While ionization energy can be taken from the literature, partition functions (as depending on temperatures above 7000 K) are not easy to obtain. Plasma temperature and electron density under practical analytical conditions have been estimated by Surkyn and Adams[11] to about 13,000 K, whereas Haas et al.[12] obtained the best approximation of their data with T = 8000 K. Both groups refer to electron densities of $n_e = 10^{18}/cm^{-3}$.

The model can be used to predict relative sensitivity factors under various experimental conditions (see Figure 3), if, for instance, plasma temperature varies with variation of the laser power. This requires, however, that reasonable guesses can be made as to partition functions and electron density. For an explicit explanation of how to obtain numerical data applicable to this method the reader is referred to the literature (e.g., Andersen and Hinthorne,[10] De Galan et al.,[22] Drawin and Felenbok,[23] Herzberg,[24] Moelwyn-Hughes,[25] and Thorne[26]).

Such "theoretical" sensitivity factors can be compared with experimentally determined data. Figures 4 and 5 demonstrate that, at least in selected material, theoretical RSFs as predicted by the LTE model agree rather well with those determined experimentally. It must be emphasized, however, that the LTE approach is far from always being an adequate concept for the quantification of LAMMA analyses. As already mentioned, ionization processes occurring in the low irradiance domain are considered not to fulfill the conditions of a local thermal equilibrium. Furthermore, with reduction of the analyzed volume of material (at spatial resolutions coming close to diffraction-limited spot sizes) one might be doubtful whether or not there is still enough material and time available in order to attain a local thermodynamic equilibrium. Matrix effects, the chemical state of the elements under investigation, inhomogeneities of energy deposition due to the sample and/or beam geometry may, in addition, limit accuracy.

In conclusion then, quantitation in LAMMA analysis requires careful evaluation of the analytical conditions and, even in the best case, is limited by the "weakness of the methodology".[11] Nevertheless, we would dare to say that with further improvements regarding the stability of laser emission, precision of focusing, specimen preparation techniques, and

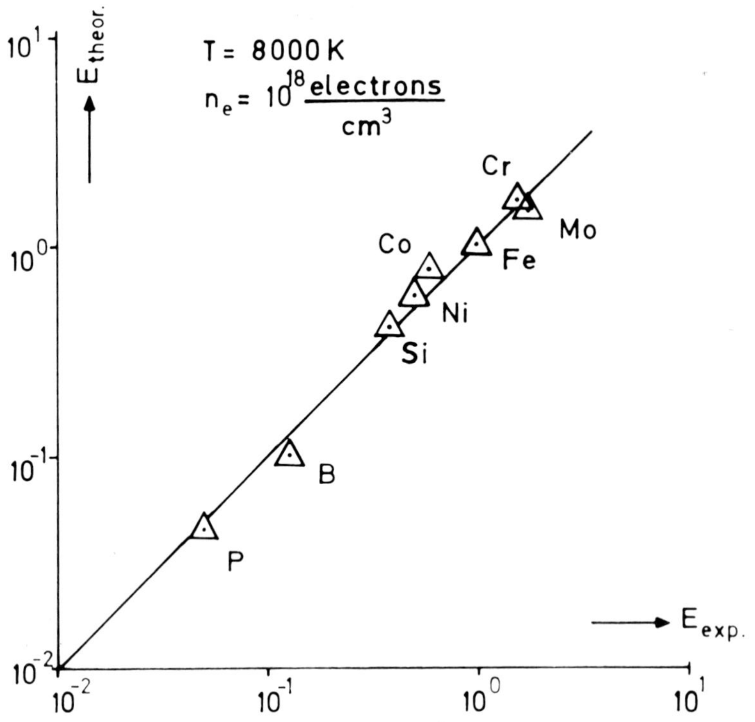


FIGURE 4. RSFs as predicted by the LTE model ($E_{theor.}$) vs. experimentally determined RSFs in a nickel-iron alloy containing boron, silicon, phosphorus, chromium, cobalt, and molybdenum.

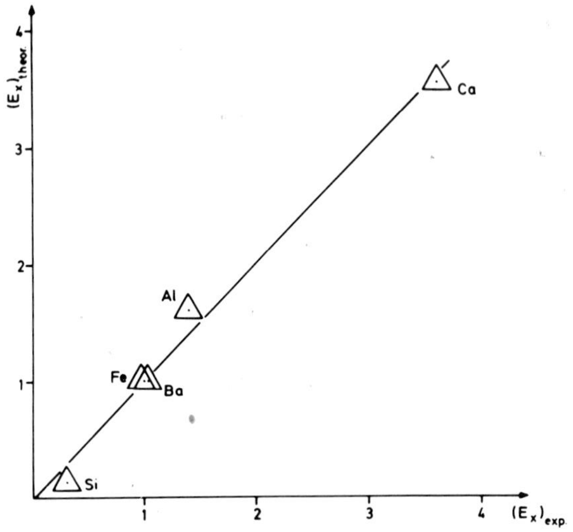


FIGURE 5. RSFs as predicted by the LTE model vs. experimentally determined RSFs in NBS K309 standard glass material, assuming T = 8000 K, $n_e = 2 \times 10^{17}/cm^3$.

signal processing one will end up with a quantitative accuracy comparable to that of the ion microprobe, say 10 to 20% SD for relative, and 20 to 50% SD for absolute quantitation.

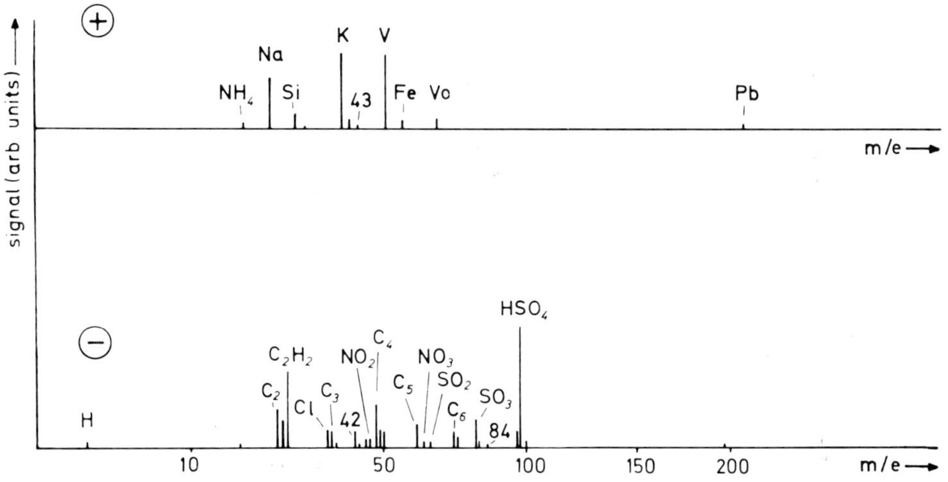


FIGURE 6. LAMMA spectra of positive and negative ions obtained from 0.5-μm atmospheric dust particles. Laser power three times threshold for ion production.

*3. Analysis of Molecules, Laser Desorption Mass Spectrometry*

With suitably prepared samples, two modes of operation, at least in principle, can be distinguished: laser pyrolysis and laser desorption (LD). For laser pyrolysis enough energy must be deposited within the sample volume in order to evaporize and ionize the sampled target volume (e.g., a particle as a whole). Molecular species originating under these conditions (mainly in the mass spectrum of negative ions such as $OH^-$, $C_2H^-$, $NO_2^-$, $NO_3^-$, $SO_2^-$, $SO_3^-$ and $HSO_4^-$ (see Figure 6) give valuable information on the chemical status of inorganic components but rather limited or no information as to potential organic constituents of the particle.

Strangely enough, this might be possible by employing low laser irradiances. There, ions begin to appear with no visible alteration or damage of the sample. In this mode, successive spectra can sometimes be recorded with great reproducibility from the same sample site (or particle). These so-called laser desorption (LD) spectra (see Figure 7) show striking similarities with spectra obtained by field desorption (FD),[27] fission fragment-induced desorption (FFID),[28] and to a lesser extent, secondary ion mass spectrometry (SIMS)[29] — the other most important methods that have been successfully applied in the mass spectrometry of nonvolatile, thermally labile (bio)organic compounds. The characteristics of these spectra are

- Substantial, often high, yield of parent molecule ions.
- Almost exclusively even electron ions (no radicals). In the positive spectrum the cathionized species $(M + Na)^+$ and $(M + K)^+$ more often than not dominate the protonated $(M + 1)^+$ ones. No cathionized fragment ions have so far been identified, however. In the negative spectrum the deprotonated $(M - 1)^-$ ion often appears as the base peak. Anionized ions are rare.
- The spectra show specific and reproducible fragmentation patterns that depend on the molecular structure in a complex way that is not yet fully understood. Negative spectra are often simpler, that is, they show fewer lines than the positive ones.
- The more strongly polar the molecules are, the easier it is to obtain desorption spectra.

Similar spectra have also been obtained by several groups that have used CW- and pulsed $CO_2$-lasers at 10.6 μm wavelength[30-32] or free run and Q-switched Nd lasers at 1.06 μm

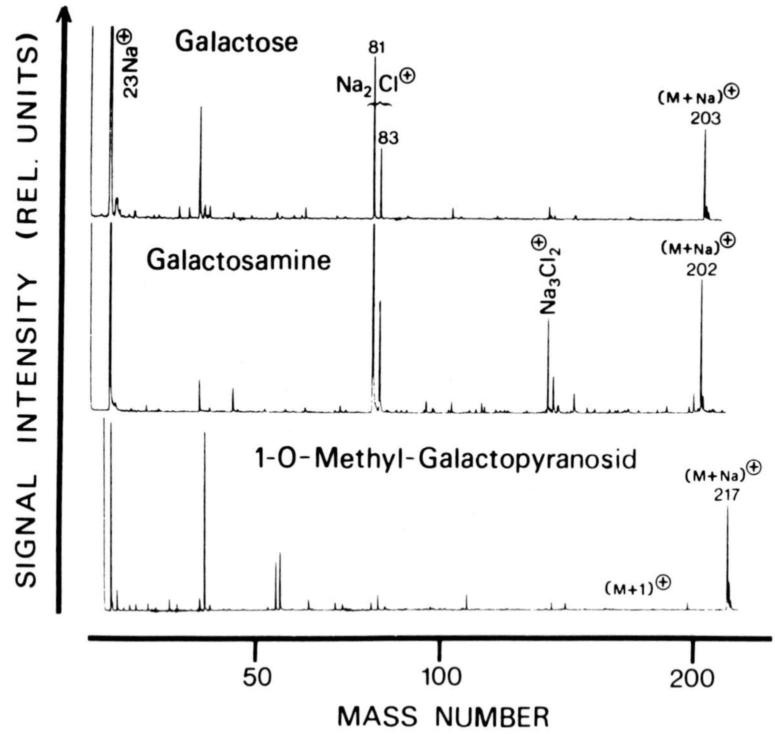


FIGURE 7. Typical laser desorption spectra obtained in thin layers of galactose and two of its derivatives. Note the strong signal of quasimolecular ions related to the (cationized) parent molecule.

wavelength,[33] all with laser-irradiated sample areas of 0.1 to 1 mm in diameter. Apparently all the laser and nonlaser desorption methods induce transition of molecules from the solid state into the gas phase that, at least locally, have a strong nonequilibrium aspect and, thereby, lead to the formation of large molecules or ions that would not be stable thermally at temperatures at which their vapor pressure is large enough for them to be generated in significant amounts by thermal evaporation in equilibrium. The details of these processes, as well as the common and/or different features for the various methods, are not yet understood, with the exception of FFID,[15] and will certainly be subjected to intensive investigation in the future.

Under certain practical conditions typically found or chosen for LAMMA of larger particles, there might be a "mixed" situation leading to mass spectra which show contributions from both types of ionization processes mentioned above. As shown in Figure 8, particulated deposits of the disaccharide sucrose (also containing small amounts of NaCl) were analyzed under LD conditions as indicated by the perforations seen in the SEM micrograph. The corresponding mass spectrum of positive ions contains signals of both the cationized parent molecule (M + Na = 365) and of the cationized monosaccharide glucose (m = 203), which are indicative for laser desorption mass spectra of this (nonvolatile) organic compound. At the same time, many fragments of lower mass number appear in the region of m = 1 to 80 which typically occur in the laser pyrolysis mode applied to organic constituents. It is assumed that due to the intensity distribution in the laser focus (see Section II.D.4), laser desorption occurs in outer regions of the analyzed area whereas pyrolysis conditions are found in the central part of that area only.

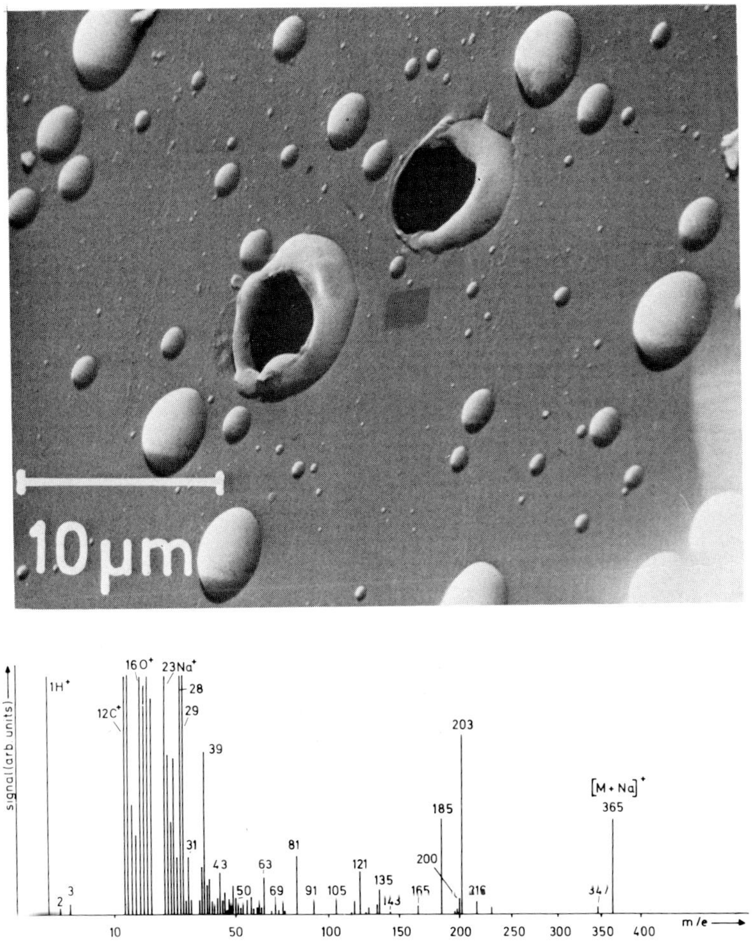


FIGURE 8. (Upper) Scanning electron micrograph of small sucrose particles deposited on a thin carbon film. Two perforations indicate sites of LAMMA at about four times the threshold for ion formation. (Lower) LAMMA spectrum of positive ions obtained in a sucrose particle as shown above. See text for further explanation.

## D. Samples and Sample Preparation

### *1. Preparation Techniques*

Since the LAMMA 500 instrument has been initially developed for thin sections of histological material, the transmission geometry of optical components and ion optics requires thin (transparent) film supports for particles to be analyzed in the actual version of the LAMMA instrument (see Figure 9). Such supporting films can be made in the same way as in electron microscopy, i.e., by coating metallic grids with, for instance, Formvar®, collodion, Mylar®, or carbon. In a recent version of the LAMMA technique a "reflection geometry" allows the analysis of bulk specimen or particles deposited on a nontransparent supporting surface.[34]

For particle deposition on either thin films or solid supports, a wide selection of techniques is described in the literature.[35-43] Size-dispersive devices such as the spiral duct centrifuge[44] or inertial impactors are the most favorable devices for preparing an aerosol specimen. Other kinds of particulate material may be deposited, for instance, by sedimentation or spraying techniques or simply by adhesion. Optimal distribution density of the particles is in the range of $10^6/cm^2$.

FIGURE 9. SEM micrograph of atmospheric dust particles as directly impacted (size dispersive) on the surface of a EM grid coated by a thin supporting Formvar® film.

Adhesive forces are not always sufficiently strong to avoid loss of particles during analysis, when the impact of laser energy on the sample creates a repelling wave which spreads over a certain area of the supporting foil. Although this phenomenon is usually restricted to only one mesh of the supporting grid, it sometimes may cause intolerable losses of particles or size selection. Sandwiching the particles between two thin films was found to remedy this problem at the expense of a slightly higher background in the recorded mass spectra.

*2. Particle Size*

There must obviously be an upper limit of particle diameter beyond which the particle can no longer be totally evaporated. With the 32 × objective this limit should be around 1 μm, if one relies on the laser beam geometry in the focus. However, if test particles of known size or volume were analyzed in the LAMMA instrument, the upper limit at which signed intensities are proportional to the volume of the particle was found to be only in the range of 0.5 μm (with the 32 × objective) by Surkyn and Adams[11] employing K 961 glass microspheres, whereas Kaufmann et al.[21] found a limit of ≃1.2 μm for the 32 × objective and ≃0.3 μm for the 100 × objective in a size-dispersive specimen of NaCl microcrystallites.

If particle size significantly exceeds this upper limit, selective or only partial evaporization and, eventually, redeposition of material must be taken into account (see Figure 10). Further disadvantages under these circumstances are saturation effects on the multiplier for signals related to leading constituents, and peak deformations due to larger time and energy spreads of the produced ions.

*3. Laser Focusing*

It is well known that the depth of focus in a microscopic objective of high numerical

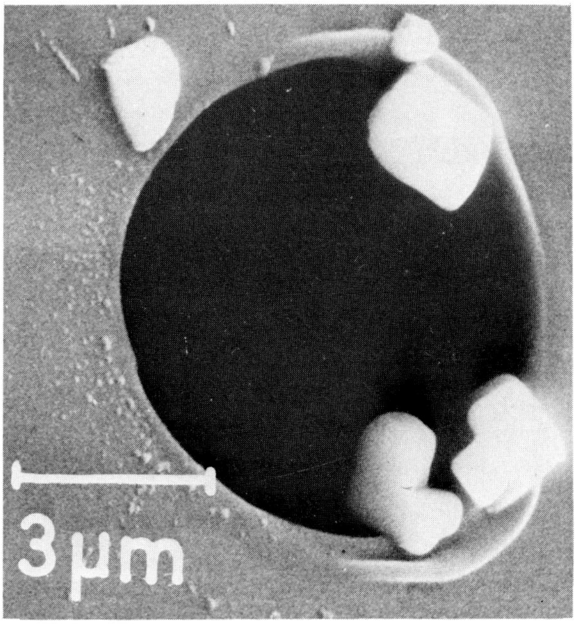


FIGURE 10. SEM micrograph of residues after laser impaction on a NaCl microcrystallite of about 10 µm in diameter deposited on a thin Formvar® foil.

aperture is rather small (≃2 µm and 0.5 µm in a 32 × and 100 × immersion objective, respectively). Consequently, field strength or irradiance in a laser beam focused through such an objective is strongly dependent on depth (z-axis of the microscope). Taking into account the nonlinear dependence of energy deposition processes occurring in the domain of laser pyrolysis (see Section II.A.1 and 2), it is easy to recognize that the exactness of focusing is an important prerequisite in obtaining reproducible LAMMA spectra in an actual analytical situation. This is particularly true in the analysis of particles the size of which come close to or exceed the upper limit for complete evaporation. Selective evaporation from either the lower or the upper side of the particle can have pronounced effects on the appearance of the respective LAMMA spectra. For further useful information on this subject the reader may refer to Wurster et al.[45] and Surkyn and Adams.[11] With particles well below the upper limit, size reproducibility can attain 10 to 20% S.D. if laser power and focusing conditions are carefully controlled (see also Figure 11).

*4. Intensity Distribution in the Laser Focus*

It has been found that the distribution of power density in the laser focus is another important determinant influencing the kind and reproducibility of mass spectra obtained in both the laser desorption and the pyrolysis mode. The ideal case, of course, would be a rectangular distribution of power density across the laser focus. With microscope objectives of high numerical aperture and an assumed Gaussian-like intensity profile in a $TEM_{oo}$-mode laser beam, such a case cannot be attained. Instead, the laws of optics predict some sort of intensity distribution in the form of an Airy-pattern, with concentric intensity maxima of exponentially decaying power density. The situation is more complex if the laser is not running in the $TEM_{oo}$ mode. Multimode emission, however, leads to rather undefined intensity distributions featuring ''hot spots'' within the focus which, in turn, render the ionization process less reproducible. Therefore, it is advisable to use array photodetectors to control the actual intensity distribution during practical LAMMA work in order to improve stability of the working conditions.

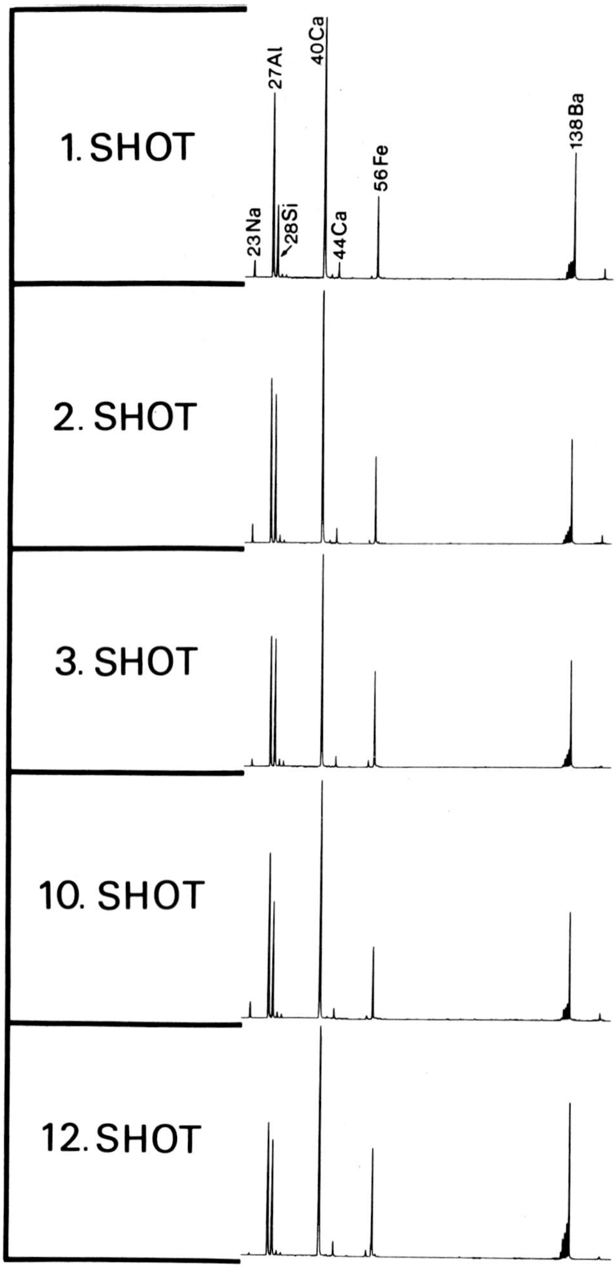


FIGURE 11. Reproducibility test in a large (diameter ~12 μm) NBS K 309 glass microsphere. With a number (N = 24) of shots on the same particle (λ = 265 nm, 32 × Ultrafluar objective, laser power five times above threshold for $Ca^+$ occurrence) the average signal intensities and standard deviations obtained were (relative units) 23 $Na^+$: 9 ± 9.2 (non-certified trace impurity), 27 $Al^+$: 54 ± 6.5, 28 $Si^+$: 36 ± 7,1, 40 $Ca^+$: 84 ± 4.7, 56 $Fe^+$: 30 ± 5.3, 138 $Ba^+$: 39 ± 10.2.[21]

### E. Comparative Evaluation of the LAMMA Technique as Applied to Particulate Materials

With respect to the special requirements for particle analysis, LAMMA offers a number of advantages if compared with alternative techniques (see also Kaufmann and Wieser[46]):

- As far as atomic composition is concerned all elements of the periodic table can be detected, many of them at sensitivities in the ppmw or even sub-ppmw range (see Table 3). This usually allows a high sensitivity for the detection of many trace contaminants in single particles.
- Since the LAMMA technique separates isotopes, isotopic ratio studies can be performed, at least within the accuracy limits given by the dynamic range of the transient recorder ($\simeq$ 6 bits). Thus, stable isotopes may be used in isotope dilution studies of kinetic processes in aerosol systems, or changes in natural isotopic abundances during the history of particle formation may be detected.
- For the chemical characterization of aerosol particles with the aim of establishing their origin, history, and ecological significance, element analysis is usually insufficient. In this regard, one needs as much information as possible about the molecular constituents (both organic and inorganic). Certainly an ideal case would be the possibility to detect, e.g., the presence of carcinogenic or toxic compounds. Since one of the unique features of LAMMA is that one obtains not only classical organic fingerprint spectra, but under favorable conditions and appropriate instrument setting also very characteristic and simple ''desorption'' spectra of major organic constituents, there is some hope that this could provide for a considerable step ahead in the organic microprobe analysis of particulate materials.
- If microprobe techniques are applied to single-particle analysis, usually the problem of distribution statistics is raised. If the measuring procedure, or the time required to perform one analysis, exclude the possibility of analyzing larger populations of individual particles there is always a loss of important statistical information. Therefore, the fact that with the LAMMA instrument the time needed for the analysis is limited only by data processing or writing the recorded mass spectra (at the time, usually 30 to 60 sec), a large number of analyses can be executed in a reasonable period of time.
- In many studies one wants to know particle morphology (size and shape), the distribution of particles over the support, or the exact position of individual particles chosen for analysis, either as independent information or in correlation with the analytical data. This requires a suitable imaging technique. In the LAMMA instrument we can obtain such structural information with the ultimate resolution a light microscope can give (about $^1/_2$ wavelength, which is $\simeq$0.13 μm at UV transillumination microscopy). In addition, all techniques of optical pattern analysis can easily be fitted to the optical part of the LAMMA instrument (see, e.g., McCrone and Delly[47] and Morton and McCarthey[48]). If the structural elements of the particles, or the whole particle itself, are below the spatial resolution of the optical microscope there is still the possibility to use size-dispersive devices for controlled particle deposition (e.g., Cadle[35] and Mercer[41]), or preferably, to make a scanning electron microscope picture of the particle(s) to be analyzed prior to LAMMA.

There are also some limitations and disadvantages of the LAMMA technique either based on physical or instrumental conditions:

- LAMMA is essentially destructive. This, for instance, excludes the possibility of recording the spectra of positive and negative ions from the same particle. Furthermore, optimizing the operational conditions of the instrument (e.g., adjusting the optimal power density) usually requires some test shots with the unavoidable loss of a corresponding number of particles.
- The fact that (up to $\sim$2 μm particle size) the whole particle is lost at once may sometimes turn into a disadvantage, particularly if one aims at surface analysis. With larger particles, however, it has been found that LAMMA allows for some modest kind of depth profiling (at least under favorable conditions).

- As in most other microprobe techniques, quantitative analysis poses many problems. However, approaches to semiquantitative analysis are possible on the basis that (1) matrix effects on ion yield are much less important in LAMMA than they are in SIMS, (2) intensity-concentration plots for elements are linear over several orders of magnitude at least in standard specimen, and (3) reference to standards is possible whenever appropriate reference materials are available.
- With every laser-induced evaporation of an individual particle, part of the supporting film is also evaporated. This material, of course, contributes to the recorded LAMMA spectrum and inevitably creates some kind of a "background". The ratio of these background signals to mass signals originating from the particle itself depends on the size of the particle, the thickness of the supporting foil, and the chemical nature of both materials. Since Formvar®, collodion, or carbon foils can be made as thin as some 100 Å there is usually no background problem encountered with particles in the range down to about 0.5 μm in size. If particles become smaller or, if trace constituents are to be analyzed, the foil-induced background can no longer be neglected. However, since the LAMMA spectra of the supporting foil are usually well-defined and highly reproducible, methods of background subtraction can be successfully applied.
- The need to bring the specimen into vacuum will create problems when the analysis of liquid or volatile particles is of interest. However, it should not be too difficult to design a suitable cold stage for LAMMA analysis of particles partially or fully in the liquid phase (see Figure 12).

## III. LAMMA IN PARTICLES

### A. Reference Particles

Many recent studies applying the LAMMA technique to particulate materials have, for obvious reasons, concentrated on reference particles of known and, hopefully, homogeneous chemical composition. Kaufmann and Wieser[46] used particles such as $(NH_4)_2SO_4$, $KMnO_4$, ammonium tartrate, kaolinite ($Al_2(OH)_4(Si_2O_5)$), talcum ($Mg_3(OH)_2(Si_4O_{10})$), as well as particulate glasses made from various NBS glass material.[21,46] Particulate and fibrous glass materials made from NBS reference glasses have also been investigated by Surkyn and Adams.[11] As demonstrated in Figures 13 and 14, the LAMMA spectra obtained from such reference particles are rather simple and can easily be related to the chemical composition of the particle under investigation. It is obvious that — on a qualitative basis — an immediate and unambiguous characterization of the chemical nature of the particle is possible based on the complementary information contained in the spectra of positive and negative ions. This is particularly true for the spectra of $(NH_4)_2SO_4$ and $NH_4C_4H_4O_5$ particles. Although in the former the appearance of the $^{96}SO_4$ ion is a little bit surprising according to the rules of classical ionization mechanisms, its assignment in laser-induced mass spectra is not questionable (as cross-checked by LAMMA analysis of $Na_2SO_4$ particles equally featuring a leading peak at mass number 96 which can truly be attributed to $SO_4$ because, in the latter case, the alternative $NH_2SO_3^-$ fragment, possibly involved in the $(NH_4)_2SO_4$ particles, is excluded).

The measurements further revealed that in particles of, presumably, both homogeneous intraindividual and interindividual chemical (inorganic) composition:

- A fairly satisfactory analytical reproducibility can be achieved (see Figure 15).
- Conditions can be found under which signal intensity ratios become independent on laser intensity such that relative sensitivity factors can be worked out (see Table 4) which apply at least for materials of similar composition.
- Absolute quantitation might be possible if the particle volume can be determined and does not exceed $^2/_3$ of the diameter of the laser focus applied (see Section II.E.).

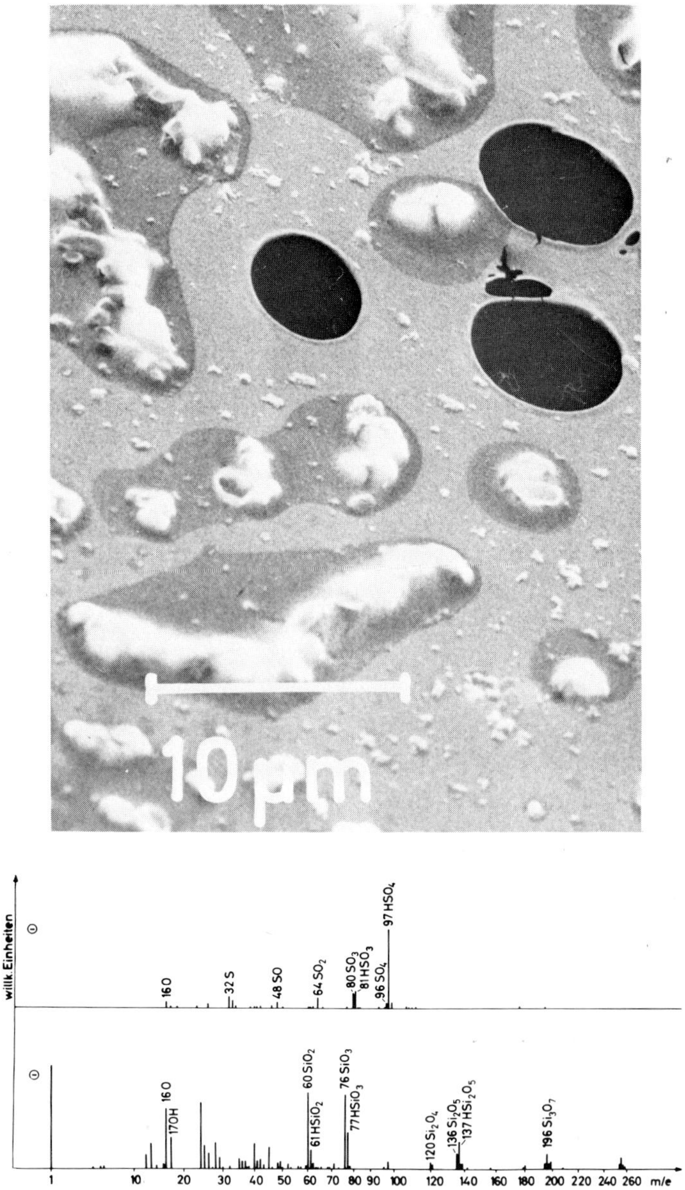


FIGURE 12. (Upper) SEM micrograph of atmospheric dust particles bearing a liquid coat as indicated by the darker area surrounding particles. Deposition was made on a thin carbon foil. Three holes indicate sites of LAMMA. (Lower) Typical LAMMA spectra of negative ions as obtained from liquid-coated aerosol particles (a) such as shown above, and of solid aerosol particles (b) with equal "aerodynamic size".

In order to determine detection limits and RSFs in a practical situation it would be most desirable to have reference particles at hand with easily adjustable qualitative and quantitative properties with respect to the (unknown) particles under investigation. To this end, an investigation was initiated to search for an appropriate matrix material which might be employed for this purpose. First attempts concentrated on highly porous silica particles used as filling material in HPLC columns. For instance SPHERISORB® particles (Phase Separation Ltd., U.K.) have a diameter of 3 µm and are fairly homogeneous with respect to size

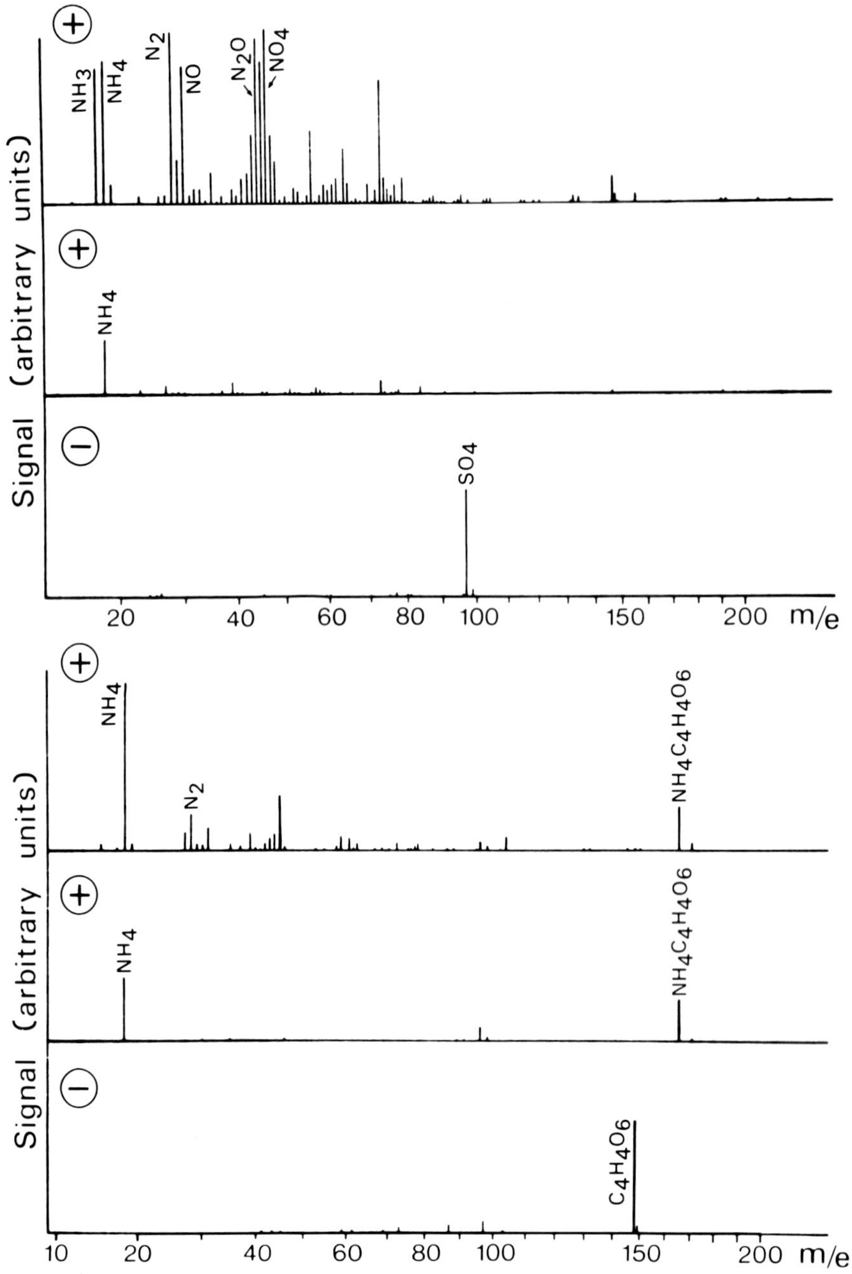


FIGURE 13. LAMMA spectra of positive and negative ions obtained from $(NH_4)_2SO_4$ (top three spectra) and $(NH_4)_2C_4H_4O_6$ (bottom three spectra) reference particles. The spectra 1 and 4 (top to bottom) were recorded at about 10 times the threshold for ion formation, whereas the 4 other spectra were taken at 3 times threshold laser intensity.[51]

and shape. A specific surface of this highly porous material amounts to 220 $m^2/g$. These microspheres can be easily loaded with various elements by soaking them in hydrous solutions of appropriate (in)organic salts. Final element concentration can be adjusted over a wide concentration range via the salt concentration employed in the incubation medium. A LAMMA spectrum of positive ions recorded from SPHERISORB® particles loaded, for instance, with mercury (Hg nitrate) and cadmium (Cd acetate) is shown in Figure 16. The amount of Hg

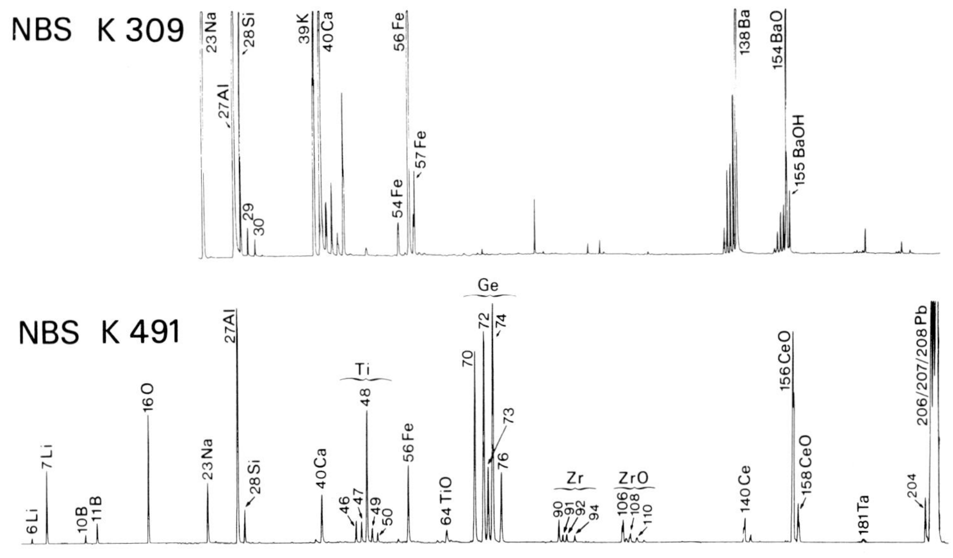


FIGURE 14. LAMMA spectra of positive ions recorded from glass particles (diameter ~1 μm) made of NBS K 309 (upper panel) and K 491 (lower panel) standard reference glass materials.[21]

in this case is about $10^{-16}$ g per particle, which corresponds to a concentration of 20 ppm. The amount of Cd is about $10^{-18}$ g per particle.

### B. Asbestos and Mineral Fibers

LAMMA spectra of asbestos fibers (Figure 17) were obtained by Spurny et al.[49] and by Adams et al.[50] They were well reproducible and characteristic for each type of asbestos. For instance, a clear-cut differentiation between serpentines and amphiboles or between amosite and crocidolite was possible. Thus LAMMA spectra can clearly discriminate different asbestos minerals on the basis of a single-fiber analysis.

In agreement with the results in the literature, a significant chemical instability of chrysotile fibers in environmental samples, in acids, and in living organisms (human lung tissue and animal tissues of lung and spleen) was found. Magnesium was relatively quickly leached out, but leaching of iron and silicon ions was also observed. In strong acids, the entire (MgOH)-sheet of chrysotile was stripped away, leaving a silica skeleton behind.

Results of analysis of about 300 chrysotile fibers (human lung tissue and UICC chrysotile after extraction in acids) examined by LAMMA indicated that the degree of corrosion (leaching of Mg, Fe, and Si) in single fibers is statistically distributed in a larger fiber population. While some chrysotile fibers showed only insignificant chemical changes after extraction with acids or after long residence time in human lung tissue, other fibers had exclusively lost their alkali elements to a rather large extent such that only Mg, Fe, and/or Si were left behind. Analytical measurements have shown that the leaching was faster in thin fibers than in fibers with a diameter $\geqslant 1.5$ μm.

In some favorable cases, where analytical profiling along an extended fiber was possible, LAMMA analysis revealed that the Mg leaching was rather nonhomogeneous along the fiber. Previous investigations with LAMMA had already demonstrated that the chemical composition of untreated chrysotile and amphibole fibers is not constant along the fiber axis.

Not only asbestos fibers (AF), but also manmade mineral fibers (MMMF), primarily glass fibers, can be changed chemically under different environmental conditions and after long residence in animal tissue. Experimental results with glass fibers have shown that these inorganic fibers undergo similar chemical and physical changes as do chrysotile fibers. They

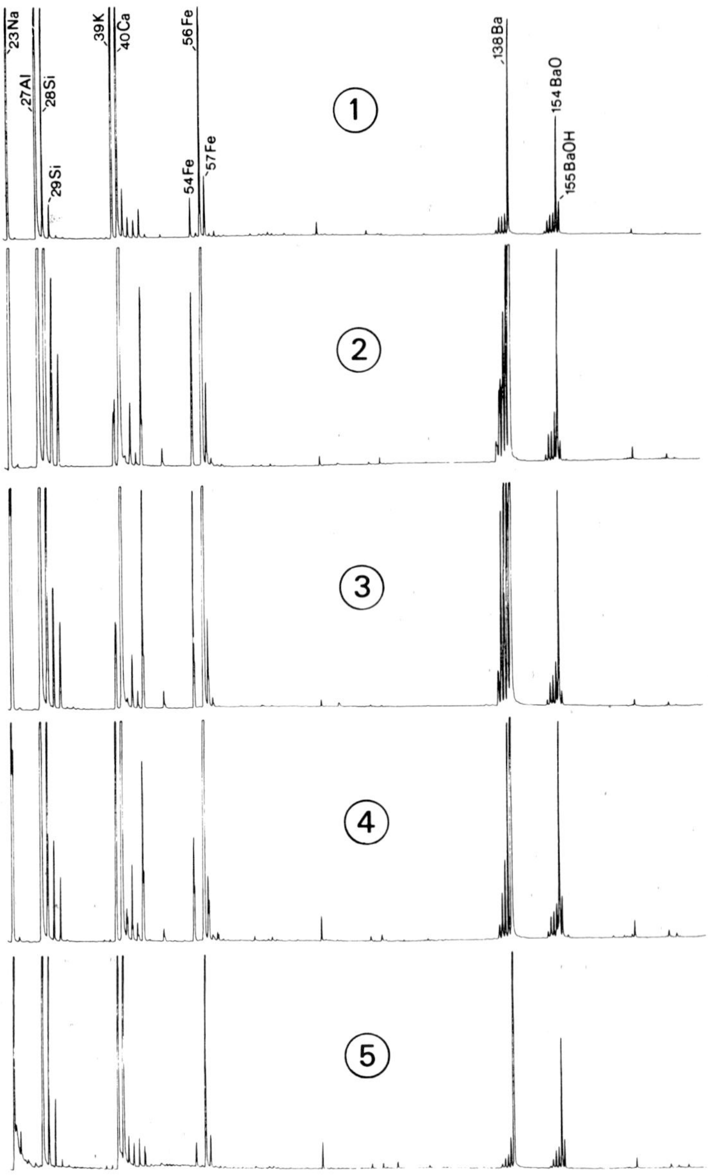


FIGURE 15. Reproducibility of LAMMA analysis in five different glass particles of NBS K 309 glass material.[21]

are not resistant to basic solutions, they lose some elements (e.g., Na, K, Ca, Zn) upon exposure to acids, and their surface becomes corroded. Changes of chemical composition have been found in glass fibers after 5-years deposition in animal tissue (instillation experiments done on rabbits). LAMMA (Figures 18a and b) revealed that a preferential leaching of elements also occurs in glass fibers; the original chemical composition, as well as the leaching of elements in glass fibers, are inhomogeneous. Analyses done along the fiber axis demonstrate that their chemical composition was different at different locations of the same fiber. Similarly, the chemical leaching of glass fibers was not constant for all fibers, but differed from one fiber to another. In any case, glass fibers seemed to be more resistant to acids and biological environments than are chrysotile fibers.

LAMMA was also successfully performed for microchemical characterization of various

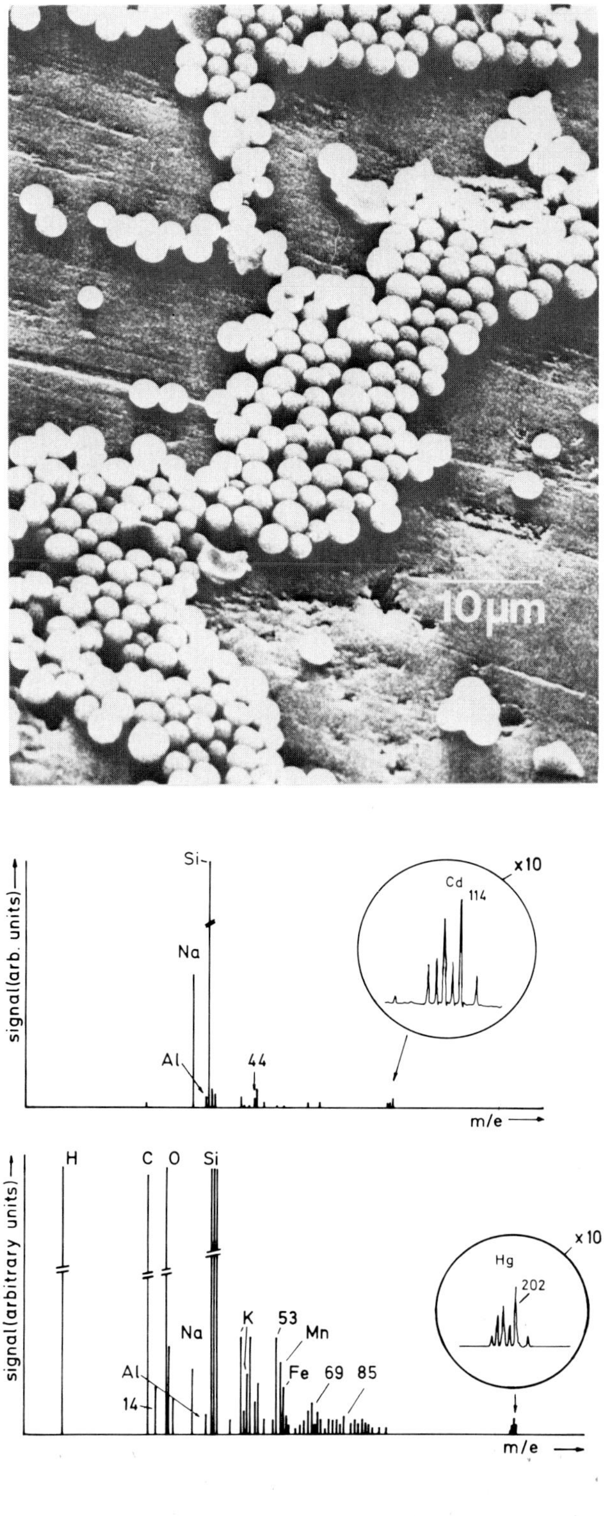


FIGURE 16. (Upper) SEM micrograph of highly porous silica particles (SPHERISORB®). (Lower) Typical LAMMA spectrum obtained from a SPHERISORB® particle loaded with trace amounts of Hg nitrate and Cd acetate.

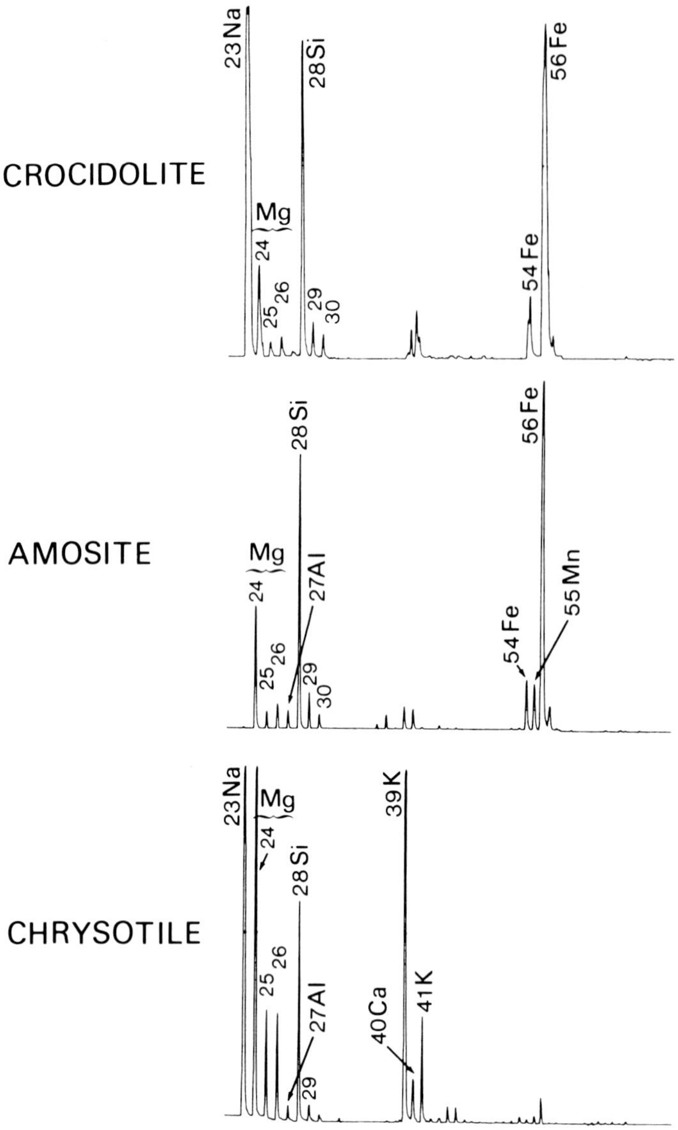


FIGURE 17. LAMMA spectra of positive ions obtained from UICC asbestos fibers.[49]

fibrous or nonfibrous zeolithes such as chabazite or erionite. As demonstrated by the LAMMA spectra of Figures 19 and 20, a large variety of (inhomogeneously distributed) trace contaminants were found, including rather exotic rare earth elements such as Zr, Gd, Tb, or Ce.

## C. LAMMA in Aerosols

Applications of the laser microprobe mass analyzer in aerosol research have been reported by Kaufmann and Wieser,[46] Kaufmann et al.,[21] Wieser et al.,[51] and Adams et al.[50] The main topics of these investigations were (1) the assessment of the chemical state of sulfur in various particle populations of different aerodynamic size, (2) the abundance of chemically "hot" particles in polluted particle populations, (3) tracing the sources of emissions in polluted aerosols, and (4) source tracing of unpolluted aerosols.

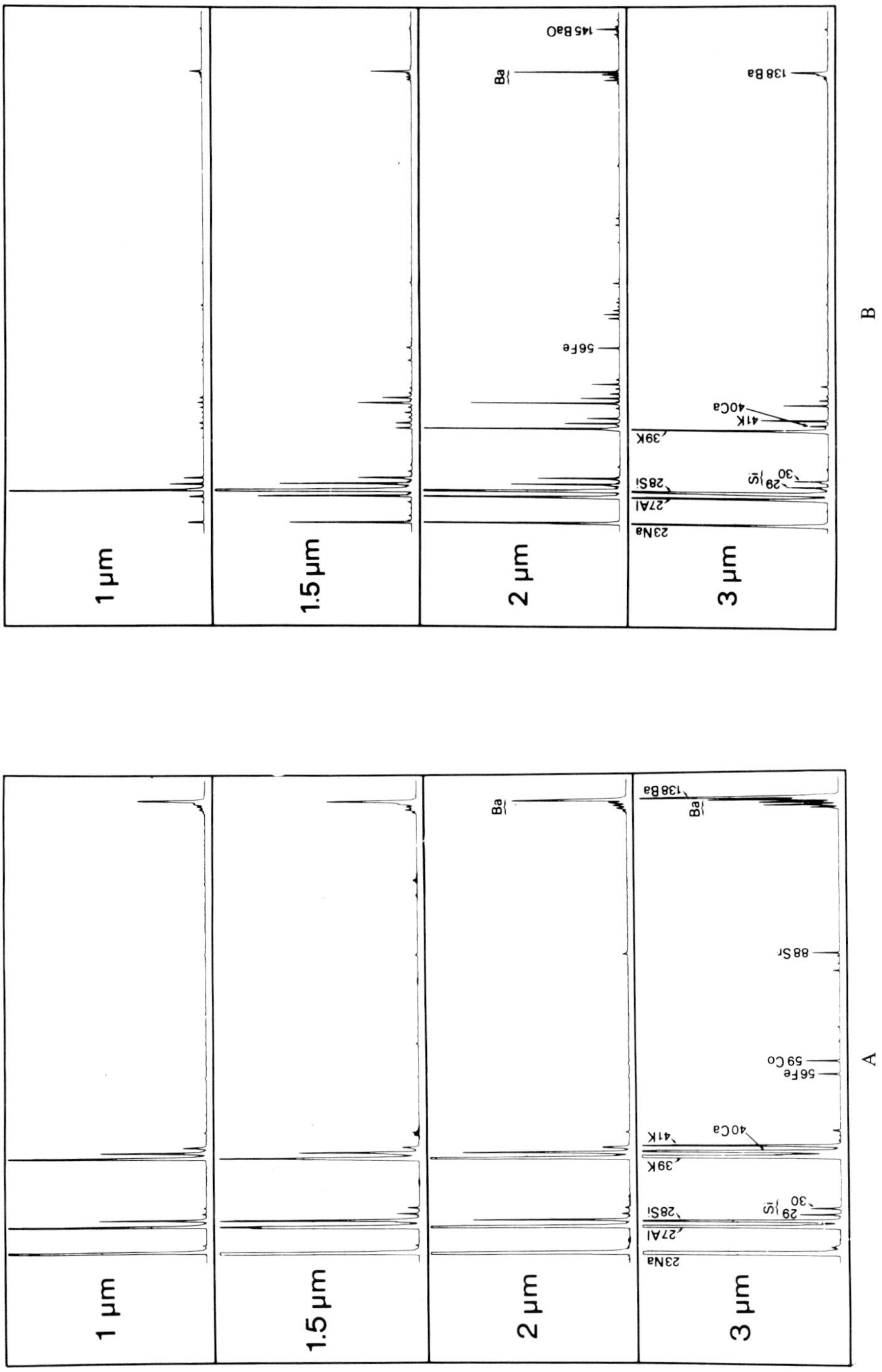


FIGURE 18. (A) LAMMA spectra obtained from manmade glass fibers (IM 100 glass) of various fiber diameter.[49] (B) LAMMA spectra obtained from the same type of glass fibers as in A after 1 week exposure to 1/100 $H_2SO_4$. Note the diameter-dependent leaching effect concerning all elements except Si.[49]

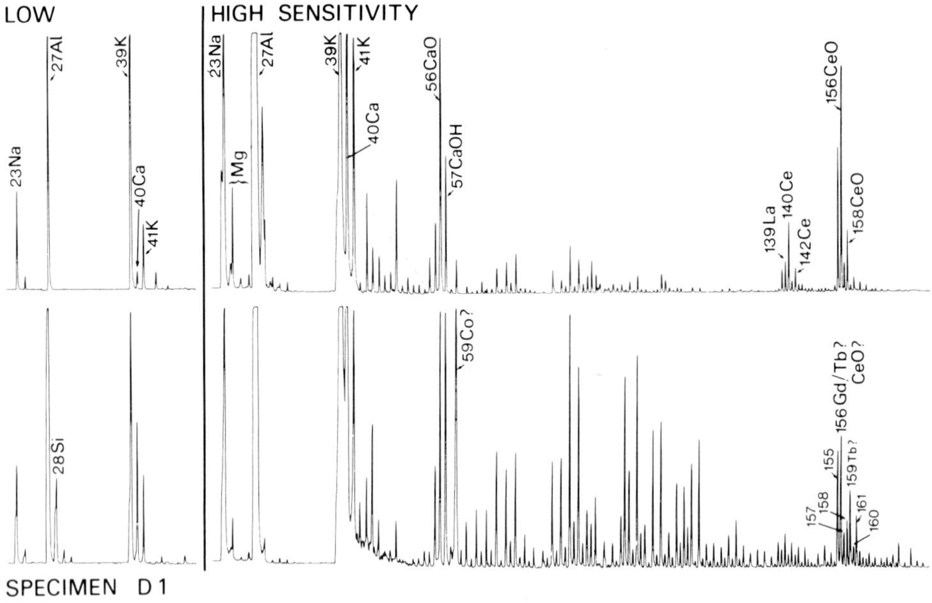


FIGURE 19. LAMMA spectra for positive ions obtained from a nonfibrous chabazite (~1 to 2 μm).[49]

With respect to sulfur chemistry, which is one of the most important subjects concerning the mechanisms of particle formation and alteration in the airborne state, it was found that in particles of the ≤0.2-μm-diameter size range sulfur appears in the LAMMA spectra of negative ions usually as $(HSO_4)^-$ and $(SO_4)^-$. These LAMMA studies confirm the dominant role of the gas to particle conversion ($H_2S$ and $SO_2$ form $(NH_4)_2SO_4$ and $NH_4HSO_4$ particles) in this size range.[52-55]

In contrast, larger particles (in the size range of about 1 to 3 μm diameter) preferentially contain sulfur in the various oxidation states. Sulfur appears in the LAMMA spectra of such negative ions as $(HSO_4)^-$, $(SO_4)^-$, $(HSO_3)^-$, $(SO_3)^-$, $(SO_2)^-$, and $(S)^-$ documenting a different origin and chemistry of sulfur in these particle populations, although the determination of the chemical state is very difficult.

Another important aspect in aerosol research is the chemical heterogeneity of individual particles ("internal mixture") or of particle populations ("external mixture"). In the latter case, each particle is a more or less pure compound, yet the population consists of different groups of such compounds.

A study of the transition metals and heavy metals appearing in the various particle groups of the atmospheric aerosol gave the following results: with the exception of the standard element combination of Cr, Mn, Fe, Ni, and Co all the others generally appear as an extremely inhomogeneous external mixture. As a consequence, we believe that knowledge of the average composition of the particulate matter of the atmospheric aerosol in polluted air is insufficient to predict and evaluate, for example, the biological and ecological effects caused by such materials. Examples demonstrating the heterogeneity of industrial aerosol samples are given in Figure 20. LAMMA studies of particles containing chlorine revealed that most of the chlorine is present in the form of halogenated hydrocarbons.

Wieser et al.[56] performed a LAMMA study of particulate pollutants from automobile exhaust. Automobile exhaust particles emitted from gasoline engines were collected by means of the spiral duct centrifuge and analyzed in the LAMMA instrument. The use of tetraethyl- and tetramethyl-lead as additives in gasoline is largely responsible for the presence of lead in the urban atmosphere.[57-59] The lead compounds in the exhaust gases of spark ignition engines have recently been identified in laboratory studies by X-ray powder diffraction.[60]

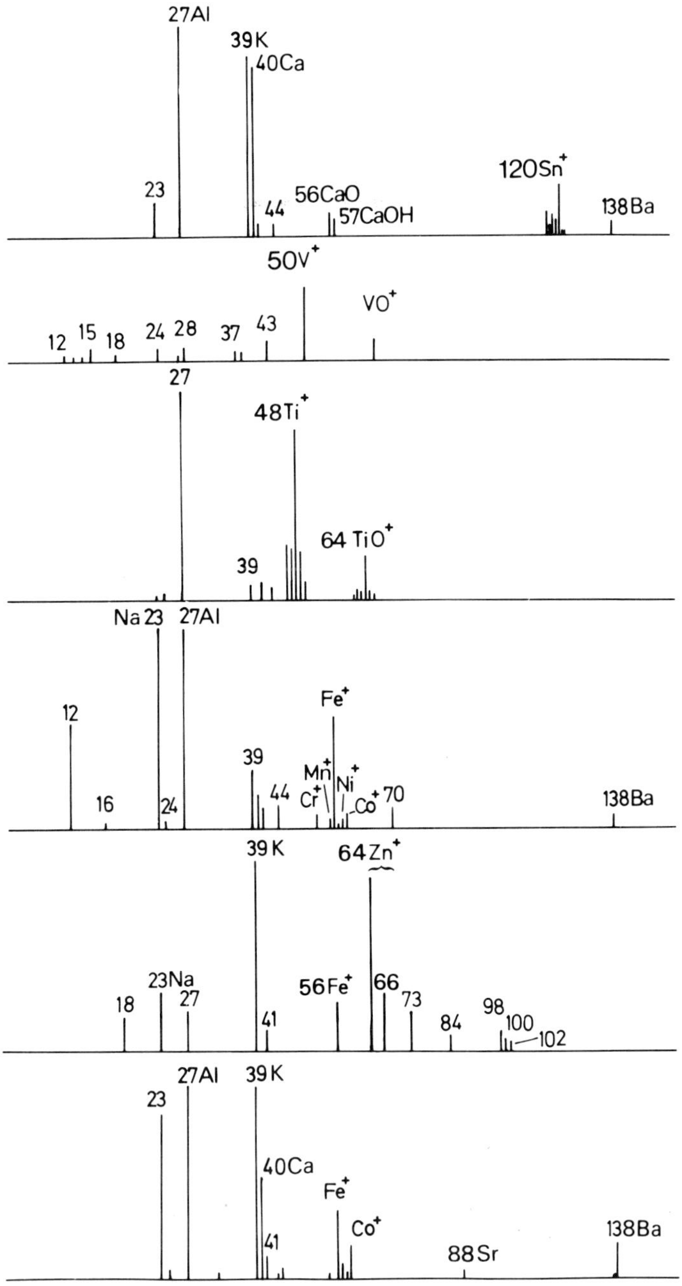


FIGURE 20. LAMMA spectra of positive ions obtained in a sample of urban dust collected near the ground. Particle size is about 1 μm. Note the variety of spectra each indicative for typical industrial sources.

It was shown that lead bromochloride (PbBrCl) is the principal lead compound in exhaust emissions of leaded gasoline containing the scavengers 1,2-dichloroethane and 2,3-dibromoethane. Information on the types of lead compounds in the atmosphere has also been collected through an analysis of individual particles with the electron microprobe.

The LAMMA results agree with the data given in the literature. Lead-containing particles occur predominantly in the smaller-size fraction. In the range of 0.1 to 0.2 μm diameter all

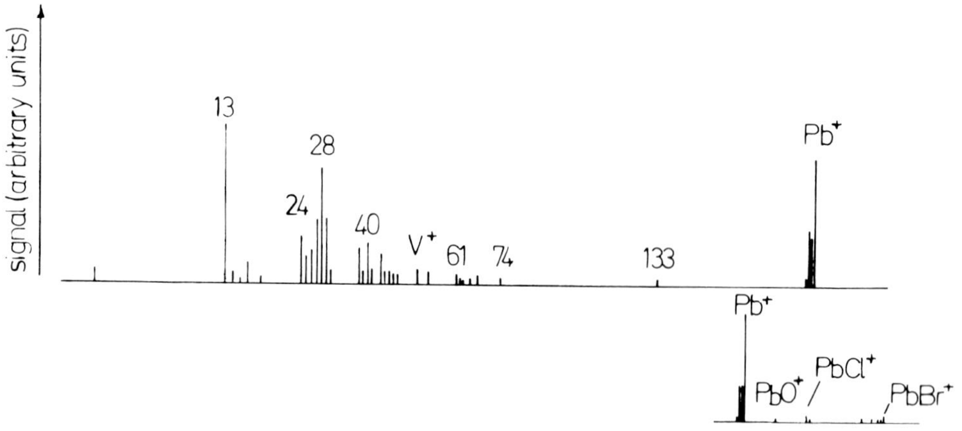


FIGURE 21. LAMMA spectrum of positive ions obtained from a small particle (0.2 μm) emitted from a combustion engine burning leaded gasoline.

collected particles contain lead (Figure 21), whereas, e.g., in the range of 1 to 2 μm diameter leaded particles had an abundance of only 5%.

LAMMA for source tracing of natural aerosols or industrial pollutants included the identification of particles originating from the Mount St. Helens volcano in a fallout over the southern part of Germany several weeks after the eruption of the volcano.[56] It traced industrial sources that emitted heavy metal-contaminated particles,[50] and assessed long-range pollution transport mechanism in either polluted and unpolluted aerosols.[50] One of the most advantageous features of the LAMMA technique emphasized by all authors, is the possibility to detect pollutant particles occurring at low abundances ($\leq$1%) in a population of aerosol particles. Contrary to other microprobe techniques, LAMMA allows the analysis of several thousands of particles per day such that statistically meaningful chemical histograms of a sample can be established.

For the possible implications of applying the LAMMA technique to current problems of aerosol research, reference is given to a paper by Adams et al.[50] dealing (among other problems) with LAMMA for source tracing of unpolluted aerosols. These authors have included the following description in the above paper:

> Within the last decade, attempts to assess pollution transport have shown a need for established reference levels of aerosol composition in remote areas of the world. In order to understand long-range transport processes and to identify sources responsible for anomalous enrichment of trace elements, baseline measurements must be made in both coastal and continental regions.
>
> When selecting a background monitoring station for marine aerosols, the question arises whether the site is affected by long-range transport of pollutants from anthropogenic sources. For continental sampling sites, the additional question arises as to whether there is a significant influx of sea-derived particulate matter. LAMMA is a quick and sensitive technique for answering such questions. The following two examples clearly illustrate this point, and also show the fingerprint spectra obtained also allow specific characterization of minor components in the particulate collection.
>
> 1. Samples were obtained during February 1981 at remote locations in Tasmania (Australia). The aerosol particles at the Cape Grim baseline monitoring station consist primarily of NaCl particles derived from sea spray (Figure 22a). For the impactor stage which collects the smallest particles there is a component with an abundance of considerably less than 1% of the particles whose mass spectrum is shown in Figure 22b. The particles contain a number of inorganic components which indicate a pollution-derived origin (e.g., Pb). The isolation of the site implies that the source of these is situated across the Bass Straight in the Melbourne metropolitan area at a distance of 500 km.
>
> Samples taken at a central location (Pencil Pine River) of the same island show the sea spray particles (Figure 22b), several hitherto unidentified components (probably of natural dispersion origin), and an extremely low particle concentration of needle-like particle (Figure 22c) whose mass spectrum corresponds closely to the amosite mass spectrum illustrated in the asbestos mass spectra of Figure 17.

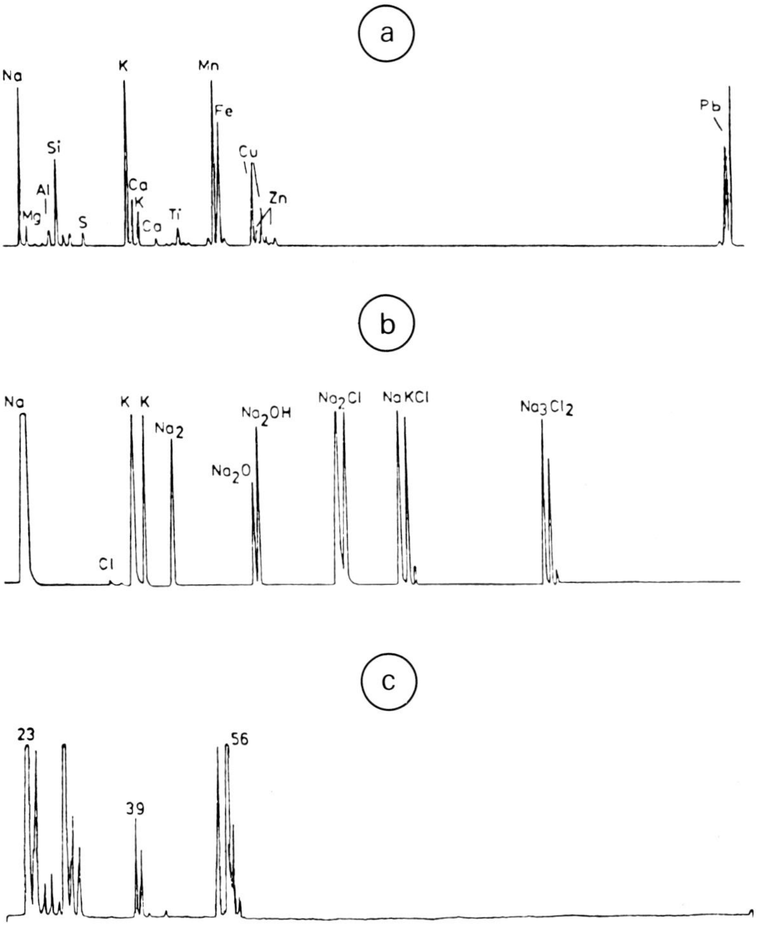


FIGURE 22. Positive LAMMA spectra of particles collected in Tasmania: (a) pollution-derived particle, (b) typically sea-salt particle, and (c) mineral (asbestos?) fiber.[50]

In both examples the virtue of LAMMA is clearly illustrated. The extremely rapid and straightforward analysis allows the ready detection of compounds which are present at minor and even trace concentrations.

2. A large number of aerosol samples were collected at Chapiquina, about 150 km east of Arica, Chile. The site is located on the slopes of the Western Cordillera at an altitude of 1800 m and is remote from any human activity. It is shielded on the western side from the Pacific and the agricultural town of Arica by the Atacama Desert, and on the east side by the Andes Mountains and the altiplano. Though the site is certainly remote enough for baseline air chemistry studies there is the question of whether there is any influence from marine atmospheric aerosols.

Microanalysis with LAMMA shows the presence in the aerosol of particles with a high concentration of Na. However, the mass spectra obtained are different from the ones shown in Figure 22b in that they do not contain elemental ions of the Cl isotopes in the negative spectrum and molecular fragments of Na and Cl in the positive or negative spectrum. The mass spectra closely correspond to that of $NaNO_3$ as appears from Figures 23 and 24. The Na-rich particles are apparently continentally produced and have a natural origin (wind-blown dust) from the $NaNO_3$-rich sediments within the area. The LAMMA investigation also provides a clue to the abnormal enrichment of sodium in the coarse particle aerosol observed by energy-dispersive X-ray fluorescence analysis and proton-induced X-ray emission in particulate matter collected in this area and on the altiplano at Chacaltaye near La Paz, Bolivia.

## D. LAMMA of Coal Mine Dust Particles

With the increasing demand for natural energy sources, the problem of coal miners' disease has attracted new attention. Since the pathogenesis of pneumoconiosis is still far

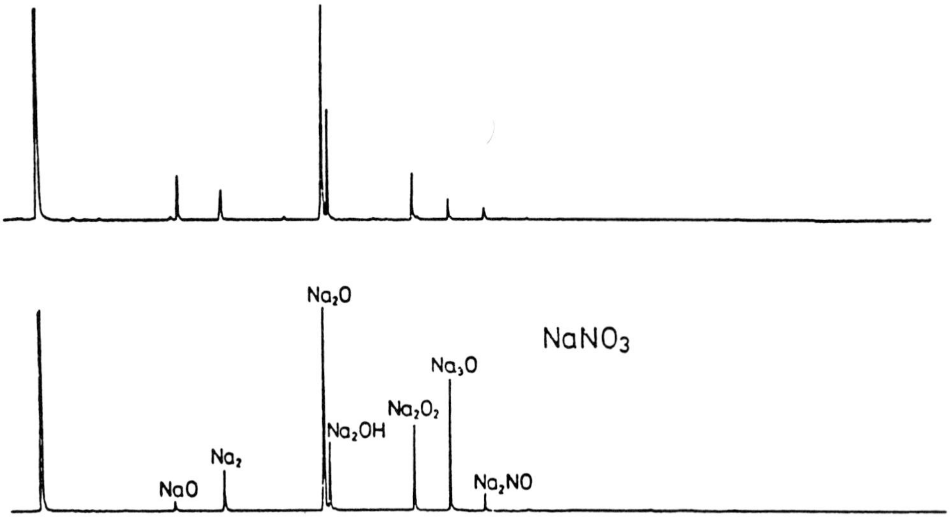


FIGURE 23. (Top) Typical positive LAMMA spectra of Na-containing particle sampled at Chapiquina, Chile, and (bottom) of an $NaNO_3$ standard particle.

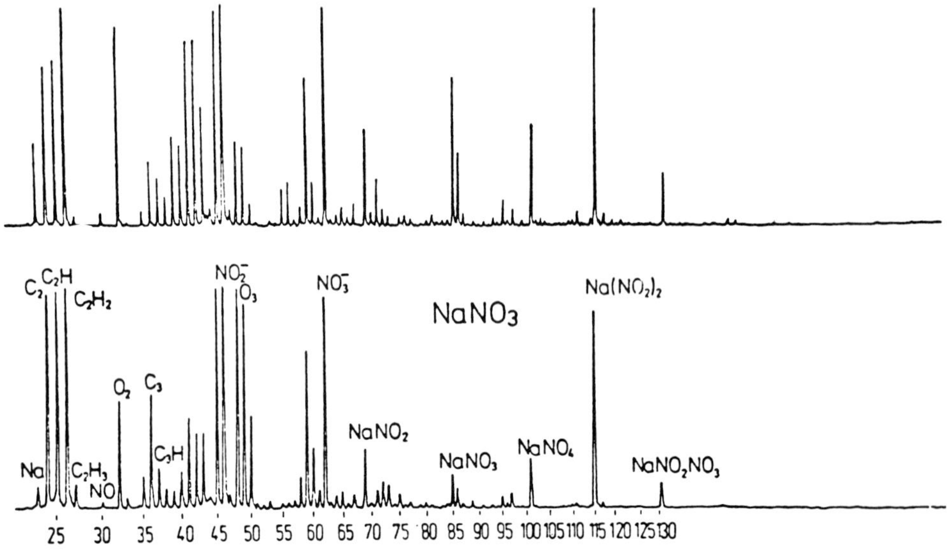


FIGURE 24. (Top) Typical negative LAMMA spectra of Na-containing particle sampled at Chapiquina, Chile, and (bottom) of an $NaNO_3$ standard particle.

from being fully understood, this problem requires new approaches for both particle characterization and possible mechanism of long range cyto- and/or histotoxicity.

In a current investigation, Kaufmann et al.[61] have used the LAMMA technique to characterize samples of coal mine dust particles (0.5 to 2 μm) collected at the sites of 12 European coal mines. In each sample, 500 particles were analyzed and characterized according to the occurrence of leading (and trace) elements in the form of "chemical histograms" such as shown in Figure 25. By means of cluster analysis, a group classification was established according to common leading elements, and sample histograms based on group abundances were formed. The data obtained by LAMMA were then correlated with the cyto- and histotoxicity data determined for each sample. As to be expected, a strong

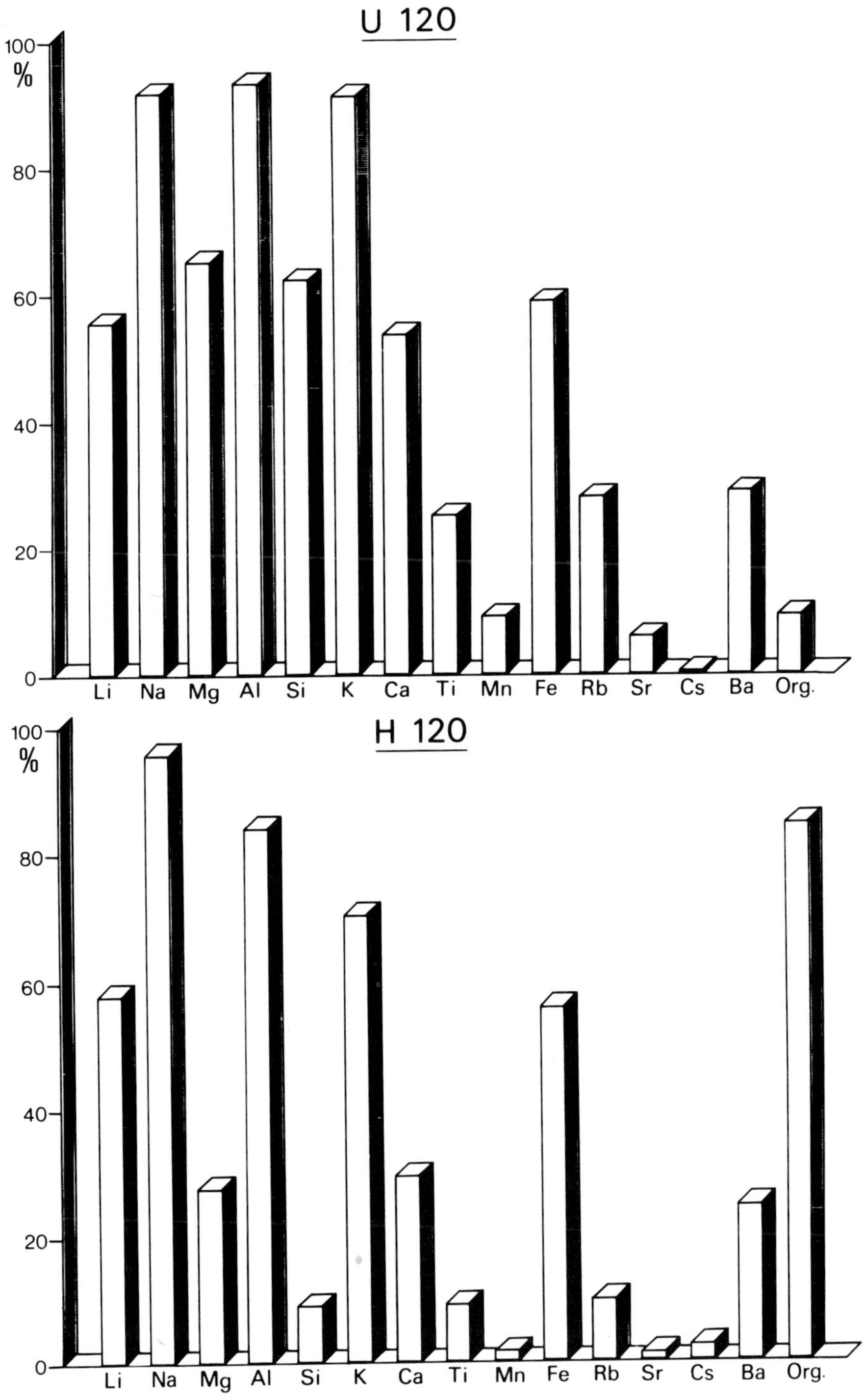


FIGURE 25. "Chemical histogram" obtained by elemental LAMMA in two samples of coal mine dust particles (0.5 to 2 μm) collected at the sites of two European coal mines. The height of the columns indicates the abundance (%) by which an element has been found within a particle population of 500 particles analyzed for each sample. The column labeled Org. indicates the abundance of signals indicative for organic constituents typically found in pit coal.

correlation was found between cytotoxicity (using pure quartz as a reference) and the abundance of particles containing Si as a leading element. However, in two of the samples an extremely low abundance of Si-containing particles was found in combination with a very high cytotoxicity. In these two samples, rather strong signals of beryllium were found, but in only 0.5 to 0.6% of the analyzed particle population. Bearing in mind the strong cytotoxicity of Be, one might tentatively explain the toxicity of these two samples to a low abundance of ''hot'', i.e., Be-containing, particles.

## IV. CONCLUSIONS

In the short time after its introduction into particle and aerosol research, the LAMMA technique has demonstrated its capability to provide many unique insights into the chemical nature of particulate materials unattainable by alternative techniques. While sensitivity allows for the detection of trace contaminants in single particles, analyzing time is fast enough to obtain statistically meaningful data in a comparatively short period of time.

Chemical histograms obtained in this way, for particle populations in the size range of about 0.1 to 2 μm, give important indications as to the nature, source, history, environmental, and biomedical significance of particulate materials. This is clearly a step toward a more reliable evaluation of particle-induced effects in, for example, biology, medicine, meteorology, and other fields related to human safety and health. The applicative scope of the technique can and certainly will be widened by further improvements in quantitation, optical classification, liquid phase analysis (cold stage), and organic mass spectrometry.

## ACKNOWLEDGMENT

The authors would like to thank Dr. Roland Wurster, who contributed all the scanning electron micrographs.

## ALPHABETICAL SYMBOLS

C = Concentrations of the elements x and of the reference element R
d = Flight path
D = Power density
FD = Field desorption
FFID = Fission fragment induced desorption
h = Planck's constant
I = Measured peak intensity
LD = Laser desorption
LTE = Local thermodynamic equilibrium
m = Mass
$m_e$ = Electron mass
$n_e$ = Electron density
$n_x^0$ = Number of densities of neutral atoms
$n_x^+$ = Number of densities of positively charged ions
$Q_x^0$ = Partition functions of atoms
$Q_x^+$ = Partition functions of ions
RSF = Relative sensitivity factor
SEM = Secondary electron multiplier
SIMS = Secondary ion mass spectrometry
t = Traveling time
T = Plasma temperature
TOF = Time of flight

eU = Energy
V = Velocity
W = Ionization energy
x = Isotope

## GREEK SYMBOLS

$\alpha_x$ = Degree of ionization
$\Delta W$ = Coulomb depression

## REFERENCES


1. **Harding-Barlow, I.,** in *Laser Applications in Medicine and Biology,* Vol. 2, Wolbarsht, M. L., Ed., Plenum Press, New York, 1974, 133.
2. **Kovalev, I. D., Maksomov, G. A., Suchkov, A. I., and Larin, N. V.,** Analytical capabilities of laser-probe mass spectrometry, *Int. J. Mass Spectrom. Ion Phys.,* 27, 101, 1978.
3. **Conzemius, R. J. and Capellen, J. M.,** A review of the applications to solids of the laser ion source mass spectrometry, *Int. J. Mass Spectrom. Ion Phys.,* 34, 197, 1980.
4. **Knox, B. E.,** Laser ion source analysis of solids, in *Trace Analysis By Mass Spectrometry,* Vol. 2, Ahearn, H. J., Ed., Academic Press, New York, 1972, 61.
5. **Moenke, H. and Moenke, L.,** *Laser Micro-Spectrochemical Analysis,* Adam Hilger, London, 1973.
6. **Etz, E. S. and Blaha, J. J.,** Scope and limitations of single particle analysis by Raman microprobe spectroscopy, in *Characterization of Particles,* NBS spec. publ., U.S. Government Printing Office, Washington, D.C., 1980.
7. **Hillenkamp, F. and Kaufmann, R.,** Laser microprobe mass analysis (LAMMA): a new approach in biomedical microanalysis and analytical mass spectrometry, in *Laser Applications in Medicine and Biology,* Vol. 4, Wolbarsht, M. L., Ed., Plenum Press, New York, 1982.
8. **Ready, J. F.,** *Effects of High-Power Laser Radiation,* Academic Press, New York, 1971.
9. **Krokhin, O. N.,** Generation of high-temperature vapors and plasmas by laser radiation, in *Laser Handbook,* Arecchi, F. T. and Schulz-DuBois, E. O., Eds., North-Holland, Amsterdam, 1972, 1371.
10. **Andersen, C. A. and Hinthorne, J. R.,** Thermodynamic approach to the quantitative interpretation of sputtered ion mass spectra, *Anal. Chem.,* 45, 1421, 1973.
11. **Surkyn, P. and Adams, F.,** Laser microprobe mass analysis of glass microparticles, *J. Trace Microprobe Tech.,* 1, 79, 1982.
12. **Haas, U., Wieser, P., and Wurster, R.,** A quantitative interpretation of LAMMA spectra based on a local thermodynamic equilibrium (LTE) model, *Fresenius Z. Anal. Chem.,* 308, 270, 1981.
13. **Fürstenau, N.,** Ion formation from organic solids, in *Springer Series in Chemical Physics,* Vol. 24, Bennighoven, A., Ed., Springer, Heidelberg, 1983.
14. **Fürstenau, N.,** Investigation of laser induced damage, evaporation and ionization with homogeneous inorganic target foils, *Fresenius Z. Anal. Chem.,* 308, 201, 1981.
15. **Krueger, F. R.,** Fast ion induced and collective electronic perturbation at the surface, *Surf. Sci.,* 86, 246, 1979.
16. **Kaufmann, R., Hillenkamp, F., and Wechsung, R.,** The laser microprobe mass analyzer (LAMMA): a new instrument for biomedical microprobe analysis, *Med. Progr. Technol.,* 6, 109, 1979.
17. **Hillenkamp, F., Unsöld, E., Kaufmann, R., and Nitsche, R.,** A high sensitivity laser microprobe mass analyzer, *Appl. Phys.,* 8, 341, 1975.
18. **Schröder, W. H.,** Quantitative LAMMA analysis of biological specimens, *Fresenius Z. Anal. Chem.,* 308, 212, 1981.
19. **Kaufmann, R., Hillenkamp, F., Wechsung, R., Heinen, H. J., and Schürmann, R.,** Laser microprobe mass analysis: achievements and aspects, *SEM II,* 279, 1979.
20. **Kaufmann, R., Heinen, H. J., Schürmann, M., and Wechsung, R.,** Recent advances of laser microprobe mass analysis as applied to biological and technical specimen, in *Microbeam Analysis,* Newbery, D. R., Ed., San Francisco Press, San Francisco, 1979, 63.
21. **Kaufmann, R., Wieser, P., and Wurster, R.,** Application of laser microprobe mass analyzer LAMMA in aerosol research, *SEM II,* 607, 1980.

22. **DeGalan, L., Smith, R., and Winnefordner, J. D.,** Electronic partition functions of atoms and ions between 1500K and 7000K, *Spectrochim. Acta,* 23B, 521, 1968.
23. **Drawin, H. W. and Felenbok, P.,** Data for plasmas, in *Local Thermodynamic Equilibrium,* Gauthier-Villars, Paris, 1965.
24. **Herzberg, G.,** *Molecular Spectra and Molecular Structure,* Vol. 1 to 3, D Van Nostrand, New York, 1966.
25. **Moelwyn-Hughes, E. A.,** *Physikalische Chemie,* Georg Thieme, Stuttgart, 1970.
26. **Thorne, A. P.,** *Spectrophysics,* Chapman & Hall, London, 1974.
27. **Krueger, R. F. and Schueler, B.,** Organic mass spectra obtained by fission-fragment- and pulsed laser-induced desorption, *Adv. Mass Spectrom.,* 718, 8, 1981.
28. **Schueler, B. and Krueger, F. R.,** Comparative study of pulsed laser and fission fragment induced desorption with nucleotic compounds, *Org. Mass Spectrom.,* 15, 295, 1980.
29. **Eicke, A., Sichtermann, W., and Benninghoven, A.,** Secondary ion mass spectrometry of nuclic acid components: pyrimidines, purines, nucleosides and nucleotides, *Org. Mass Spectrom.,* 15, 289, 1980.
30. **Posthumus, M. A., Kistemaker, P. G., Meuzelaar, H. L. C., and Ten Noever de Brauer, M. C.,** Laser desorption mass spectrometry of polar nonvolatile bio-organic molecules, *Anal. Chem.,* 50, 985, 1978.
31. **Kistemaker, P. G., Lens, M. M. J., Van der Peyl, G. J. Q., and Boerboom, H. A. J.,** Laser-induced desorption mass spectrometry, *Adv. Mass Spectrom.,* 928, 1980.
32. **Stoll, R. and Röllgen, F. W.,** Laser desorption mass spectrometry of thermally labile compounds using a continuous wave $CO_2$ laser, *Org. Mass Spectrom.,* 14, 642, 1979.
33. **Heresch, F., Schmidt, E. R., and Huber, J. F. K.,** Repetitive laser desorption mass spectrometry for nonvolatile organic compounds, *Anal. Chem.,* 52, 1803, 1980.
34. **Schueler, B., Nitsche, R., and Hillenkamp, R.,** Possibilities for a laser-induced micro-mass analysis of bulk surfaces, *SEM II,* 597, 1980.
35. **Cadle, R. D.,** *The Measurement of Airborne Particles,* John Wiley & Sons, New York, 1975.
36. **Davies, C. N.,** *Aerosol Science,* Academic Press, New York, 1966.
37. **Dennis, R.,** *Handbook on Aerosols,* Tech. Inf. Cent., U.S. Department of Energy, Washington, D.C., 1976.
38. **Fuchs, N. A.,** *The Mechanics of Aerosols,* Pergamon Press, Oxford, 1964.
39. **Zebel, G. and Hochrainer, D.,** Zur Messung der Größenverteilung des Feinstaubes mit einem verbesserten Spektralimpaktor, *Staub, Reinhaltung, Luft,* 3, 91, 1972.
40. **Liu, B. Y. H.,** *Fine Particles,* Academic Press, New York, 1976.
41. **Mercer, T. T.,** *Aerosol Technology in Hazard Evaluation,* Academic Press, New York, 1973.
42. **Mercer, T. T., Morrow, P. E., and Stöber, W.,** *Assessment of Airborne Particles,* Charles C Thomas, Springfield, Ill., 1972.
43. **Strauss, W.,** *Air Pollution Control,* Vol. 3, John Wiley & Sons, New York, 1978.
44. **Stöber, W. and Flachsbart, H.,** Size separation precipitation of aerosols in a spinning spiral duct, *Environ. Sci. Technol.,* 3, 1280, 1969.
45. **Wurster, R., Haas, U., and Wieser, P.,** Electron microscopical investigations of laser irradiated foils and particles, *Fresenius Z. Anal. Chem.,* 308, 206, 1981.
46. **Kaufmann, R. and Wieser, P.,** Laser microprobe mass analysis (LAMMA) in particle analysis, in Characterization of Particles, Heinrich, K. F. J., Ed., NBS Spec. Publ. 533, U.S. Government Printing Office, Washington, D.C., 1980.
47. **McCrone, W. and Delly, J. G.,** *The Particle Atlas,* Ann Arbor Scientific Publications, Ann Arbor, Mich., 1975.
48. **Morton, R. R. A. and McCarthey, C.,** The omnicon pattern analysis system, *Microscope,* 23, 239, 1975.
49. **Spurny, K. R., Schörmann, J., and Kaufmann, R.,** Identification and microanalysis of mineral fibers by LAMMA, *Fresenius Z. Anal. Chem.,* 308, 274, 1981.
50. **Adams, F., Block, P., Natusch, D. F. S., and Surkyn, P.,** Microscopical analysis for source identification in air chemistry and air pollution, Proc. Int. Conf. Environ. Pollution, Thessalonica, Greece, September 20 to 24, 1981.
51. **Wieser, P., Wurster, R., and Seiler, H.,** Identification of airborne particles by laser induced mass spectroscopy, *Atmos. Environ.,* 14, 485, 1980.
52. **Charlson, R. J., Covert, D. S., Larson, T. V., and Waggoner, A. P.,** Chemical properties of tropospheric sulfur aerosols, *Atmos. Environ.,* 12, 39, 1978.
53. **Novakov, T., Chang, S. G., and Harker, A. B.,** Sulfates as pollutions particulates: catalytic formation on carbon (soot) particles, *Science,* 186, 259, 1974.
54. **Cox, R. A.,** Photochemical oxidation of atmospheric sulfur dioxide, in *Pathways of Pollutants in the Atmosphere,* Sudgen, T. M., Ed., Royal Society, London, 1979.
55. **Rasool, S. I.,** *Chemistry of the Lower Atmosphere,* Plenum Press, New York, 1973.

56. **Wieser, P., Wurster, R., and Haas, U.**, Application of LAMMA in aerosol research, *Fresenius Z. Anal. Chem.*, 308, 260, 1981.
57. **Perera, F. P. and Ahmed, A. K.**, *Respirable Particles*, Ballinger Publishing, Cambridge, Mass., 1979.
58. **Wesolowski, J. J., John, W., and Kaifer, R.**, Lead source identification by multi-element analysis of diurnal samples of ambient air, in *Trace Elements in the Environment*, Kothny, E. L., Ed., American Chemical Society, Washington, D.C., 1973.
59. **Committee on Biological Effects of Atmospheric Pollutants, Airborne Lead in Perspective, National Academy of Science-National Research Council, Washington, D.C., 1972.**
60. **Biggins, P. B. E. and Harrison, R. M.**, Atmospheric chemistry of automotive lead, *Environ. Sci. Technol.*, 13, 558, 1979.
61. **Kaufmann, R., Barthes, G., Schmitz, G., Spurny, K., and Bruch, J.**, LAMMA-studies in mineral fibres and coal mine dust particles, Proc. 12th Annu. Symp. Anal. Chem. of Pollutants, Amsterdam, April 14 to 15, 1982.

Chapter 3

# THE ANALYSIS OF PARTICLES BY ELECTRON MICROSCOPY AND SPECTROSCOPY

**David L. Davidson**

## TABLE OF CONTENTS

## I. INTRODUCTION

The purpose of this chapter is to discuss the analysis of particles as captured from fluid streams by analytical methods which use electrons. The electrons may be used either as the illumination source or as the information carrying secondarily excited, and subsequently detected, radiation. The chapter is meant to be a survey of the analytical techniques rather than a review of the results. Each of the analytical techniques is, in itself, complex and requires extensive description and explanation of numerous factors related to both the instrumental technique itself and specimen preparation required of particulate matter for that technique. It is not the purpose of this chapter to go into these details.

All of the techniques discussed here are meant to be related to particles captured from fluids. They might well also apply to particles found imbedded in solids; however, this chapter concentrates on unimbedded particles. Some examples of the types of particles to which these techniques may be applied are (1) airborne respirable particles from natural or industrial processes, (2) ingestible particles found in drinking water supplies or particles found in body fluids, and (3) particles of metals used in preparing powder metallurgy parts. The latter particles are formed from the liquid metal and rapidly quenched in gas streams, from whence they are recovered.

There are a number of steps involved in obtaining information from any particulate matter which in general, include the following: (1) collection, (2) sample selection, (3) sample preparation, (4) electron microscopy or spectroscopy, and (5) interpretation of results.

It must constantly be kept in mind that all of these steps affect the final interpretation of the microscopy or spectroscopy data collected. The technique used, together with the care used in applying that technique in each of the steps leading to (4), the actual data collection, is extremely important and it is usually necessary to plan the experiments with a verification procedure for each of steps 1 through 3. The purpose of this verification procedure is to ensure that the data collected are not artifacted by the specifics of the techniques used in sample collection, selection, and preparation. For example, if the surface chemical composition of airborne particles is of interest, there should be verification that the collection technique is not altering the chemical nature of the surfaces of the particles being collected. What might be used as a verification procedure in this case would be to collect the particles by two different techniques, if that is possible, and examine the effect of those collection techniques on the surface chemistry of the particles. Another verification procedure might be to collect particles of both unknown and known surface chemistry by the same technique and determine the effect of that technique on subsequent analysis.

Interpretation of results published in the literature is often made very difficult due to inadequate information on each of the steps required to get to the interpretation of the results; this is particularly true for the collection, sample selection, and preparation steps. Thus, the importance of adopting sampling techniques which may be verified cannot be overemphasized.

It must be noted, although it is obvious, that all the techniques utilizing electrons require vacuum in the specimen chamber to prevent scattering of the electrons by gas molecules. Many particulates contain volatile liquids and dissolved gases which are an important part of their chemical characteristics. Insertion into a vacuum system will usually drive these out of the specimen, or at least markedly alter their concentration in the particles. Thus, the effect of vacuum on the particulates to be studied by electron microscopy and spectroscopy must be seriously considered.

## II. TECHNIQUES

The observational techniques discussed in this section are given in terms of their capabilities. Application of any one of the techniques to particulate matter depends upon many

**Table 1**
**ANALYSIS TECHNIQUES**

| Technique | Abbreviation | Principal use | Ref. |
|---|---|---|---|
| Electron microscopy | | | |
| Scanning | SEM | Surface structure, size, shape of individual particles down to 100 Å | 18—20 |
| Transmission | TEM | Interior structure and crystallography | 29—32 |
| Scanning Transmission | STEM | of individual particles which occur as electron transparent or have been sliced to be transparent | 33 |
| Electron stimulated spectroscopy | | | |
| X-ray | | | |
| Energy dispersive | EDS | Bulk[a] elemental composition of individual particles or aggregates from | 1—17 |
| Wavelength dispersive | WDS | atomic no. 20 (ca.) and above; see Note (1) | |
| Diffraction | XRD | Compound identification of the bulk of an aggregate of particles; see Note (2) | |
| Scanning Auger | SAM | Surface[b] elemental composition of individual particles or aggregates, all elements except H and He | 34—37 |
| Energy loss | EELS | Bulk elemental composition of electron transparent particles, complementary to EDS, very sensitive to low at. no. elements | |
| Photoelectron spectroscopy | | | |
| X-ray excited | XPS, ESCA | Surface[b] element and compound identification on aggregates of particles | 34—36, 38—39 |

*Notes:* (1)These spectroscopies may be used in conjunction with SEM, TEM, or STEM for individual particle analysis or with X-ray or radioactive sources for the composition of aggregates; (2) XRD is most often used with an X-ray source and therefore it is not an electron-related technique — it is included here as not being covered elsewhere in this book and is the only technique listed here which is not necessarily applied in a vacuum.

[a] Entire particle is yielding information; particles are up to 10 to 20 μm average dimension.
[b] Top two to ten atomic layers yielding information.

factors such as the size; atomic number or composition of the particulate; whether structural information, bulk chemistry, surface structure, surface chemistry, chemical compound, or elemental chemistry information is desired; and whether the required information is needed on each individual particle or on agglomerates of those particles. Information averaged over many particulates is usually referred to as information from an aggregate. This term is used to differentiate between the term "bulk" which is used to refer to information from whole particles, in contrast to the term "surface" which refers to the top two to ten atomic layers of material.

The analysis techniques referred to in this chapter are listed in Table 1 along with common abbreviations and a brief description of the principal use for each technique. For example, to obtain surface compositional information from individual particles an investigator must use scanning Auger spectroscopy (SAM), whereas for bulk composition on aggregates of particles, EDS, WDS, or XRD would probably be used.

For each of the techniques shown in the table, references are given to which the interested investigator can go for further information both on application of that technique to particles and on the details of the specimen preparation technique used. These references are taken mainly from microscopy journals and proceedings where technique is emphasized, but each reference has other references related to the specific relevant application of the technique from spectroscopy, biological, or metallurgical journals and books.

## III. AUTOMATION OF TECHNIQUES

Without automated techniques of sample analysis, any investigation will sample very few particles because of the labor involved in making each analysis. This is particularly true of quantitative analysis. It is not surprising, therefore, that considerable effort has gone into methods of automated analysis. The references list several extensive and detailed descriptions of both the hardware and computer software related to the automation of spectroscopic particle examination.

## IV. SPECIMEN PREPARATION

For any spectrographic work, specimen preparation is an important factor, usually requiring considerable effort. This is especially true of particles. As outlined in Section I, there are essentially five steps in the analysis of particles. The results obtained may depend strongly on the details of specimen collection, selection, and preparation. Excellent references exist[23-28] which discuss in detail the options, alternatives, and correct and incorrect techniques for handling particles. It should be noted that there are some unique problems encountered in handling particles; the techniques used are different than those associated with bulk specimens. In preparing an investigation of particles, sufficient budget must be assigned to specimen preparation to allow for these special problems.

Preparation of particles for TEM or STEM may require extraordinary effort if they are not already transparent to the electron beam. If some form of thinning or slicing is required, it is especially difficult for particles. Ion thinning might be useful for some particles, most likely very hard, glass-like ones — and microtoming, such as is used in biological thin-section preparation, may be useful for softer particles, including metal powders. Each investigation will require its own technique development.

## REFERENCES

The surveyed journals, books, and proceedings have been separated into categories in order to assist in the finding of an area of interest. These categories are listed, followed by the reference numbers related to that category. An explanation of the abbreviations for each category is also shown. References published later than mid-1982 should be sought especially in *Microbeam Analysis* published annually by the San Francisco Press, 547 Howard Street, San Francisco, Calif., 94105 and *Scanning Electron Microscopy/(Year),* SEM Inc., P.O. Box 66507, AMF O'Hare, Ill., 60666. The following reference notations are categorized by category and reference numbers (in parentheses).

SEM/EDX: (1-17); SEM: (18-20); Spec. Prep.: (21-28); TEM: (29-32); STEM: (33); SAM/XPS: (34-37); XPS: (38-40); Auto: (41-43); Stds. and Technique: (44-47).

**Categories:**

| | |
|---|---|
| SEM/EDX | Scanning electron microscopy together with energy dispersive X-ray spectroscopy |
| SEM | Scanning electron microscopy alone |
| Spec. Prep. | Techniques of specimen preparation |
| TEM | Transmission electron microscopy, including selected area electron diffraction |
| STEM | Scanning transmission electron microscopy |
| SAM | Scanning Auger microprobe |
| XPS | X-ray photoelectron spectroscopy, also sometimes called ESCA |
| Auto | Automated techniques using electron spectroscopies |


1. **Fleming, E., Tucker, P., and Hersh, S.,** Characterization of cotton dust by elemental analysis, *Proc. Microbeam Analysis Soc.,* San Francisco Press, San Francisco, 1976, 15A.
**SEM/EDS**
2. **Heinrich, K. F. J., Ed.,** Characterization of Particles, NBS Spec. Publ. 533, National Bureau of Standards, Washington, D.C., 1980.
**SEM/EDS**
3. **Mosley, W. C.,** Automated electron microprobe identification of minerals in stream sediments for national uranium resources evaluation programs, *Microbeam Analysis — 1979,* San Francisco Press, San Francisco, 1979, 101.
**SEM/WDS**
4. **Small, J. A., Newbury, D. E., and Myklebust, R. L.,** Analysis of particles and rough samples by FRAME P, a ZAF method incorporating peak-to-background measurements, *Microbeam Analysis — 1979,* San Francisco Press, San Francisco, 1979, 243.
**SEM/EDS**
5. **Alden, G. D. and Buseck, P. R.,** Rapid quantitative analysis of individual particles by energy-dispersive spectrometry, *Microbeam Analysis — 1979,* San Francisco Press, San Francisco, 1979, 254.
**SEM/EDS**
6. **Armstrong, J. T.,** Rapid quantitative analysis of individual microparticles using the α-factor approach, *Microbeam Analysis — 1980,* San Francisco Press, San Francisco, 1980, 193.
**SEM/EDS**
7. **Brownlee, D. E.,** Terrestrial and extraterrestrial pollution in the stratosphere. *Microbeam Analysis — 1980,* San Francisco Press, San Francisco, 1980, 199.
**SEM/EDS**
8. **Alden, G. D. and Buseck, P. R.,** The chemical composition of individual submicron sulfate particles from an aerosol near a power plant, *Microbeam Analysis — 1980,* San Francisco Press, San Francisco, 1980, 203.
**SEM/EDS**
9. **Bradley, J. P. and Buseck, P. R.,** Fibrous particles from a copper smelter aerosol, *Microbeam Analysis — 1980,* San Francisco Press, San Francisco, 1980, 209.
**SEM/EDS**
10. **Armstrong, J. T. and Buseck, P. R.,** Application of individual particle quantitative analysis in the determination of ambient and source aerosol heterogeneity, *Microbeam Analysis — 1980,* San Francisco Press, San Francisco, 1980, 213.
**SEM/EDS**
11. **DeNee, P. B. and Carpenter, R. L.,** Elemental analysis of aerodynamically separated respirable fly ash particles, *Microbeam Analysis — 1980,* San Francisco Press, San Francisco, 1980, 217.
**SEM/EDS**
12. **Mah, M. and Boatman, E. S.,** Scanning and transmission electron microscopy of new and used asbestos-cement pipe utilized in the distribution of water supplies, *Scanning Electron Microscopy/1978/I,* SEM Inc., AMF O'Hare, Ill., 1978, 85.
**STEM/EDS**
13. **Barbi, N. C., Giles, M. A., and Skinner, D. P.,** Estimating elemental concentrations in small particles using X-ray analysis in the electron microscope, *Scanning Electron Microscopy/1978/I,* SEM Inc., AMF O'Hare, Ill., 1978, 193.
**SEM/EDS**

14. **Hayes, T. L., Pawley, J. B., and Fisher, G. L.,** The effect of chemical variability of individual fly ash particles on cell exposure, *Scanning Electron Microscopy/1978/I,* SEM Inc., AMF O'Hare, Ill., 1978, 239.
**SEM/EDS**

15. **Basu, S. and Ferriss, S.,** A refined collection technique for rapid search of gunshot residue particles in the SEM, *Scanning Electron Microscopy/1980/I,* SEM Inc., AMF O'Hare, Ill., 1980, 375.
**SEM/EDS**

16. **Ferrell, R. E., Paulson, G. G., and Walker, C. W.,** Evaluation of an SEM-EDS method for identification of chrysotile, *Scanning Electron Microscopy/1975/II,* Illinois Institute of Technology Research Institute, Chicago, 1975, 537.
**SEM/EDS**

17. **Perrins, N. M., Alben, K., and Shpirt, E.,** SEM/EDX analysis of granular activated carbon from a water treatment facility, *Microbeam Analysis – 1981,* San Francisco Press, San Francisco, 1981, 222.
**SEM/EDS**

18. **DeNee, P. B., Rebar, A. H., DeNicola, D. B., and Pickrell, J. A.,** Application of SEM backscattered electron imaging to the study of the effects of fibrous glass and asbestos on lung tissue, *Microbeam Analysis — 1979,* San Francisco Press, San Francisco, 1979, 11.
**SEM**

19. **DeNee, P. B.,** Measurement of mass and thickness of respirable size dust particles by SEM backscattered imaging, *Scanning Electron Microscopy/1978/I,* SEM Inc., AMF O'Hare, Ill., 1978, 741.
**SEM**

20. **Mitchell, M. and Ludi, P.,** SEM and chemical evaluation of coal ash and related erosion and corrosion from some Western coals, *Microbeam Analysis — 1981,* San Francisco Press, San Francisco, 1981, 167.
**SEM**

21. **Millette, J. R., Clarke, J. P., and Pansing, M. F.,** Sizing of particulates for environmental health studies, *Scanning Electron Microscopy/1978/I,* SEM Inc., AMF O'Hare, Ill., 1978, 253.
**Spec. Prep.**

22. **Armstrong, J. T.,** Methods of quantitative analysis of individual microparticles with electron beam instruments, *Scanning Electron Microscopy/1978/I,* SEM Inc., AMF O'Hare, Ill., 1978, 455.
**EDS, Spec. Prep.**

23. **DeNee, P. B.,** Collecting, handling and mounting of particles for SEM, *Scanning Electron Microscopy/1978/I,* SEM Inc., AMF O'Hare, Ill., 1978, 479.
**Spec. Prep.**

24. **DeNee, P. B.,** Mine dust characterization using the SEM and dust samples, *Scanning Electron Microscopy/1971,* Illinois Institute of Technology Research Institute, Chicago, 1971, 209.
**Spec. Prep.**

25. **Johari, O. and DeNee, P. B.,** Handling, mounting and examination of particles for SEM, *Scanning Electron Microscopy/1971,* Illinois Institute of Technology Research Institute, Chicago, 1971, 249.
**Spec. Prep.**

26. **Brown, J. A. and Teetsov, A.,** Some techniques for handling particles in SEM studies, *Scanning Electron Microscopy/1976/I,* Illinois Institute of Technology Research Institute, Chicago, 1973, 385.
**Spec. Prep.**

27. **Huggins, F. E., Kasmak, D. A., Huffman, G. P., and Lee, R. J.,** Coal mineralogies by SEM automatic image analysis, *Scanning Electron Microscopy/1980/I,* SEM Inc., AMF O'Hare, Ill., 1980.
**Spec. Prep., Auto**

28. **Corn, M. and Esmen, N. A.,** Field sampling of airborne particulates, *Am. Lab.,* 10, 13, 1978.
**Spec. Prep.**

29. **Sprys, J. W. and Short, M. A.,** Quantitative elemental analysis of transparent particles in the TEM, *Proc. Microbeam Analysis Soc.,* San Francisco Press, San Francisco, 1976, 9A.
**TEM/EDS**

30. **Beaman, D. R. and Baker, H. M.,** Improvement in asbestos analysis using microdiffraction, *Microbeam Analysis Society — 1980,* San Francisco Press, San Francisco, 1980, 221.
**TEM**

31. **Lee, R. J.,** Basic concepts of electron diffraction and asbestos identification using SAD. I. Current methods of asbestos identification using SAD. II. Single crystal and SAD, *Scanning Electron Microscopy/1978/I,* SEM Inc., AMF O'Hare, Ill., 1978, 677; 687.
**TEM**

32. **Goodman, P.,** Setting up convergent beam diffraction for materials analysis using a TEM, *Scanning Electron Microscopy/1980/I,* SEM Inc., AMF O'Hare, Ill., 1980, 53.
**TEM**

33. **Steeds, J. W.,** Convergent beam electron diffraction, in *Introduction to Analytical Electron Microscopy,* Hren, J. J. et al., Eds., Plenum Press, New York, 1979, 387.
**STEM**

34. **Gay, A. J., vonRosenstial, A. P., and vanDuin, P. J.,** Investigation of fly ash particles with respect to the morphology and surface enrichment of matrix and trace elements, *Microbeam Analysis — 1981,* San Francisco Press, San Francisco, 1981, 229.

**SEM/EDS, AES, XPS**

35. **Davidson, D. L. and Gause, E. M.,** XPS analysis of fly ash changes due to guinea pig alveolar macrophage exposure, *Microbeam Analysis Society – 1979,* San Francisco Press, San Francisco, 1979, 237.

**XPS, AES**

36. **Linton, R. W., Loh, A., Natusch, D. F. S., Evans, L. A., and Williams, P.,** Surface predominance of trace elements in airborne particles, *Science,* 191, 852, 1976.

**AES, XPS**

37. **Wikjord, A. G., Doern, F. E., Hachkowski, L., and Baird, J.,** Electron beam microanalysis of radioactive particulates from the disintegration of Russian satellite Cosmos 954, *Scanning Electron Microscopy/1980/I,* SEM Inc., AMF O'Hare, Ill., 1980, 355.

**SEM/EDS, AES**

38. **Novakov, T., Mueller, P. K., Alcocer, A. E., and Otros, J. W.,** Chemical composition of Pasadena aerosol by particle size and time of day. III. Chemical states of nitrogen and sulfur by photoelectron spectroscopy, *J. Colloid Interface Sci.,* 39, 225, 1972.

**XPS**

39. **Chang, S. G. and Novakov, T.,** Formation of pollution particulate nitrogen compounds by NO-soot and $NH_3$-soot gas-particle surface reactions, *Atmos. Environ.,* 9, 495, 1975.

**XPS**

40. **Grieger, G. R.,** Electron spectroscopy for chemical analysis of airborne particulates, *Am. Lab.,* 8, 77, 1976.

**XPS**

41. **Lee, R. J. and Kelly, J. F.,** Overview of SEM-based automated image analysis, *Scanning Electron Microscopy/1980/I,* SEM Inc., AMF O'Hare, Ill., 1980, 303.

**Auto**

42. **Kelly, J. F., Lee, R. J., and Lentz, S.,** Automated characterization of fine particles, *Scanning Electron Microscopy/1980/I,* SEM Inc., AMF O'Hare, Ill., 1980, 311.

**Auto**

43. **Hanna, R. B., Karcich, K. J., and Johnson, D. L.,** Determination of particle identities via a computer assisted SEM-EDXA system, *Scanning Electron Microscopy/1980/I,* SEM Inc., AMF O'Hare, Ill., 1980, 323.

**Auto**

44. **Small, J. A., Heinrich, F. K. J., Fiori, C. E., Myklebust, R. L., Newbury, D. E., and Dilmore, M. F.,** The production and characterization of glass fibers and spheres for microanalysis, *Scanning Electron Microscopy/1978/I,* SEM Inc., AMF O'Hare, Ill., 1978, 445.

**Analytical Standards**

45. **Statham, P. J. and Pawley, J. B.,** Methods of quantitative analysis of individual microparticles with electron beam instruments, *Scanning Electron Microscopy/1978/I,* SEM Inc., AMF O'Hare, Ill., 1978, 469.

**EDS**

46. **Fritz, G. S., McCarthy, J. J., and Lee, R. J.,** Interactive software for automated particulate analysis, *Microbeam Analysis — 1981,* San Francisco Press, San Francisco, 1981, 57.

**EDS**

47. **Solberg, T. N., Abrecht, J., and Hewitt, D. A.,** Graphical procedures for the refinement of electron microprobe analysis of fine-grained particles, *Microbeam Analysis — 1981,* San Francisco Press, San Francisco, 1981, 160.

**EDS**

# *Part II*
# *Characterization in Various Fields of Application*

Chapter 4

# CHARACTERIZATION OF PHARMACEUTICAL PARTICULATE MATERIALS

**John A. Hersey**

## TABLE OF CONTENTS

# I. INTRODUCTION

The pharmaceutical industry is dependent upon the physiological activity of chemical entities called drugs which may either be derived from natural sources, principally of vegetable origin, or synthesized in the laboratory. It is not surprising, therefore, that the characterization of fine drug particles is largely directed toward their chemical structure and purity. For this purpose, the whole armamentarium of the analytical chemist is freely available. Thus the range extends from the various chromatographic procedures of thin-layer, gas-liquid, and high performance liquid chromatography to a variety of spectrophotometric methods, including ultraviolet, infrared, and mass spectroscopy. Such sophistication as the use of differential thermal analysis coupled with thermogravimetric analysis, where the effluent gases are subjected to gas chromatographic separation and identified by mass spectroscopy, is not unknown.

Initially, physical characterization also involved physiological activity with the realization that different stereoisomers of the same drug have different activities. Thus, the manual separation in 1848 of the sodium ammonium racemate of tartaric acid into its optical enantiomorphs became a classical experiment.[1]

The fact that other physical parameters such as particle size and the related surface area might have physiological influence was realized relatively recently.[2] The dissolution rate of sparingly soluble drugs became recognized as an important property affecting the bioavailability and, hence, physiological and clinical efficacy of the medication.[3] The relationship between the dissolution rate and surface area had been established very early[4] and methods were already available in pharmaceutical technology laboratories for particle characterization. This was due to the recognition that particle characteristics have considerable influence on the processing of these drugs into dosage forms. Thus, equipment and techniques for the physical characterization of fine drugs were already established and available for this new role. The characterization of fine drugs, however, remains predominantly chemical, but there is a considerable expertise available for the characterization of physical parameters. The former knowledge is largely contained in official compendia such as the *U.S. Pharmacopeia* and *National Formulary*[5] while the latter remains mostly in the province of undisclosed commercial "know-how".

# II. PARTICLE SIZE AND DISTRIBUTION

## A. Standardization

The official methods for characterization of particle size and shape of drugs are quaint by modern standards. The official compendia rely largely on sieving techniques introduced when drugs were derived solely from vegetable sources. Then, the only means of quantifying the strength of the extract of the vegetable drug was to specify the drug, its condition (dryness, particle size, etc.), and the extracting medium and method. Larger particles would be less well extracted than finer ones, leading to a more dilute product with a decreased physiological effect.

The *U.S. Pharmacopeia*[5] lists tables of sieves (Table 1) and descriptions of powders in terms of those sieve sizes (Table 2), and the ability of the powder to pass through a coarse screen and be retained at some percentage level on a finer screen. The various known problems of sieving are given sparse mention, including the use of worn or damaged sieves, optimum time of sieving, sample size, screen area, amplitude of vibration, etc. The terminology used is also wide and imprecise. To add to this confusion, there is a considerable overlap in definitions as has been shown for those in the *British Pharmacopoeia*[7] which have similar definitions to those in the U.S. version. Thus, a given powder may be defined within more than one single category (Figures 1 and 2).

**Table 1**
**SIEVE SIZES OF THE *U.S. PHARMACOPEIA***

| Sieve no. | Aperture size mm | Aperture size μm |
|---|---|---|
| 2 | 9.52 | 9520 |
| 3.5 | 5.66 | 5660 |
| 4 | 4.76 | 4760 |
| 8 | 2.38 | 2380 |
| 10 | 2.00 | 2000 |
| 20 | 0.84 | 840 |
| 30 | 0.59 | 595 |
| 40 | 0.42 | 420 |
| 50 | 0.297 | 297 |
| 60 | 0.250 | 250 |
| 70 | 0.210 | 210 |
| 80 | 0.177 | 177 |
| 100 | 0.149 | 149 |
| 120 | 0.125 | 125 |
| 200 | 0.074 | 74 |
| 230 | 0.063 | 63 |
| 270 | 0.053 | 53 |
| 325 | 0.044 | 44 |
| 400 | 0.037 | 37 |

**Table 2**
**DESCRIPTION OF THE FINENESS OF POWDERS USED BY THE *U.S. PHARMACOPEIA***

| | Vegetable and animal drugs | | Chemicals | |
|---|---|---|---|---|
| Powder classification | Sieve no.[a] (all passing) | Sieve no.[a] (% passing) | Sieve no. (all passing) | Sieve no. (% passing) |
| Very coarse | 8 | 60/20 | | |
| Coarse | 20 | 60/40 | 20 | 40/60 |
| Moderately coarse | 40 | 80/40 | 40 | 60/60 |
| Fine | 60 | 100/40 | 80 | 80/100 |
| Very fine | 80 | 80/100 | 120 | 120/100 |

[a] Refers to sieves in Table 1.

Alternative nomenclature[8] has not been adopted by the industry (Table 3). Nevertheless, this official sieving method and its accompanying definitions appear to serve the requirements of regulatory authorities for both vegetable and synthetic drugs. Even so, there are a few minor exceptions in the *British Pharmacopoeia* which do not carry over into the U.S. counterpart, e.g., there is the requirement for the anthelmintic, bephenium hydroxynaphthoate, to have a surface area not less than 7000 $cm^2/g$ using an air permeability method. In this case the Fisher Sub-Sieve Sizer® is suggested and the density of the drug given. Another example is the microscopic description of insulin zinc suspension (crystalline) as consisting of rhombohedral crystals having a dimension between 10 and 40 μm, whereas the amorphous form has particles of no uniform shape and rarely exceeds 2 μm in maximum

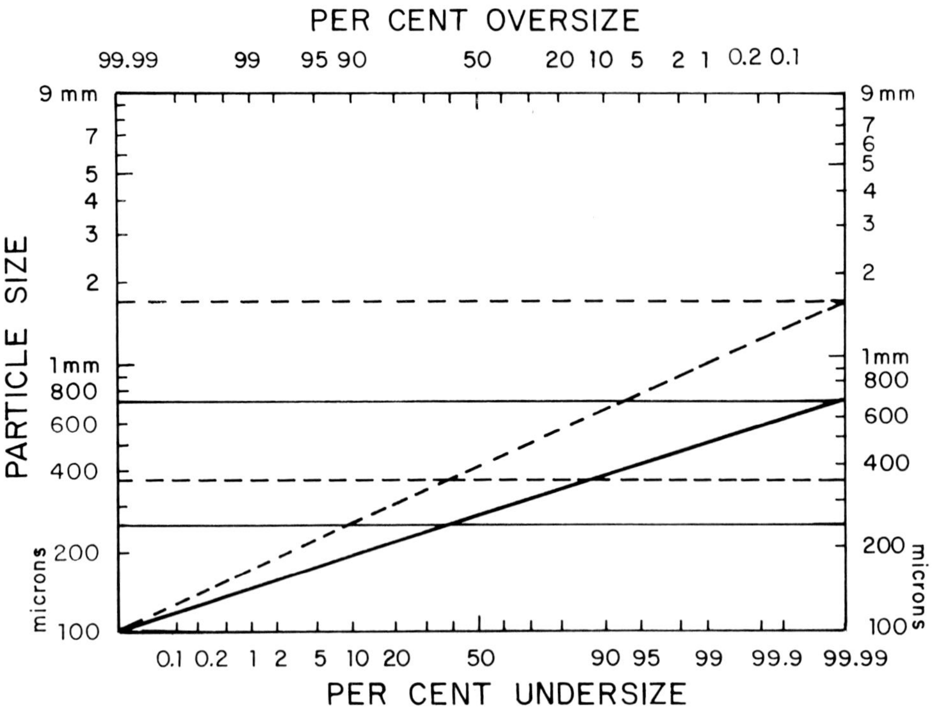


FIGURE 1. Log-normal size distributions of powders. Any straight line within the broken boundaries satisfies the requirements of the definition of a coarse powder. Similarly any straight line within the unbroken boundaries satisfies the requirements of the definition of a moderately coarse powder.

dimension. The control on bephenium hydroxynaphthoate is a result of the necessary difference between efficacy against the parasite and toxicity in the host animal. In the case of the insulin zinc suspensions, variation of the crystal size and form affects the release rate of the drug as it slowly dissolves into the bloodstream. An intermediate effect can be shown by mixing the two systems.

**B. Contamination Control**

Contamination is controlled in a variety of pharmaceutical preparations. Most solutions, especially those for parenteral use, are required to be clear and free from particulate matter. However, further guidance on this point is not clear except in the case of large-volume transfusion fluids where specific limits of particulate contamination and appropriate methods are applied.

The *U.S. Pharmacopeia*[5] applies a well-defined membrane filtration system to the examination of particulate contamination in single-dose large-volume injection solution. The whole procedure must be carried out under clean room conditions. Alternative methods may be employed provided they are correlated with the described method. The result is not valid if a control count on a blank gives more than five particles having a linear dimension in excess of 25 μm. The test requirements are that not more than 50 particles/mℓ should be larger than 10 μm and not more than 5 particles/mℓ are larger than 25 μm.

The *British Pharmacopoeia*[7] uses an automatic instrument operating on the electrical zone-sensing principle, such as the COULTER® Counter, in order to measure particulate contamination in large-volume parenterals. A disadvantage of this system is that electrolyte infrequently has to be added and sometimes the viscosity has to be reduced. A blank determination is made and subtracted from the actual data. The test requirements here are

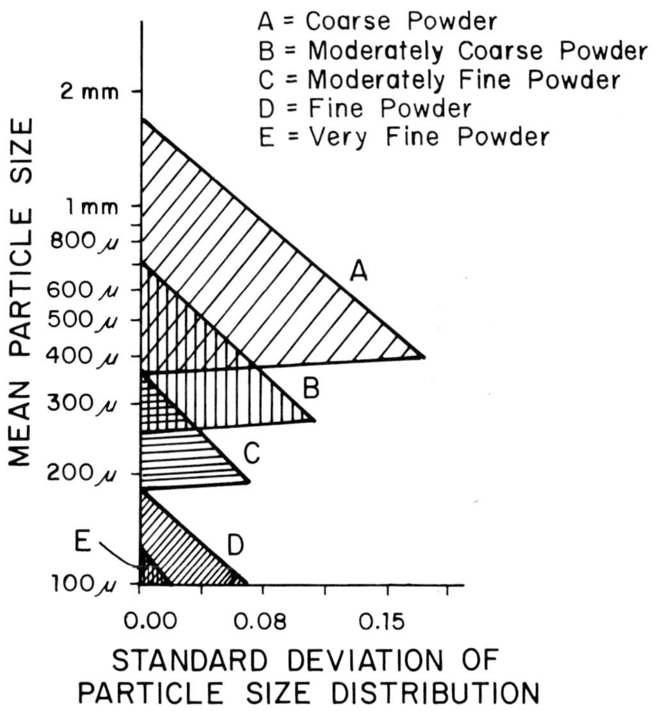


FIGURE 2. The overlap of definitions of different powder systems in the *British Pharmacopoeia*. The illustration is limited to particles above 100 μm and is valid only for log-normal distributions.

that there should not be more than 1000 particles/mℓ exceeding 2 μm and not more than 100 greater than 5 μm.

The alternative use of a light-blockage instrument, e.g., the Royco or Hiac, is allowed although the test requirements in these circumstances are reduced to not more than 500 particles/mℓ exceeding 2 μm and not more than 80 particles/mℓ exceeding 5 μm. In Australia, the light-blockage method appears to be preferred.

These techniques as applied to large-volume parenterals have not been as successfully applied to small-volume products such as ampoules and vials. This is partially due to the small sample volumes that would need to be considered, but is also due to the extra contamination which may occur in opening the dosage form.

A second dangerous situation exists where contaminant metal particles remain in metal tubes due to their manufacturing process and where such tubes are used for ointments and creams, etc. This is particularly so in the case of ophthalmic ointments. Thus, the *U.S. Pharmacopeia*[5] limits to an unobjectionable level the number of such discrete metal particles having a maximum dimension larger than 50 μm. The technique employed is to liquify a sample of the ointment by melting and allow the bigger, heavy particles to sediment so that they can be counted and sized microscopically. No tube is allowed to contain more than eight such particles and the average number per tube should be less than five.

The *British Pharmacopoeia*[7] requires only that all collapsible tubes should be as free as possible from contaminants. There are, however, limits on the particle size of ingredients dispersed in eye ointments. In any sample corresponding to 10 μg of the solid base, there should be not more than 20 particles with a maximum dimension of 25 μm, not more than 2 above 50 μm, and none above 90 μm when scanned microscopically. This is more to test adequate dispersion than to limit contamination. Other dispersions will be considered later.

**Table 3**
**TERMINOLOGY PROPOSED BY THE INSTITUTE OF CHEMICAL ENGINEERS WORKING PARTY COMPARED WITH THAT USED OFFICIALLY BY THE *BRITISH PHARMACOPOEIA*[7]**

| Particle size | 100 mm | 10 mm | 1 mm | 100 μm | 10 μm | 1 μm | 0.1 μm |
|---|---|---|---|---|---|---|---|
| Terminology proposed by working party | Individual component | Lump | Grain | Granule | Granular particle | Superfine particle | Ultrafine particle |
| | | | **Granular Material** | | | | |
| | | | **Bulk solid** | | | **Fine powder** | |
| | | Coarse broken solid | Broken solid | Granular solid | Granular powder | Superfine powder | Ultrafine powder |
| | | | **Coarse Powder** | | | | |
| Official Terminology | | No terminology | | m.c.p.[a]<br>m.f.p.[b] | | | |
| | | | | | | Fine powder<br>Very fine powder<br>Ultrafine powder | |

[a] m.c.p. = moderately coarse powder.
[b] m.f.p. = moderately fine powder.

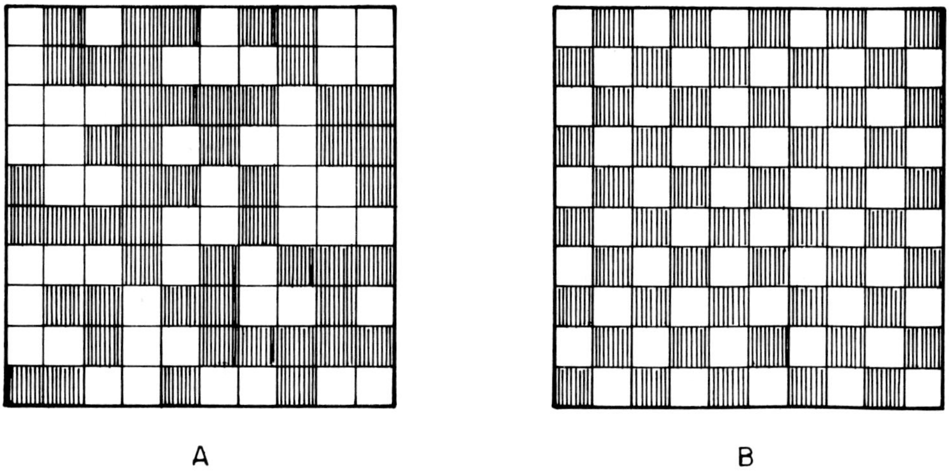


FIGURE 3. Diagrammatic representations of (A) a random mixture of particles, and (B) an ordered mix.

In general, all forms of contamination are detrimental to pharmaceutical products and should be avoided. Prevention of contamination during pharmaceutical processing is largely controlled by the various codes of good manufacturing practice which are legally enforced throughout the world. In the U.S this is under the jurisdiction of the Food and Drug Administration.[9]

### C. Technological Aspects

Pharmaceutical processing has been considerably involved with fine particles, and the mortar and pestle represents a symbol of pharmaceutical processing which dates back into antiquity. Vegetable drugs had to be ground before they would efficiently yield their active principles. Then came the era of the synthetic chemist, who preferred to prepare larger crystals due to the better purity these offered. In order to mix drugs with other solid excipients, it was necessary to reduce their particle size by comminution processes. When size is reduced, the importance of shape differences can largely be eliminated and the effects of density differences are considerably reduced. Particles of equal size could be readily mixed if free-flowing, according to Lacey's statistical interpretation of powder mixing, to give a random mix.[10,11] An alternative mixing approach, that of cohesive mixing of very fine particles to give an ordered mixture, was later suggested by Hersey.[12,13] The latter approach provided the capability of forming more homogeneous systems by using the intrinsically cohesive nature of fine particles to distribute themselves throughout a mass of carrier particles without subsequent segregation problems. The two systems are diagrammatically represented in Figure 3.

Powder mixing was studied by Train,[14] who demonstrated the importance of fine particles in pharmaceutical systems in order to produce dose uniformity of the highly potent drugs used in medicine today. This work subsequently led to the establishment of dose uniformity testing, particularly in the area of tablets, but also in other areas. The *U.S. Pharmacopeia*[5] now has limits on the dose variation of all tablets containing 50 mg or less of active drug.[15]

The dose variability in other dosage forms, particularly suspensions[17] and ointments,[18] was subsequently questioned[16] and investigated. It is clear from this survey that fine particles are necessary for achieving dose uniformity in pharmaceutical systems. These fine particles have to be controlled, and a variety of techniques have been used for this purpose including microscopic techniques, sedimentation or centrifugation methods, air-permeability as a batch quality control measure, B.E.T. adsorption, light blockage, or electrical zone sensing methods. In short, all available methods have been used from time to time somewhere within the industry.

Particle size is also reduced in other systems — for example, in suspensions to prevent sedimentation and to ensure relative ease of resuspension on shaking, so that the patient once again receives the correct dose of dispersed drug. In topical dosage forms, such as ointments and creams, it is necessary to keep the size of the particle below 60 μm. Above this size a gritty feeling would be apparent. Powders which cannot be detected by touch are described as impalpable. On the other hand, powders dispersed in toothpastes may be required to give a gritty effect, similar to sandpaper cleaning the tooth enamel, although such an effect may hopefully only be illusionary.

Particle size must also be reduced in the preparation of suspensions for injection, simply in order that they may be able to pass through the needle when administered. For these reasons, and others, the industry is aware of the need to monitor the particle size of its drugs. It is also becoming increasingly aware of the need to monitor the particle size of nonactive ingredients or excipients used in the manufacture of the dosage form. For example, pigment color is dependent on particle size and dispersion. Lubrication by solid tablet lubricants is similarly influenced by particle size and dispersion of the excipient. Tablet disintegration and subsequent dissolution of the active drug may also be dependent upon particle size and dispersion of the disintegrating agent throughout the tablet matrix.

However, the reduction of surface area for sparingly soluble drugs is frequently carried out in order to increase the rate of dissolution of the drug. In these cases, drug dissolution may be the rate limiting step in the absorption process. By increasing the rate of dissolution, then, blood levels and therapeutic effects will be demonstrably higher. Griseofulvin[19] and digoxin[20] are notable examples of the effect of such processing, however, they must not be regarded as unique. Quality control tests are now official for many tablets and capsules in the *U.S. Pharmacopeia* and require the rate of drug dissolution from the tablet to exceed certain given levels. Dissolution testing replaced the less pertinent disintegration testing for such tablets.

Particle size reduction and its control is not a panacea in pharmaceutical practice. Increasing the surface area also increases the likelihood of deterioration through interaction with other excipients, or by oxidation or hydrolytic mechanisms. Thus, a compromise often has to be achieved between drug stability and the need to increase uniformity and/or bioavailability.

Frequently a particle size increase is necessary, as in the granulation step preceding tabletting or encapsulation. Here a twofold advantage can be obtained in that the volumetric fill of the tablet die can be unified by the easier-flowing, larger granulate particles and the degree of mixing obtained with the fine powders can be maintained during the handling necessary in the subsequent tabletting operation. During this mixing process, adjuvants aiding the drug may be used to manufacture suitable tablets possessing the requirements of stability, strength to withstand handling, packaging, etc. as well as the necessary bioavailability of the drug in order to produce the desired effect.

It is to be expected that particle size will be controlled in many monographs of the *Excipient Codex* currently under preparation through the auspices of the American Academy of Pharmaceutical Sciences.[21]

## III. MONITORING OTHER PARTICLE CHARACTERISTICS

Apart from particle size and distribution, few other particle characteristics have been monitored in pharmaceutical technology. Where specific problems arise, these individual materials are adequately dealt with using the techniques available. In this section, examples will be given for individual drug problems.

Particle density is frequently required to be known in order to monitor particle size. True

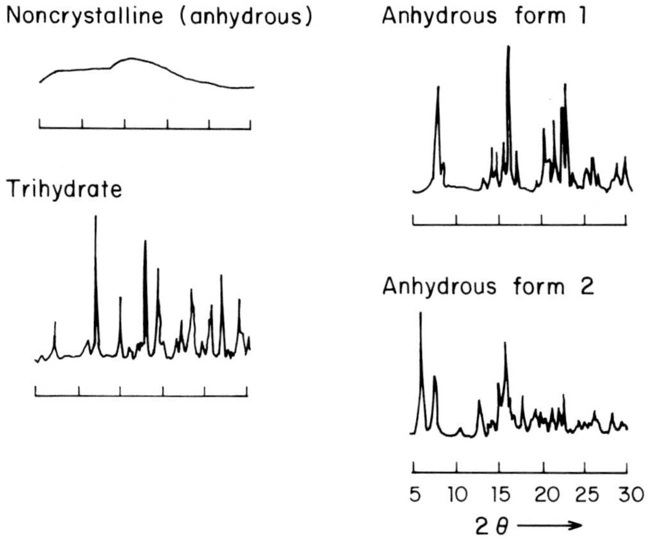


FIGURE 4. Typical X-ray powder patterns obtained for four solid phases of the drug ampicillin.

density is best measured using either air or helium displacement methods, since it is virtually impossible to expel air included in all the pores of fine particles, thus rendering low results when density is measured using liquid displacement pycnometry. Bulk density is more readily measured, and is a useful guide for tabletting operations and as an indication of package volumes necessary for powdered products.

Porosity has infrequently been measured as a research tool, but has little to offer in routine quality control. The increased pore volume adds significantly to surface area, thus increasing the rate of dissolution or reaction of the drug.

Color is also an important characteristic, but this is more usually a problem with blends, maintenance of stability of the colored products, batchwise color reproduction, and mottling effects. Various instrumentation is available to match colors using X, Y, and Z color coordinates or using the visible range of the spectrophotometer in the reflectance mode.

The occurrence of a particulate drug in more than one polymorphic or solvated form has considerable implications in both the technology of subsequent processing and the bioavailability of the drug in the tablet form which is produced. Thus, monitoring of polymorphic transitions using X-ray crystallography or thermochemical techniques is commonly performed. The *U.S. Pharmacopeia*[5] cites the example of ampicillin and gives the necessary X-ray crystallographic detail in order to separate the various forms (Figure 4). Differential thermal analysis offers a viable alternative solution to the identification of polymorphic forms, hydrates, or solvates (Figure 5).

Particle morphology has been less studied in the pharmaceutical area. The normal practice is to reduce particle size, thus substantially eliminating differences due to particle shape. This overcomes difficulties, for example, in the segregation of differently shaped particles on mixing or handling. The deliberate manufacture of reentrant lactose particles to act as a carrier for drugs in an ordered system is a unique example of the use of particle morphology in pharmaceutical processing.[22]

Surface properties of particles have been monitored by measurement of their adhesion and cohesion. The effect of contact angle with various liquids is also an indication of this parameter as a guide to the dispersion of the solid or its dissolution in the medium.[23]

While the strength of particulates would be a useful parameter to measure in order to evaluate further processing potential, e.g., tabletting, the irregular size and shape of the

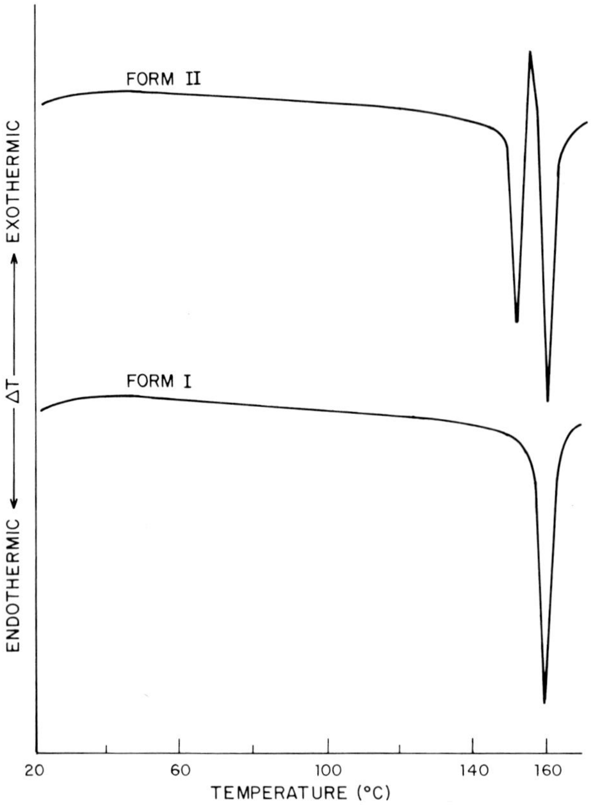


FIGURE 5. Differential thermograms of stable amylobarbital polymorphs I and II.

particles encountered makes such measurement difficult. This has been overcome to some extent by the use of a friability index determination,[24] in which the particles are placed in a sieve well capable of their support. They are then shaken with plastic balls and their breakage rate determined. This index can then be used in assessing performance of particular granulations made by different methods or using different binding agents.[25]

## IV. CONCLUSIONS

Pharmaceutical drugs are characterized chiefly for their physiological effect. Thus chemical properties largely predominate, and these are effectively controlled in official compendia. Physical attributes of fine particles are also compendium controlled, but in a less sophisticated manner and only where they are likely to influence therapeutic effect. Physical properties, however, are well monitored in the area of particle size analysis where they have a considerable influence on processing and technology. Apart from particle size considerations, other physical parameters are only characterized where this has been demonstrated as being necessary because they affect the biological performance of the drug or subsequent processing.

## REFERENCES

1. **Finar, I. L.,** *Organic Chemistry: The Fundamental Principles,* Vol. 1, Longmans, London, 1967, 434.
2. **Newman, A. C. C. and Axon, A.,** *Fine Powders and the Pharmaceutical Industry, Monogr. 14,* Society for Chemical Industry, London, 1961, 291.
3. **Parrott, E. L., Wurster, D. E., and Higuchi, T.,** *J. Pharm. Sci.,* 44, 269, 1955.
4. **Noyes, A. A. and Whitney, W. R.,** *J. Am. Chem. Soc.,* 19, 930, 1897.
5. The U.S. Pharmacopeia XX and National Formulary XV, U.S. Pharmacopeial Convention, Washington, D.C., 1980.
6. **Hersey, J. A.,** *Pharm. J.,* 200, 125, 1967.
7. British Pharmacopoeia, Her Majesty's Stationery Office, London, 1980.
8. **Richards, J. C., Ed.,** *The Storage and Recovery of Particulate Solids,* Institution of Chemical Engineers, London, 1966.
9. Good Manufacturing Practice Requirements, Federal Register, September 29, 1978, U.S. Government Printing Office, Washington, D.C.
10. **Lacey, P. M. C.,** *Trans. Inst. Chem. Eng.,* 21, 53, 1943.
11. **Lacey, P. M. C.,** *J. Appl. Chem.,* 4, 257, 1954.
12. **Hersey, J. A.,** *Powder Technol.,* 11, 41, 1975.
13. **Hersey, J. A.,** *Powder Technol.,* 15, 149, 1976.
14. **Train, D.,** *Pharm. J.,* 185, 129, 1960.
15. **Cooper, J. and Hersey, J. A.,** *Cron. Farm.,* 5, 3, 1970.
16. **Hersey, J. A. and Cook, P. C.,** *J. Pharm. Pharmacol.,* 26, 126, 1974.
17. **Orr, N. A. and Hill, E. A.,** *Pharm. J.,* 224, 547, 1980.
18. **Orr, N. A., Hill, E. A., and Smith, J. F.,** *J. Pharm. Pharmacol.,* 32, 766, 1980.
19. **Atkinson, R. M., Bedford, C. B., Child, K. J., and Tomich, E. G.,** *Nature (London),* 193, 588, 1962.
20. **Greenblatt, D. J., Smith, T. W., and Koch-Weser, J.,** *Clin. Pharmacokinet.,* 1, 36, 1976.
21. **Cooper, J. and Chowan, Z.,** *Codex of Pharmaceutical Excipients,* American Academy of Pharmaceutical Sciences, Washington, D.C., in preparation.
22. **Staniforth, J. and Rees, J. E.,** *Powder Technol.,* 28, 3, 1981.
23. **Fell, J. T. and Efentakis, E.,** *Int. J. Pharm.,* 4, 153, 1979.
24. **Ho, T. and Hersey, J. A.,** *J. Pharm. Sci.,* 1, 47, 1979.
25. **Krycer, I., Pope, D. G., and Hersey, J. A.,** in preparation at the time of Dr. Hersey's death.

Chapter 5

# FRACTAL DESCRIPTION OF FINEPARTICLE SYSTEMS

**Brian H. Kaye**

## TABLE OF CONTENTS

## I. INTRODUCTION

Mandelbrot[1] made an extensive study of the mathematical properties of rugged curves. In particular, he studied a class of curves which had no differential function, such as the curve shown in Figure 1. This particular curve, known as Koch's triadic island, has the property that it confines a finite area within an infinite boundary. This apparent paradox of finite and infinite properties of a non-Euclidean curve challenges the thinking of traditionally trained mathematicians. Adequate description of this type of curve requires a reorientation in mathematical strategy from calculus to Cantorian sets and non-Euclidean geometry. Stages in the construction of Koch's triadic island are shown in Figures 1a, 1b, and 1c. To create the 2nd order Koch's triadic island, equilateral triangles are constructed on the sides of the 1st order triangle, as shown in Figure 1b. Further subsidiary triangles are drawn on the 2nd order Koch's triadic island to generate the 3rd order island shown in Figure 1c. When this process is repeated ad infinitum, the infinitely long boundary of the ideal Koch's triadic island is achieved. When crystals grow upon existing crystals the growth of the resultant cluster of crystals is not unlike the growth of Koch's triadic island illustrated in the sequence of Figures of 1a through 1c. This analogous appearance suggests that it may be useful to explore the possibility that the mathematical techniques developed by Mandelbrot for describing curves such as Koch's triadic island may be applicable to the problem of describing crystal structure. In 1977, the writer made this intuitive step after comparing pictures of computer-drawn islands having infinite coastlines shown in Mandelbrot's book with those of carbonblack profiles which the writer was attempting to describe. The possibility that Mandelbrot's theoretical techniques for describing rugged boundaries could be used to describe the boundaries of rugged fineparticle profiles was explored in a series of papers.[2-5]

In these papers, it has been shown that Mandelbrot's mathematics form the basis of useful shape description factor which is known as the fractal index of ruggedness. This fractal index is the magnitude of a numerical dimension between 1 and 2 which describes the space-filling capacity of the profile boundary.

Before exploring the concept and the meaning of the term "fractal" it is interesting to note that fineparticle specialists had, in their attempt to develop useful geometric shape factors, already discovered the paradox of infinite boundaries associated with fineparticles before the publication of Mandelbrot's book. For example, one dimensionless shape factor which had been suggested for use in shape description was the circularity defined as

$$C = \frac{\text{Measured perimeter}}{\text{Perimeter of circle of equal area}}$$

The problem with this shape factor was that every time the resolution of the microscope used to characterize the fineparticle was increased, the measured perimeter increased. This shape factor tended toward infinity for all profiles. Thus the paradoxical situation was reached in which all the profiles of highly textured fineparticles were infinite, but some were more infinite than others. We will discuss the paradox again after exploring the use of fractals to describe rugged fineparticle boundaries.

To describe the boundaries similar to those of Koch's triadic island, Mandelbrot suggested allocating a number between 1 and 2 to the boundary to describe its space-filling ability. Mandelbrot described this number between 1 and 2 as a fractal dimension, a term which has been shortened by common usage to fractal. Thus, if a rugged line had a fractal of 1.6 it would be more efficient at filling space than a rugged line of fractal 1.3. Some scientists initially reject the concept of a fractal dimension as being impossible to picture conceptually. Mandelbrot anticipated this difficulty in his book where he discusses at length the arbitrary

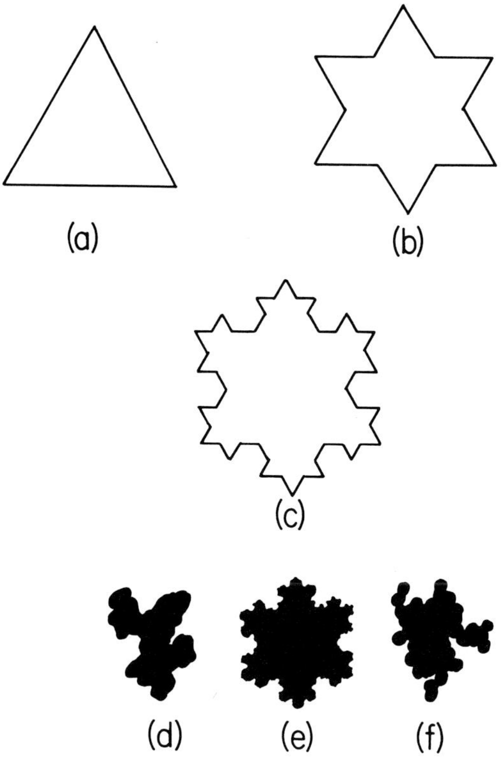


FIGURE 1. Koch's triadic island, constructed by building smaller and smaller equilateral triangles on top of each other according to the pattern illustrated in (a) to (c), has finite area and infinite perimeter: (a) first order Koch's triadic island, (b) second order Koch's triadic island, (c) third order Koch's triadic island, (d) infinite order Koch's triadic island, (e) carbonblack profile, (f) simulated agglomerate studied by Exner and Schwarz[17] and Kaye.[23]

nature of our classical assignment of dimensions to any specific geometric system. He argues persuasively that the dimensionality of an object is an operational concept which has no physical validity outside of a given physical perspective. Thus, a thin sheet of paper viewed from a distance of 20 in. is essentially a two-dimensional object, but from the point of view of bacteria trapped within a filter paper, the paper has the physical reality of a twisted tangle of stout ropes. A fractional dimension is a useful way to describe the space-filling ability of a curve and is on a par, conceptually, with the numerical listing of the average family of 2.5 children. The fractional child in such a description does not correspond to any physical attribute of a given child, but is useful as a description of the population density of a country. Thus a line of 1.3 dimension is a useful way of describing the ruggedness of a line, which immediately conveys an impression of its space-filling ability.

Mandelbrot's major contribution to the art of describing a rugged boundary was to demonstrate a technique for allocating a numerical fractal dimension to a natural boundary. To explain Mandelbrot's method for measuring the fractal dimension of a given curve it is useful to study the problem of measuring the shoreline of an island. This problem was explored experimentally by Richardson.[6] Richardson discovered empirically that if one estimated the extent of a coastline by constructing a polygon of side length $\lambda$ within the coastline boundary, then the boundary estimate based upon the perimeter of the polygon

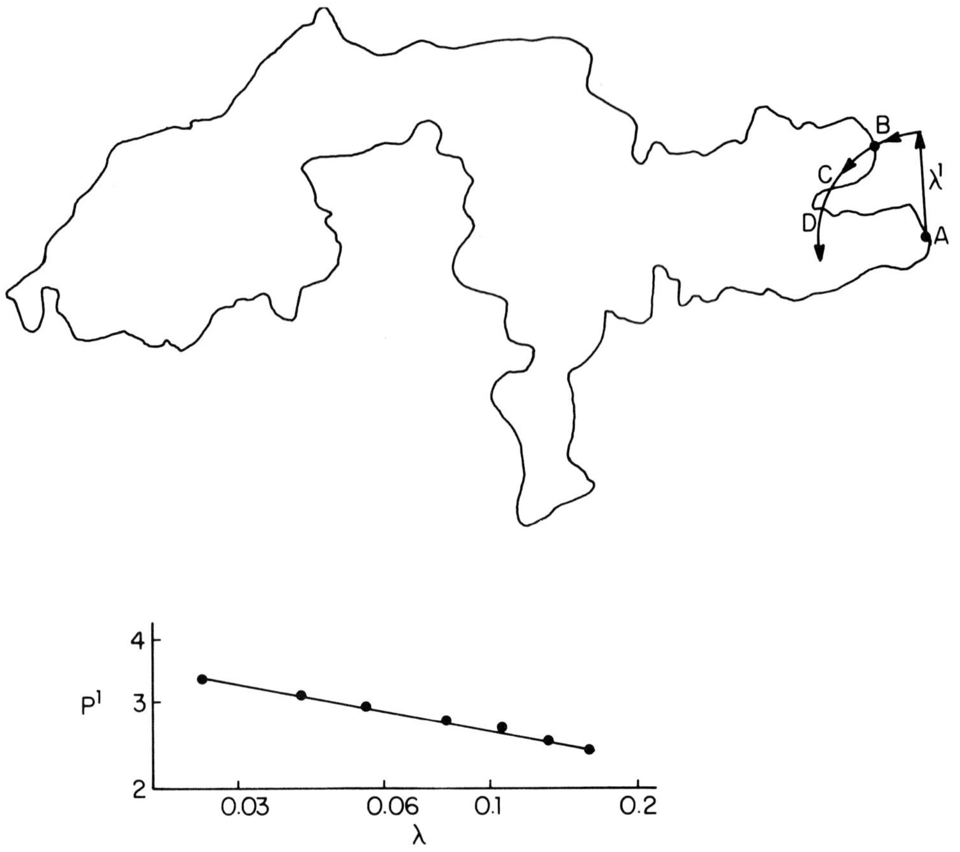


FIGURE 2. Perimeter data for Lake Ramsey illustrate Richardson's discovery that estimates of a coastline based on striding around the boundary with stride $\lambda$ tend to infinity: $P'$ = shoreline estimates normalized with respect to maximum projected length of lake; $\lambda$ = stride length normalized with respect to the maximum projected length of the lake. The slope of the line connecting the data points is $-0.18$; fractal dimension of the shoreline is 1.18.[6]

was dependent on the magnitude of $\lambda$. Constructing the polygon on the shore boundary is equivalent to striding around the boundary with a stride of magnitude $\lambda$. Richardson showed that if one plotted a graph of $[P']_\lambda$, the coastline perimeter estimate made with a stride of magnitude $\lambda$, against $\lambda$ on log-log graph paper one obtained for rugged coastlines and rugged international boundaries a straight line which tended toward infinity. What appeared to be distinctive about the various coastlines and international boundaries studied was the rate at which the boundaries tended toward infinity. Thus, we see that boundaries which are all essentially infinite can be described by the rate at which they tend toward infinity.

The technique used by Richardson can be illustrated by the data of Figure 2. The profile of Lake Ramsey (Laurentian University is built on the shores of Lake Ramsey) shown in Figure 2, is taken from a survey map of Sudbury, Ontario. The lake is approximately 7 miles long. To estimate the boundary of the lake a divider is set with points $\lambda$ apart. One then "walks" around the perimeter counting the number of strides needed to complete the walk. It is important when using this strategy, which is essentially the same as that used to measure the ruggedness of a fineparticle profile described later in the review, to practice a systematic procedure when approaching bays and promentaries on the profile. The technique developed by the author uses a swing of the divider leg into the boundary from outside taking the first point of contact with the boundary line as the definitive contact point from which one takes the next stride. Thus, in Figure 2, when operating from the point A with a stride of length λ one swings into the coastline and takes B as the next stepping-off point, not points C and D. Usually, a fraction of a step is required to complete the walk around the coastline. The simplest way to estimate the magnitude of this fractional step is to measure the length of the line from the termination of the final whole stride step to the initial starting point, which is then expressed as a fractional value of λ. It is usually convenient when making boundary estimates in this way to normalize both the perimeter estimate and the stride magnitude using the maximum projected length of the profile. Thus a stride of 0.1 in Figure 2 represents a "stride" of approximately 0.7 miles since the maximum projected length of the lake profile is of the order of 7 miles. It is important to realize that for a given size of stride λ various estimates of P′, the length of the profile, can be obtained depending on the starting point on the profile. This variation is not an experimental error, it is an inherent variation in the defined parameter — "perimeter estimate". The significance of this variation is discussed elsewhere.[7] Each data point in the graph of Figure 2 represents an average of three different estimates of the coastline using the same stride magnitude, but starting off from different, randomly selected points along the coast.

As can be seen from the data of Figure 2, the coastline estimates plotted against λ on log-log scales produce a linear relationship tending toward infinity with a slope of −0.18. Richardson discovered similar relationships for the coastlines of Australia, the west coast of Britain and the land frontiers of Portugal and Germany (pre-World War II boundary). The slopes of the various data lines differed indicating the different rates at which the boundaries tended to infinity. Richardson pointed out that if the boundaries of the various coasts and frontiers had been simple geometric figures the estimated boundary data lines would have become asymptotic to the λ axis for small values of λ. Thus, for a simple circular island the data line of the type shown in Figure 2 would become asymptotic to the value 3.14 for small λ. Richardson appears to have treated his discovery of this apparently infinite extent of natural rugged boundaries as an "empirical curiosity". Mandelbrot, however, showed that it was a widely occurring phenomenon associated with many natural rugged boundaries. Furthermore, he showed that the slope of the data line of the type shown in Figure 2 could be related to the fractional dimension, that is, the space-filling attributes of the rugged boundary. He showed that the fractional dimension was equal to 1 plus the absolute magnitude of the slope. Thus, from the data in Figure 2, the fractal of the shore of Lake Ramsey is 1.18.

A plot of a series of estimates of perimeter against the magnitude of λ, of the type shown in Figure 2, is described as a "Richardson plot". In the case of the Richardson plot of Figure 2, λ is the stride magnitude used in the "walking" estimate of the length; in other techniques for estimating fractal dimensions λ can be the length of the basic mosaic tile used in estimating the perimeter from a mosaic transformation of the profile, or the spacing of lines of a grid used to explore the perimeter of a profile (see later discussion of various techniques for measuring the fractal dimension of a profile). It should be noted that the theory developed by Mandelbrot for deducing the fractal dimension of a curve from a Richardson plot is only strictly valid for curves which are true fractal curves similar to Koch's triadic island of infinite order. The criterion for a rugged boundary to be a mathematically ideal fractal curve is that it should be self-similar. Stated in an alternative way, a rugged boundary which is a perfect fractal should exhibit the same structure at every scale of scrutiny. Thus, when looking at a photography of a fractal boundary, one should not be able to tell if it had been taken through a telescope or a microscope.[7] In practice, one cannot know before carrying out an empirical study of any natural boundary if the ruggedness of the boundary can be described by a fractal dimension. Furthermore, one cannot safely predict anything about the boundary structures beyond the scale of scrutiny used in the experimental exploration of the boundary. Thus, all that one can say about the data summarized in Figure

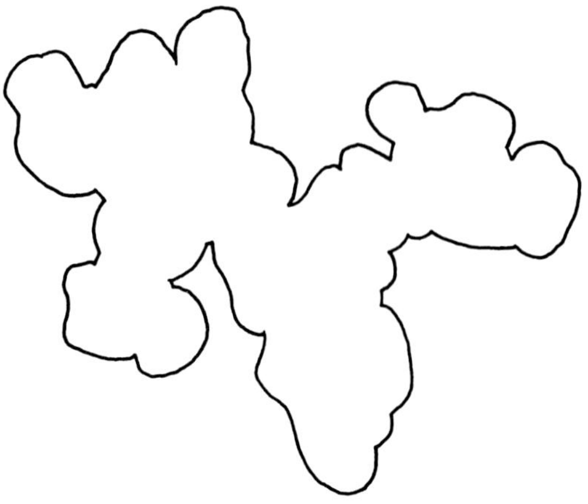

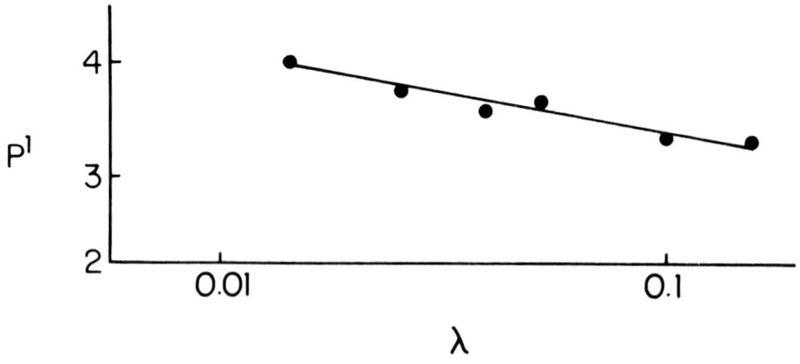


FIGURE 3. The ruggedness of a carbonblack profile can be described by a fractal 1.18 over the range of normalized λ, 0.015 to 0.15[8] (λ is normalized with respect to the longest projected length of the profile); λ = normalized stride magnitude and P′ = normalized perimeter estimate.

2 is that when examined at scales of scrutiny from $\lambda = 0.16 \rightarrow 0.025$, λ being normalized with respect to the maximum project length of the profile, the ruggedness of the boundary is describable by the fractal dimension 1.18.

The first fineparticle profile to be studied to see if it could be described with the aid of a fractal dimension was the carbonblack profile shown in Figure 3.[8] From the Richardson plot summarized in Figure 3 it can be seen that the ruggedness of the profile can be described by the fractal dimension 1.18 for the normalized values of λ in the range $0.015 \rightarrow 0.15$. Carbonblack agglomerates of the type shown in Figure 3 are formed by random collision of carbon spheres in turbulent areas of the flame used to produce the carbonblack. The rugged profile described by the dimension 1.18 is essential to the agglomerate envelope. Although it is not feasible when using the stride technique to go to smaller values of λ than those shown in Figure 3, through the use of higher-resolution technology available with automated characterization on the Quantimet image analyzer it has been shown that at small

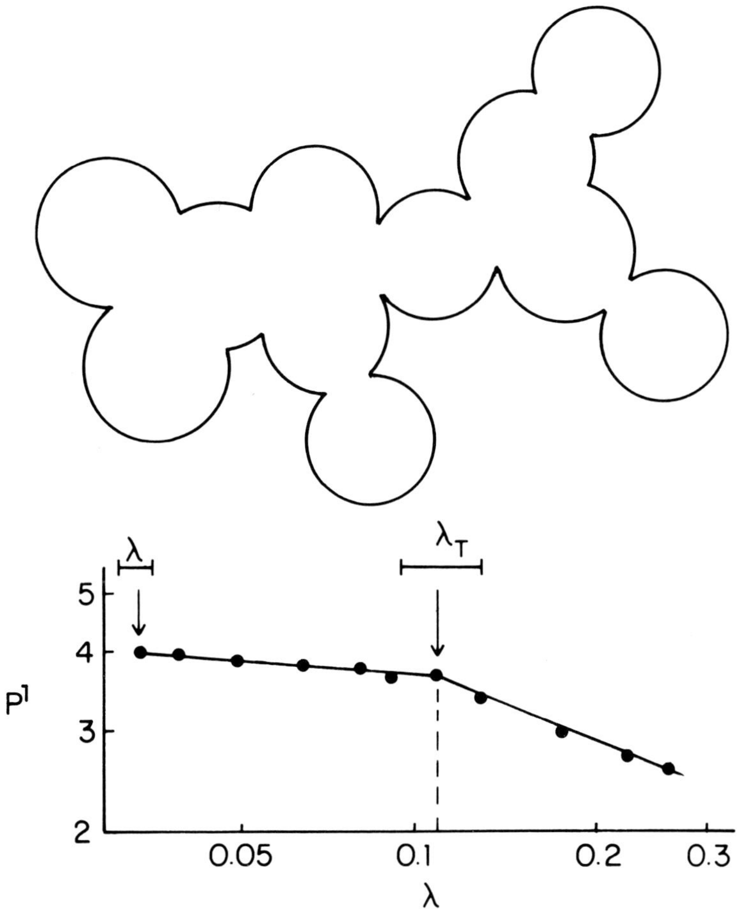


FIGURE 4. A simulated agglomerate exhibits two different fractal relationships on a Richardson plot: $P'$ = perimeter estimates normalized with respect to the longest projected length, $\lambda$ = stride magnitude normalized with respect to the longest projected length, and $\lambda_T$ = stride magnitude at transition between the two fractals. From $\lambda$ 0.28 to 0.11 fractal dimension of profile boundary is 1.39. From $\lambda$ 0.11 to 0.035 fractal dimension of profile boundary is 1.07.

values of $\lambda$ carbonblack agglomerates similar to that of Figure 3 exhibit a smaller fractal which is apparently descriptive of the packing of the carbon spheres within the agglomerate.[9,14] To illustrate how a rugged boundary can be described by different fractal dimensions at different scales of scrutiny, consider the simulated agglomerate shown in Figure 4. The Richardson plot for this profile (each data point represents the average of three perimeter estimates) contains two linear regions. The first region for $\lambda$ in the range $0.28 \rightarrow 0.11$ has a slope of $-0.39$ corresponding to a fractal dimension of 1.39. This describes the large scale ruggedness of the profile. For $\lambda = 0.11 \rightarrow 0.035$ the slope is $-0.07$ corresponding to a fractal of 1.07. It should be noted that the value of $\lambda_T$ at the transition between the two regions is 1.1, which is approximately the radius of the circles constituting the agglomerate. It seems reasonable to state that when taking strides larger than $\lambda_T$ one is exploring the gross morphology, that is the spatial structure, of the agglomerate whereas at small strides the search procedure does not look at the global structure of the agglomerate but explores the detailed packing of the circles. The physical significance of the statements made in the last sentence can be illustrated by considering the analogous exploration of an island. The large-stride exploration of the island corresponds to a survey of the gross features of the island, which establishes the existence of bays, estuaries, and promontories. The small-scale exploration of the island explores the nooks and crannies of the coast line without reference to the large-scale structure of the island.

Where a profile exhibits two linear regions on a Richardson plot of the type shown in Figure 4 it would seem useful to refer to the fractal index at low resolution search as a "structural fractal" and the one at high resolution as a "textural fractal". It is interesting to speculate on the physical significance of the two types of fractals in the case of a carbonblack profile and an island. In the case of an island, the structural fractal could probably be related to the large-scale geological forces which formed the island, i.e., tectonic plate movement, strata upthrust from subterranean stress, and wind and sea erosion. The high-scale resolution probably corresponds to the type of rocks constituting the geology of the island. It may be that fractal descriptions of islands and lakes could prove useful in analyzing satellite photographs. In the case of the carbonblack agglomerates, it is possible that the structural fractal contains information on the turbulent conditions under which the agglomerate grows by collision, whereas the textural analysis may indicate the forces holding the subunits into the structure of the agglomerate. Since the size of the stride at the position point from one type of fractal to another appears to be of the same order of magnitude as the radius of the subunits constituting the agglomerate, it may be that the size of the subunits can be derived from the transition point in the fractal plot.[9] The textural fractal of a carbonblack agglomerate may be useful in describing the chemical reactivity and absorptive properties and may also be useful in describing the strength of a bond between the agglomerate and any rubber matrix into which it is introduced. Both fractals which describe the carbonblack agglomerate may be operative in different aspects of the properties of plastic filled with a carbonblack material. It is interesting to note that the magnitude of the fractal describing the ruggedness of the carbonblack agglomerate of Figure 3 is the same as that for the outline of Lake Ramsey in Figure 2. This brings out the point that the fractal description by itself is not a complete description of the morphology of the fineparticle or any other rugged boundary. Several geometric shape factors can be used in conjunction with the ruggedness to describe in more detail the overall morphology of a fineparticle profile. For example, the writer has found it useful to use a shape index factor known as the "chunkiness". For a two-dimensional profile the chunkiness is defined as the ratio of the breadth of a fineparticle divided by the length of the fineparticle. The length is taken to be the maximum projected length and the breadth is taken to be the length of the particle enclosed by two tangents to the profile at right angles to the direction in which the length is measured. The chunkiness has the advantage that it runs from zero to 1 and corresponds to our intuitive appreciation of the fact that an equidimensional fineparticle has a chunkiness of 1.[5] It should be noted that the chunkiness is the inverse of the defined elongation.[5] The chunkiness of the lake profile of Figure 2 is 0.54, and the chunkiness of the carbonblack agglomerate of Figure 3 is 0.76. Thus $\delta = 1.18$, $C = 0.54$ is a more complete description of the morphology of the lake profile as contrasted to $\delta = 1.18$, $C = 0.76$ for the carbonblack agglomerate. Later in this chapter the utility of combinations of geometric shape factors and ruggedness in describing the properties of fineparticle systems shall be discussed.

The fact that the range over which a fineparticle profile exhibits fractal behavior contains useful information can be appreciated from the data summarized in Figures 5 and 6. In Figure 5, the profile of an aluminum fineparticle produced by atomization is shown. In an atomization process the fineparticles are produced by the breakup of a jet of molten metal. As the metal drops cool, surface tensions ensure that the surface of the fineparticle is smooth. Depending on the flow condition around the molten jet as it breaks down into droplets, and on the conditions under which the droplets are cooled, one can achieve relatively spherical fineparticles or ones of irregular structure of the type shown in Figure 5. This aluminum

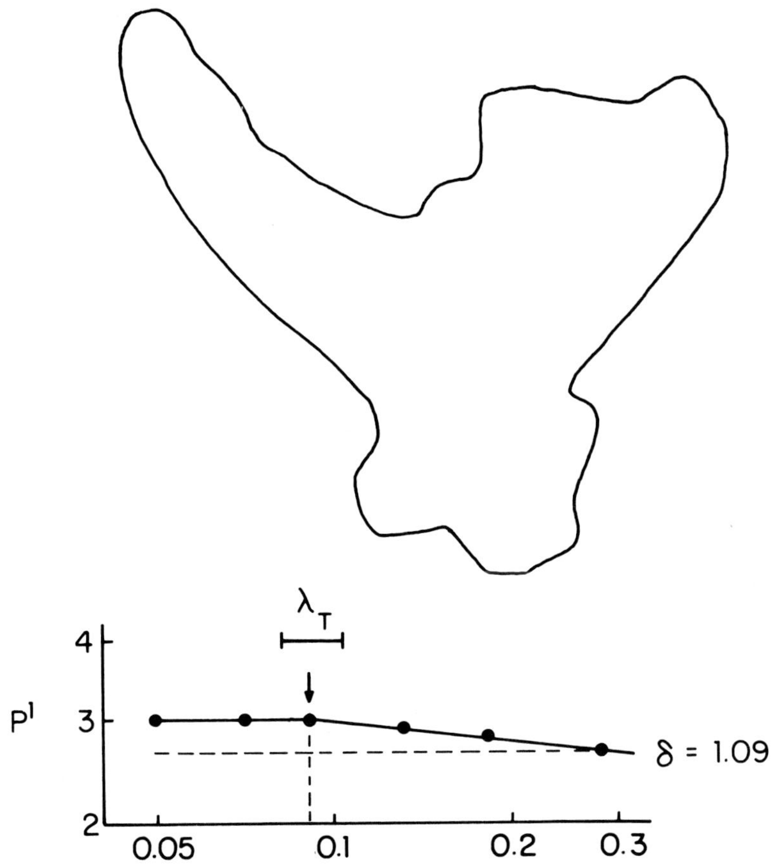


FIGURE 5. Limited-range fractals can be used to describe relatively simple structures which have smooth surfaces, as shown by the Richardson plot for an atomized aluminum fineparticle: $\lambda_T$ is the magnitude of the transition stride at changeover from a non-Euclidean to a Euclidian boundary, $\lambda$ is stride normalized with respect to maximum Feret's diameter, and $P^1$ is perimeter estimate normalized with respect to maximum Feret's diameter.

fineparticle probably cooled quickly in a turbulent air region. Because of the smooth-textured surface, the profile of Figure 5 will have a finite perimeter and a Richardson plot should exhibit an approach to a finite perimeter with the data line asymptotic to the stride axis. On the other hand, the low-level ruggedness should generate a structural fractal of limited range in an initial large-stride exploration of the profile. From the Richardson plot summarized in Figure 5, it can be seen that, indeed, the perimeter estimates tend to a finite limit of approximately 3, but a structural fractal of magnitude of 1.09 exists over the normalized stride range of 0.29 to 0.09. Again, it can be seen that the magnitude of the stride at the transition point from fractal to finite curve corresponds approximately to the magnitude of the smaller concavities and protuberances in the profile. Thus, a combination of the fractal and the transition stride at the changeover from rugged to finite boundary estimates is a measure of the ruggedness and the magnitude of indentedness that exists in the profile.

In Figure 6 we show a series of profiles for an aluminum fineparticle drawn from the same population as that of Figure 5 which has been subjected to acid erosion over a period of time. It can be seen that as the erosion proceeds the ruggedness of the profile changes, and that the stride range over which the structural fractal applies increases as erosion proceeds. The data from Figures 5 and 6 are taken from a larger study which demonstrates the utility of the fractal description of eroding fineparticles. A more extensive discussion of this work

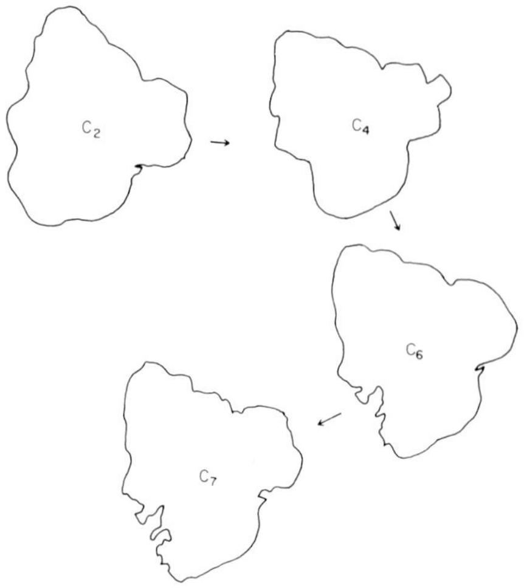


FIGURE 6. The rugged profiles generated as a fineparticle erodes in acid can be described by a series of fractals.[10]

**Table 1**
**FRACTAL DIMENSIONS OF ERODING ALUMINUM FINEPARTICLES**

| | | | |
|---|---|---|---|
| Fractal of | $C_2$ (t = 130 sec) | $\delta$ = 1.053 | $\lambda_n$ (0.30 → 0.10) |
| | $C_4$ (t = 234 sec) | $\delta$ = 1.081 | $\lambda_n$ (0.34 → 0.09) |
| | $C_6$ (t = 930 sec) | $\delta$ = 1.079 | $\lambda_n$ (0.31 → 0.05) |
| | $C_7$ (t = 1,050 sec) | $\delta$ = 1.091 | $\lambda_n$ (0.31 → 0.05) |

*Note:* Subscript to C refers to profile reference number in original studies;[10] t = time fineparticle has been immersed in the acid.

$\delta$ = fractal dimension, and $\lambda$ = stride length normalized with respect to maximum Feret's diameter of the profile.

will be published in the near future.[10] It is obvious that the use of fractals to describe the progress of acid erosion probably has applications in the general study of solution rates, corrosion studies, and chemical attack of surfaces.[16]

If a fineparticle such as that shown in Figure 5 were to be sedimented in a viscous fluid it is probable that the fractal index will be useful in estimating the amount of bound fluid carried by the fineparticle profile. Thus, the liquid trapped within the bays and estuaries of the profile is not able to participate in the flow around the profile. This factor decreases the effective density of the fineparticle. An estimate of the immobilized fluid would, therefore, be useful in investigating a relationship between the geometric configuration of the profile and its Stokes diameter.

In all of the diagrams presented so far it has become obvious that one cannot make any prediction concerning the nature of a rugged boundary beyond the smallest stride used to explore the boundary. When presented with the concept of fractal description of a real boundary some scientists raise the problem of interpreting the tendency to infinity in real physical terms. In practice, one has to recognize that even the physical meaning of the questions disappear as one proceeds to lower and lower scales of scrutiny. Thus, the concept of what is the shoreline of Lake Ramsey starts to disappear as we come down to a 1-m measurement exploration because the rise and fall of the water level with seasonal variations and wave action due to wind makes it difficult to even conceive of a finite boundary. Furthermore, if one were to proceed to a millimeter level of exploration of the shoreline one starts to travel in and out of the nooks and crannies of the rocks and the measured boundary becomes enormous. If one then proceeds to the theoretical situation in which one explored the boundary line with Angstrom-level resolution the atoms forming the rocks become indistinct to our measurement techniques. Thus the question "how long is the shoreline?" has no meaning except from a perspective in which one asks the question in such a way that one wishes to use the information generated. In the same way, if one increased the resolution with which one examined a carbonblack one would soon find that the concept of a boundary became indistinct as the scale of resolution reached that of the interspatial distance between layers of carbon atoms. Conversely, then, one avoids the problem of envisaging infinite boundaries by clearly stating that the fractal description can only be linked to a range of exploration, and that the range of exploration is essentially determined by the pragmatic conditions which lead to the need to explore the boundary in the first place.

The other main feature of the diagrams presented so far is that stride exploration of greater than 0.3 or 0.4 has not been used in the investigation of the boundary. The last real point that can exist on the stride axis is the value of 1, which corresponds to the maximum value of Feret's diameter. It obviously is meaningless to explore a rugged profile with a ruler which is as long as the profile. As one decreases the stride magnitude down to 0.6, one is still not able to explore the profile in any detail. In general, the more rugged the boundary the larger the stride that one can effectively use to explore that boundary. In practice, it appears that steps larger than 0.35 are not generally useful in the exploration of rugged boundaries. It should be emphasized that this is an empirical statement based upon experience and is not based on theroretical considerations.

In this introductory discussion of the concept and utility of fractal description of a fine-particle boundary we have restricted the discussion to complete profiles. It should, however, be pointed out that fractal description of ruggedness has also proven to be useful in describing the fracture appearance of broken composite materials and fractured metal specimens.[11-13] Current investigations at Laurentian University are seeking to link the fracture ruggedness of a pigmented plastic with the dispersion characteristics of the fineparticles loaded into the plastic.

## II. TECHNIQUES FOR MEASURING FRACTALS

In the introductory discussion of the basic concepts and utility of fractal descriptions of rugged profiles, the stride perimeter estimate technique for evaluating a fractal was discussed in some detail since it remains the simplest and probably the most efficient technique for measuring a fractal from a photograph. When investigating the structure of a small number of fineparticles the method is straightforward and adequate. Since the original introduction of fractal dimensions to describe rugged boundaries, various workers have developed different techniques for measuring the ruggedness of a profile. In particular, Flook[13,14] has programed the erosion dilation methodology of the Quantimet image analyzer to measure fractal dimensions of rugged boundaries. Coster and co-workers[11,12] have also developed algorithms for automated image analysis procedures for calculating the fractal dimensions of a profile. Schwarz and Exner[17] have programed a small computer used in conjunction with a graphical digitizer to evaluate the fractal dimensions of various profiles.

In the original publication concerned with the use of fractal dimensions to describe rugged boundaries, it was suggested that the profile could be transformed into a mosaic using a square grid and the fractal evaluated by counting elements of the mosaic which lay on the profile of the fineparticle. This is the preferred method when using a charged coupled device camera (known as a CCD camera) which generates an image in mosaic form. This image can be interfaced directly to a computer so that the counting of picture elements lying on a boundary becomes a relatively simple task. Note that it is common usage among electrical engineers processing mosaic images to talk about each mosaic element as a ''tile'' or a ''pixel''. To obtain a smaller or larger pixel mosaic (corresponding to different stride length in the perimeter walk method of evaluating the fractal dimensions) one can either change the optical magnification of the image being produced by the CCD camera or one can amalgamate the elements of a high-resolution mosaic in a systematic manner to produce lower and lower resolution moasics. The logic associated with mosaic evaluation of the ruggedness of a profile is the subject of several investigations proceeding at Laurentian University.

Reid[18] has suggested that a method for evaluating a fractal with a linescan system, such as that used in conventional electric-beam TV camera interface to digital computers, is to use triangulation procedures to calculate the perimeter of a polygon which can be drawn between the intersections created on a profile boundary by the presence of a linescan set of parallel lines.

## III. APPLICATIONS OF FRACTAL DESCRIPTION OF FINEPARTICLES

Figure 7 shows the outline of a rubber crumb produced by a process designed to facilitate the recycling of rubber.[19] The Richardson plot for this profile indicates that the rubber crumb can be described by a fractal of magnitude 1.09 for normalized strides from 0.45 to 0.09. This fractal may be useful in describing the dry powder flow of the rubber crumbs and may be related to the compressive strain which could be sustained by an assembly of the fineparticles when under pressure. To look at the packing properties of the profile, one would probably have to use, in addition to the fractal ruggedness, a geometric shape such as the chunkiness.

In Section I we noted the similarity in appearance of a Koch's triadic island to a cluster of crystals. In Figure 8, the outline of an agglomerate of barite crystals is shown.[20] It can be seen from Figure 8 that the outline of the barite crystals can be described by a fractal dimension of magnitude of 1.08 over a normalized stride range of 0.45 to 0.18. This would suggest that fractal descriptions of crystal structures may be useful in following the progress and the effect of varying conditions of a crystalizer.

In Figure 9 the profile of a fineparticle discovered in a lubricating oil is shown. This fineparticle had been generated by frictional wear in a gear system. This type of profile has been explored by Beddow and Meloy[16] using Fourier analysis of geometric signature waveforms. From the data given in Figure 9, one can see that the ruggedness of the profile can be described by a fractal dimension of 1.12 for the normalized stride range of 0.34 to 0.07. Again, this diagram is suggestive of the fact that fractal description may be useful in study of fineparticles produced by wear and friction.

In Figure 10, the profiles of several fragments of shattered plastic are shown. They are taken from a study of the cryogrinding of plastic. To generate the profiles of Figure 10 a piece of Styrofoam® was chilled in liquid nitrogen and then allowed to shatter by falling to the floor from a height of 1 m. The plastic becomes very brittle at the temperature of liquid nitrogen and, as can be seen from the fragments shown in the diagram, the resultant fineparticles are extremely rugged. In the earlier discussions presented in this review of fractal description of fineparticles we have concentrated on describing individual fineparticles. The profiles of Figure 10 show that a complete description of a population of fineparticles requires data on the different sizes, gross morphology, and ruggedness of a series of fineparticles. It has been proposed by the author that a good technique for describing such a population

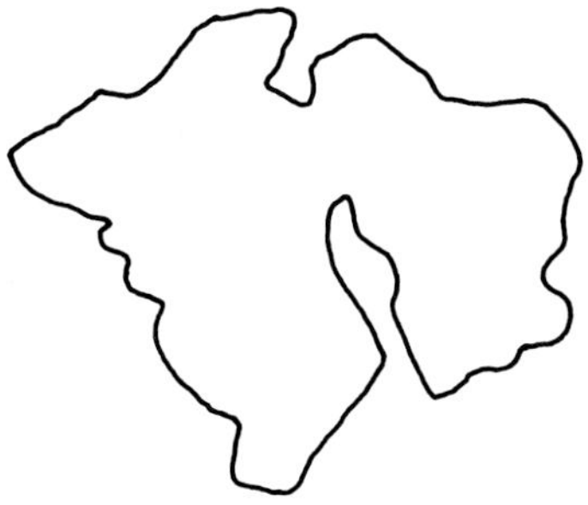

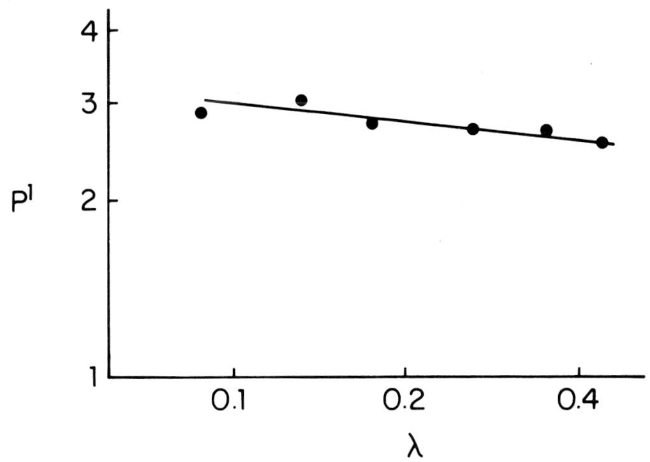


FIGURE 7. Fractal dimension of a 20-μm rubber crumb: P′ represents the perimeter, λ is the stride length normalized with respect to Feret's diameter, and δ = 1.09 (for normalized λ, 0.45 to 0.09).[19]

of fineparticles is to employ a three-dimensional graph in which the three mutually perpendicular axes represent size, a suitable geometric index of morphology, and ruggedness. Thus, for the profiles of Figure 10, a suitable size parameter would be the diameter of the circle of equal area and a useful geometric index would be the chunkiness factor discussed earlier. The ruggedness would then be described by the fractal index of the profiles. The data for many fineparticles would then be represented by a cloud of points in three-dimensional space. To calculate a parameter representative of the range of shape and size present in a population of fineparticles one could then treat the points as representing small points of mass, so that the center of gravity of the cloud of data points could serve as a representative average. In the same way, the radius of gyration of the points treated as elemental masses could serve as a dispersion index.[5]

In Figure 11, the profile of a diesel exhaust fineparticle is shown. This is essentially a soot agglomerate and appears to be built of spherical subunits. This agglomerate was described and presented by Kittleson and Dolan in a study of fumes emitted by diesel engines.[22] The Richardson plot for this diagram demonstrates that it is describable by a fractal of 1.28 over the normalized stride range of 0.35 to 0.06. Again, the fractal description will probably be useful in describing the pockets of immobilized air associated with the profile, as it falls

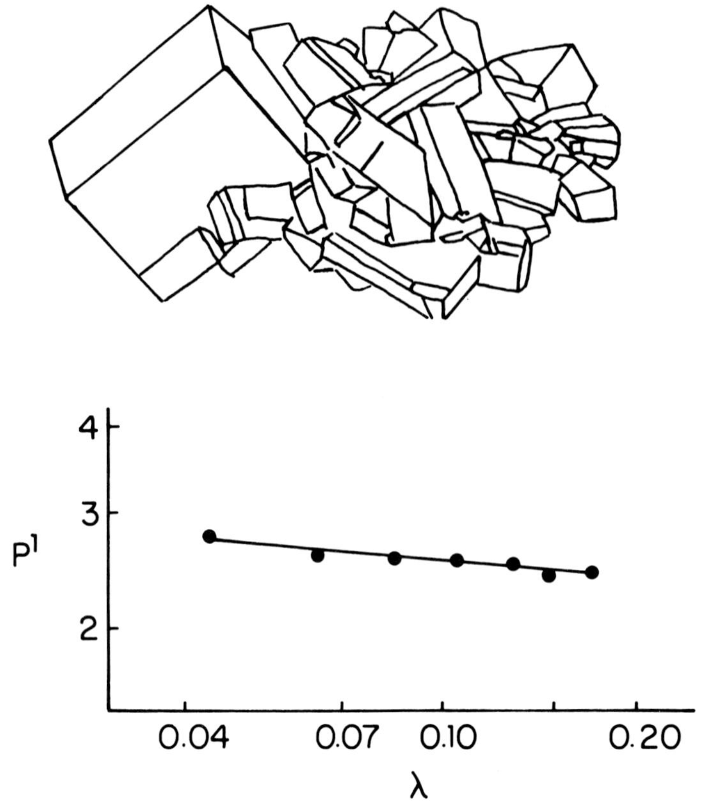


FIGURE 8. The perimeter of an agglomerate of barite crystals can be described by means of a fractal dimension: P′ is the perimeter estimate, $\lambda$ is the stride length normalized with respect to maximum, and $\delta = 1.08$.[20]

under laminar flow conditions which should enable one to start predicting a link between the aerodynamic diameter of the agglomerate and its physical dimensions. This relationship will probably prove significant in predicting the relationship between the apertures in a physical filter used to collect diesel exhaust fumes and any hazard to the lung posed by inhaled diesel emission fineparticles.

In Figure 12, the profile of a sponge iron fineparticle profile is shown. This profile is one of a series which has been studied extensively by Beddow and Meloy.[16] The Richardson plot for this profile shows that the ruggedness of the profile can be described by a fractal dimension $\delta = 1.20$ over the stride range $\lambda$ 0.28 → 0.07. This iron fineparticle is spongy in texture and it may well be that the fractal description is related to the surface energy of the fineparticle.[16,21]

In Figure 13 an interesting profile, which can also be described with the aid of a fractal dimension, is shown. It is the outline of an inclusion found inside a meteorite; this rugged inclusion has a fractal dimension of 1.48 over the stride range of 0.10 → 0.03. This demonstrates that fractal description can be useful when examining sectioned structures such as composite materials, metals, and ore specimens.

Fractal description can also be applied to profiles which are of biological origin. Thus, in Figure 14 the outline of a Purkinje brain cell is shown. In the original diagram of this cell, presented by Llinás, nerve junctions (synapses) at the lobe-type protruberances were shown.[25] For the sake of clarity, only the profile of the actual central body of the brain cell is shown in Figure 14. As can be seen from the Richardson plot of Figure 14, the ruggedness

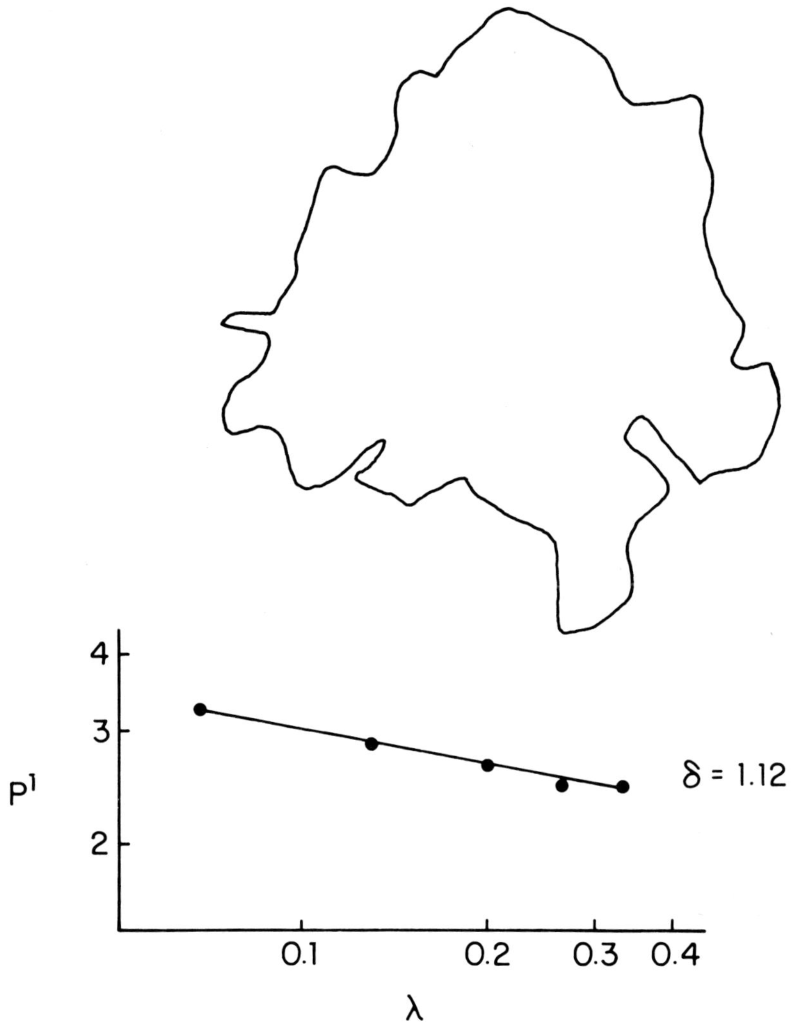


FIGURE 9. The shape of some fineparticles produced by friction and wear can be described by fractal-based ruggedness factors: P′ represents the perimeter normalized with regard to maximum Feret's diameter; λ is the stride magnitude normalized with respect to maximum Feret's diameter; and δ = 1.12 fractal dimension for λ, 0.34 to 0.07.

of the cell can be described by the fractal δ = 1.42 for the range λ 0.20 → 0.02. It has been suggested that fractals may be very useful in the description of malignant cells, which often exhibit rugged structure.[26]

## IV. ANTICIPATING FRACTAL BOUNDARIES

The possibility that a boundary can usefully be described from a given operative perspective must always be explored and established experimentally. This is, however, no different from the situation in which the investigator suspects that a given population of data is Gaussianly distributed and then proceeds to carry out tests which will determine whether or not the population is indeed described adequately by the Gaussian distribution. (Note that "adequately" establishes the operative nature of the attempt to describe a population by a Gaussian distribution.)

However, when testing for the adequacy of a Gaussian distribution, it may become clear

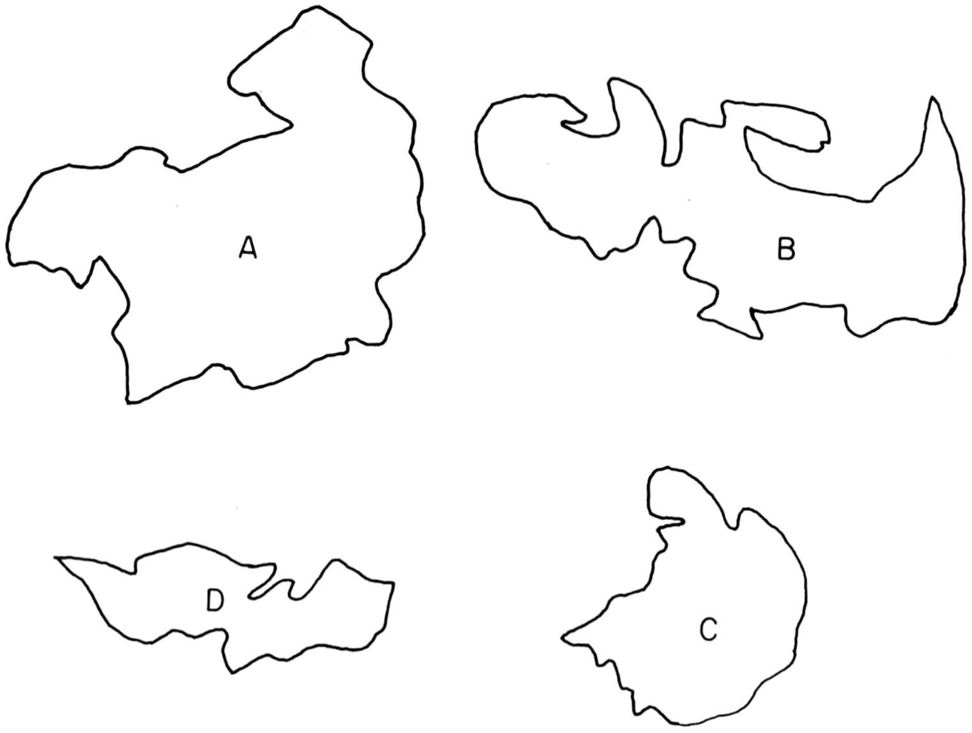


FIGURE 10. Typical fragments from a Styrofoam® cup cooled in liquid nitrogen and dropped on the floor from a 1-m height (magnification × 50). Measured fractals over a normalized stride range of 0.35 to 0.05: A, $\delta$ = 1.12; B, $\delta$ = 1.26; C, $\delta$ = 1.13; and D, $\delta$ = 1.16.

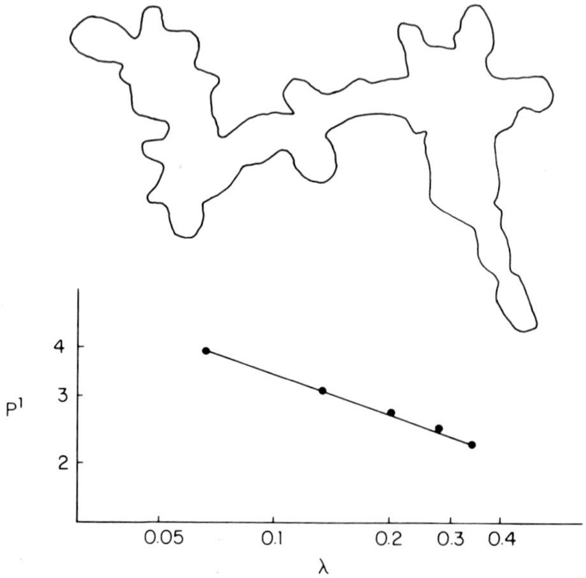


FIGURE 11. A diesel exhaust fineparticle exhibits structure ruggedness that can be described by a fractal dimension (actual size approximately 5-μm long): P′ is the estimated perimeter normalized with respect to maximum Feret's diameter; $\lambda$ is the stride length normalized with respect to maximum Feret's diameter; and $\delta$ = 1.28 for $\lambda$, 0.35 to 0.07.[22]

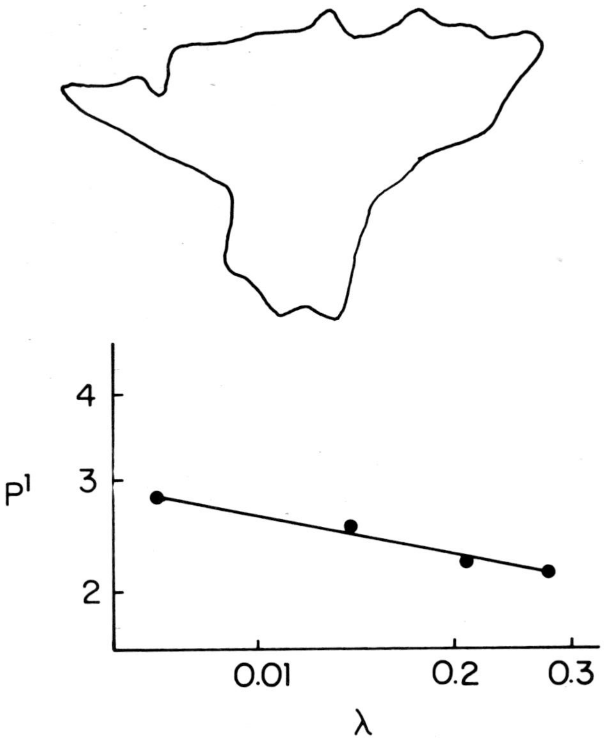


FIGURE 12. The rugged structure of a sponge iron fineparticle can be described by a fractal dimension: P′ is the perimeter estimate normalized with respect to maximum Feret's diameter; λ is the stride length normalized with respect to maximum Feret's diameter; and δ = 1.20 for λ, 0.28 to 0.07.[16,21]

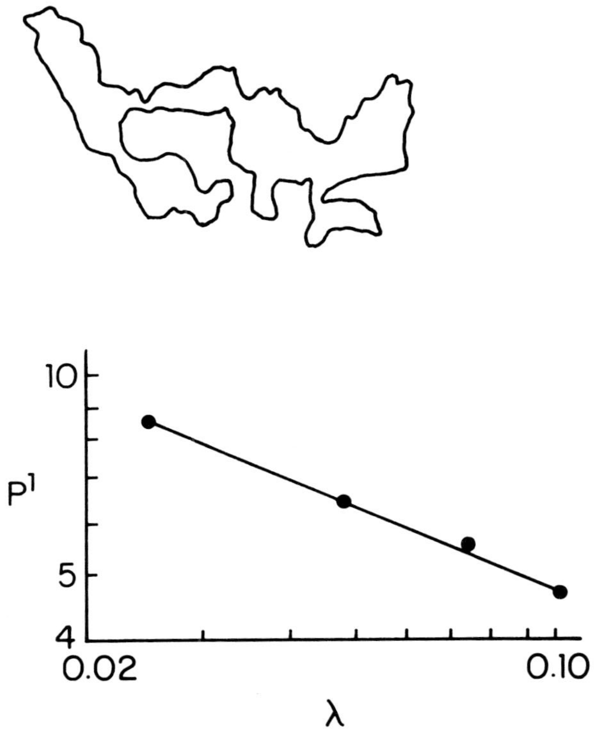


FIGURE 13. An inclusion inside a meteorite has a boundary which can be described by a fractal dimension: P′ is the perimeter estimate normalized with respect to maximum Feret's diameter; λ is the stride length normalized with respect to maximum Feret's diameter; and δ = 1.48 for λ, 0.1 to 0.03.[24]

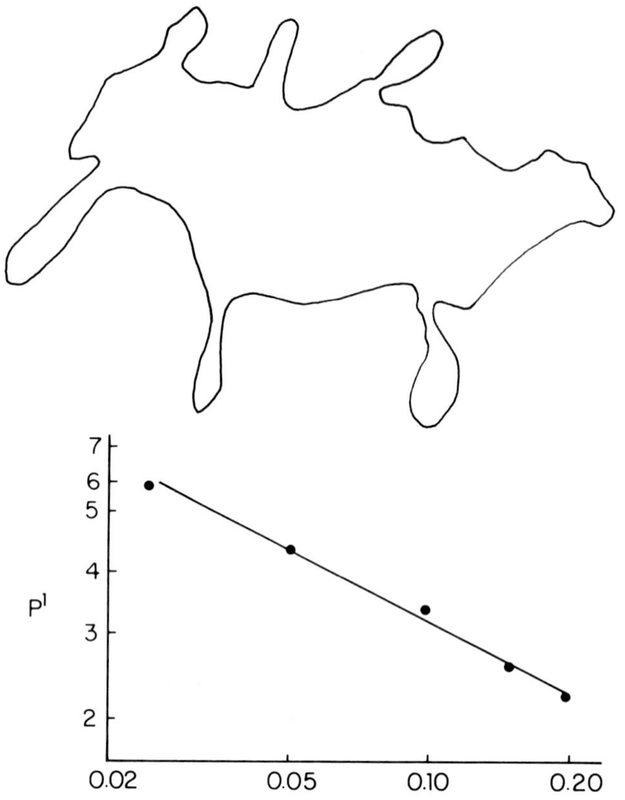


FIGURE 14. Some biological cells exhibit a ruggedness that can be described by a fractal. Thus, the ruggedness of this Purkinje cell can be studied using a Richardson plot. P′ is the perimeter estimate normalized with respect to maximum Feret's diameter; λ is the stride length normalized with respect to maximum Feret's diameter; and δ = 1.42 for λ, 0.20 to 0.02.

that it is more appropriate to describe the set of data by some other distribution function, such as the Rosin Rammler distribution function or the log-normal distribution function. Through accumulated experience, scientists have come to anticipate that a Gaussian distribution may be relevant for describing a system produced by the random interaction of many causes of similar magnitude. On the other hand, a log-normal distribution function is generated when a given value of a particular property requires the relatively rare favorable combination of the causes interacting to form an object. Thus, heights of individuals tend to be Gaussianly distributed, whereas salary levels among engineers are log-normally distributed with higher salaries requiring a favorable combination of talent, industry, and opportunity. What are the underlying interactions of physical causes which give rise to rugged boundaries which can be described by fractals? No absolute guidelines can be given, but from the case histories discussed in the review it would seem reasonable to suggest that boundaries generated by the random accumulation of effects produced by an iterative process will exhibit fractal regions on a Richardson plot. Thus, crushed materials formed by repeated crack failure should have fractal boundaries. Precipitates and condensates formed under turbulent conditions are, again, the cumulative result of iterative random processes. As more automated procedures for characterizing fractal boundaries are developed the technique should prove useful in the study of the properties of a major class of fineparticles formed by this type of random process.

## ACKNOWLEDGMENTS

The writer wishes to thank Mr. John Leblanc, Research Associate in the Institute for Fineparticles Research, who evaluated many of the fractal boundaries discussed in this review.

## SYMBOLS

$C$ = Circularity
$\lambda$ = Length of side of polygon
$[P']$ = Estimate of perimeter

## REFERENCES

1. **Mandelbrot, B. P.,** *Fractals Form, Chance and Dimension,* W. H. Freeman, San Francisco, 1977.
2. **Kaye, B. H.,** Characterization of the surface area of a fineparticle profile by its fractal dimension, in *Proc. Salford Conf. on Particle Size Analysis,* Groves, M. J., Ed., Heyden & Sons Ltd., London, 1977, 250.
3. **Kaye, B. H.,** Specification of the ruggedness and/or texture of a fineparticle profile by its fractal dimension, *Powder Technol.,* 21, 1, 1978.
4. **Kaye, B. H.,** *Fractal Dimension and Fineparticle Science,* International Powder Institute, Chicago, Ill., 1978.
5. **Kaye, B. H.,** *Direct Characterization of Fineparticles,* John Wiley & Sons, New York, 1981.
6. **Mandelbrot, B. P.,** How long is the coast of Britain? *Science,* 156, 636, 1967. (Mandelbrot discusses Richardson's work in some depth; Mandelbrot's discussion here and in Reference 1 constitute accessible information on Richardson's work.)
7. **Kaye, B. H.,** The description of natural rugged boundaries in fineparticle science, in preparation.
8. **Medalia, A.,** Carbonblack profile, Cabot Corporation, Billerica, Mass., 1978.
9. **Kaye, B. H. and Flook, A.,** Fractal description of carbonblack agglomerates, in preparation.
10. **Kaye, B. H., Leblanc, J., and Abbott, P.,** Fractal description of an eroding aluminum fineparticle, in preparation.
11. **Chermant, J. L. and Coster, M.,** Fractal object in image analysis, Proc. Int. Symp. Quantitative Metallography, Florence, 1978, 125.
12. **Coster, M. and Deschanvres, A.,** Fracture, object fractal et morphologie mathematique, *Sonderbande der Praktischen Metallorgraphie,* 8, 61, 1978.
13. **Flook, A. G.,** The use of dilation logic on the Quantimet to achieve fractal dimension characterization of textured and structured profiles, *Powder Technol.,* 21, 295, 1978.
14. **Flook, A. G.,** The characterization of textured and structured particle profiles by the automated measurement of their fractal dimensions, Proc. PARTEC Eur. Symp. Particle Characterization, Nuremberg, 1979, 591.
15. **Kaye, B. H.,** *Sequential Mosaic Amalgamation as a Strategy for Evaluating Fractal Dimension of a Fineparticle Profile,* Res. Rep. 21, Institute for Fineparticle Research, Laurentian University, Sudbury, Ontario, available from the Institute.
16. **Beddow, J. K. and Meloy, T. P.,** *Advanced Particulate Morphology,* CRC Press, Boca Raton, Fla., 1980, chap. 1.
17. **Schwarz, H. and Exner, H. E.,** The implementation of the concept of fractal dimensions on a semiautomatic image analyzer, *Powder Technol.,* 27, 207, 1980.
18. **Reid, A.,** private communication.
19. **Anon.,** *Process for Recycling Vulcanized Rubber,* Gould Inc., Cleveland, Ohio, 1978.
20. **Jansen, H.,** *Natural History,* February 1978, p. 63.
21. **Kaye, B. H.,** Fractal and shape index description of the shape and surface structure of metal powders, in Proc. 1980 Powder Metallurgy Conf., Washington, D.C., June 22 to 27, 1980.
22. **Kittelson, D. B. and Dolan, D. F.,** Diesel exhaust aerosols, Publ. No. 387, Particles Technology Laboratory, Mechanical Engineering Department, University of Minnesota, Minneapolis, 1978.

23. **Kaye, B. H.,** Trends in fineparticle characterization, *Loughborough Conf. Fineparticle Characterization,* British Institute of Chemistry, London, 1981.
24. **Grossman, L.,** The most primitive objects in the solar system, *Sci. Am.,* February 1975, p. 30.
25. **Llinás, R. R.,** The cortex of the cerebellum, *Sci. Am.,* January 1975.
26. **Martens, A.,** Bausch & Lomb, Rochester, N.Y., private communication.

Chapter 6

# BEHAVIOR OF BULK SOLIDS

**David L. O. Smith and Robert A. Lohnes**

## TABLE OF CONTENTS

# I. HISTORICAL PERSPECTIVE

Prior to the 1880s designers of bulk storage facilities for granular or particulate materials believed that such material behaved like a fluid and calculated hydrostatic pressures based on the apparent fluid density. It was realized that granular systems possess shear strength and thus the ratio of lateral to vertical stress is not equal to one; however, the strength and stress-strain behavior of these materials cannot be characterized by the models used for continuous solids. The mechanical properties of bulk particulate systems lie between those of a fluid and a solid.

Soils have been studied more extensively in terms of strength and stress-strain characteristics than any other particulate material. Theoretical and experimental work on the mechanical properties of soils dates back to the 17th century, with perhaps the most significant early contributions coming from Coulomb[1] and Rankine.[2] Late-19th century designers successfully used the theories of Rankine and Coulomb for predicting stresses in shallow bins, but these same theories were inadequate for predicting stresses in deep silos although they are still used in the design of soil retaining walls. Roberts[3,4] experimented with wheat and dried peas and found that they exerted shear stresses on the walls of silos and that lateral stresses increased nonlinearly with depth. Janssen[5,6] developed theoretical equations to predict both lateral and vertical stresses within silos. These equations, currently widely used in silo design,[7] require knowledge of the frictional characteristics of the stored material as well as its bulk density. The theory that Janssen developed for stored granular material was incorporated into soil mechanics literature when Marston[8] used the same differential equation to predict soil loads on buried conduits.

Shortly after Janssen's work, Prante[9] reported increases in lateral stresses when grain was discharged from the bottom of a bin. Some investigators[10-12] reported similar behavior, whereas others[13-15] reported no increase in lateral stress upon unloading. Takhtamishev, as cited by Turitzin,[16] measured lateral stresses during the unloading of elevators, observed the flow patterns, and concluded that increased stresses were associated with material properties, however no mechanistic explanation of the phenomenon was presented.

Various theories have been offered to predict the static and flow loadings which granular materials impose upon silos and bins.[17-22] Virtually all of these theories require data on the shear strength of the particulate material and/or the coefficient of friction of the material on the walls of the structure. As a result of the need for data on the strength of bulk solids, several studies have focused on measuring the strength or "flowability" of these materials.[23-27] A majority of studies on granular materials, other than soils, have emphasized measurements of shear strength; however, recent investigators have measured stress-strain as well as strength characteristics of granular materials.[28-30] In all cases, the equipment used to measure the mechanical properties of the materials is identical to, or a modification of, equipment used for soil mechanics.

In both theory and practice the study of the bulk mechanical properties of granular systems has been entwined with soil mechanics. Thus, the following sections on testing and theory of the strength and stress-strain characteristics of particulate materials will borrow from soil mechanics. A discussion of soil mechanics usually includes cohesive materials such as overconsolidated clays, and noncohesive or purely granular materials such as sands and gravels. Although many particulate systems exhibit cohesion under certain conditions, bulk solids behave more like granular soils so this chapter will emphasize noncohesive materials.

The preceding historical development is intended to show the chronological development of ideas relative to the mechanistic behavior of granular materials. Specific examples of how these ideas are applied to the design of storage facilities for bulk solids are described in the following sections.

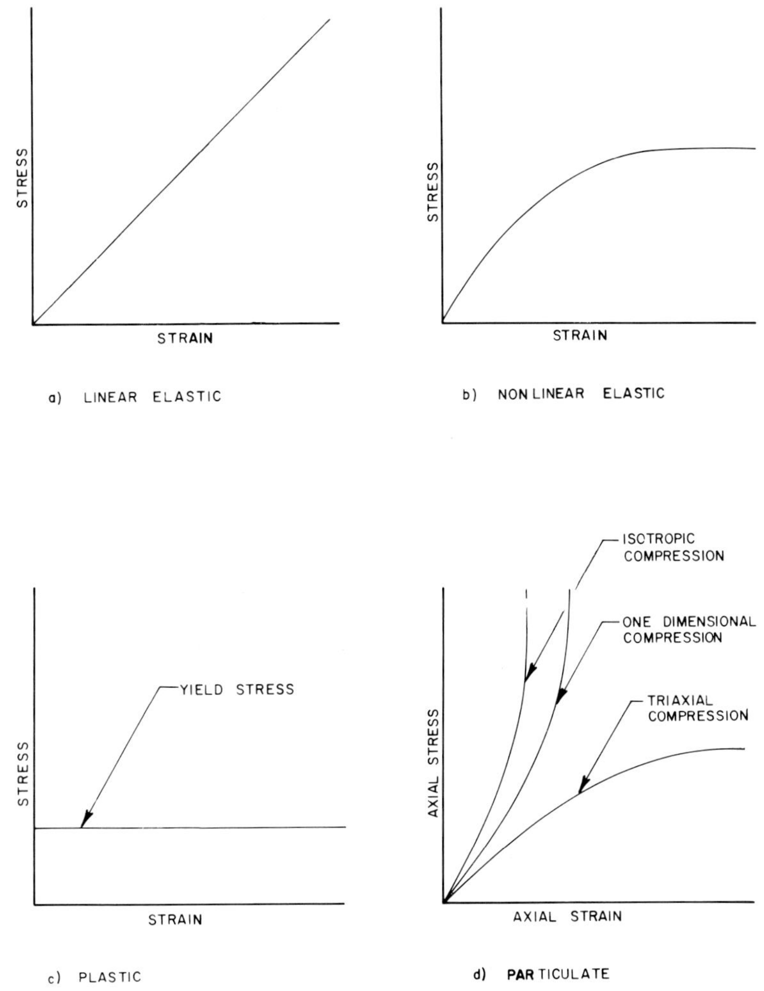


FIGURE 1. Stress-strain curves for idealized and real materials.

## II. GENERALIZATIONS ON STRENGTH AND STRESS-STRAIN BEHAVIOR

### A. Stress-Strain Characteristics

The behavior of continuous solids can be modeled by several types of idealized stress-strain relationships as shown in Figure 1. Elastic materials, which regain their initial shape once the stresses are released, may be linearly elastic (Figure 1a) or nonlinearly elastic (Figure 1b). The perfectly plastic material (Figure 1c) does not strain initially upon the application of stress, but strains continuously with no additional stress increase upon reaching a yield point. If the stress falls below the yield point the straining ceases, but the material retains the strain thus far achieved.

The behavior of particulate systems is characterized by various types of nonlinear stress-strain relationships, which depend upon the test conditions (Figure 1d). These variations in types of deformation result because the major portion of the strains in a particulate system is due to interparticle movement. Little strain results from deformation of individual particles.

Current practice in the design of storage facilities for bulk solids requires data on the limiting stresses which define shear failure within the particulate system, therefore it is useful to discuss the stress-strain behavior of these materials in shear strength tests. The upper two

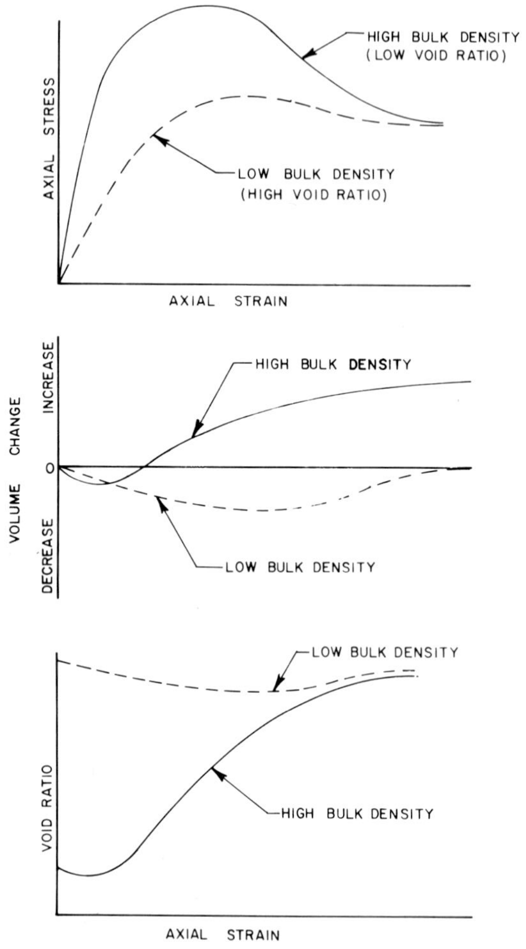


FIGURE 2. Stress-strain and volume change characteristics of bulk solid.

curves of Figure 2 are representative of stress-strain and volume change curves of a sand at two different initial bulk densities. The material, which was originally at a high bulk density, exhibits a rapid stress increase to a peak value which defines the yield stress or shear failure. As strain continues to increase beyond the peak stress, the stress decreases to a constant value. The low bulk density material shows a less rapid increase in axial stress which levels out at higher strains, approaching the ultimate stress value of the high bulk density material.

The volume change curves show that the high bulk density material initially experiences a small volume decrease followed by a volume increase, whereas the low bulk density material decreases in volume and then returns to the original volume.

It appears that a high bulk density material has an anomalous behavior because it expands in volume when compressed. This behavior is not so anomalous if the nature of the particle movement is taken into consideration. In a bulk solid, the major portion of the deformation is due to sliding, rolling, and rearrangement of the grains relative to one another. In the case of a high bulk density granular material in which the grains are tightly interlocked, it is necessary for the grains to roll up over one another in order for the bulk solid to shear. The tendency for dense materials to increase in volume when sheared is referred to in soil mechanics terminology as dilation. For a loosely packed, or low bulk density material the individual grains will move into open void spaces as the material shears and compression will be accompanied by a volume decrease. Higher stresses are required, therefore, to cause failure in dense materials as dilation occurs. In loose bulk solids, where it is not necessary for dilation to occur as the system shears, the failure stresses are lower. Thus, the shear strength of particulate material depends upon the packing or degree of interlocking of the individual grains.

If a high bulk density, granular material is sheared in such a manner that volume increase is not permitted, the only way in which failure can occur is by the deformation or breaking of individual particles. If the particulate system is composed of very soft grains, shearing of tightly packed material may occur without dilation, but with the degradation of individual particles.

If high bulk density materials dilate and low bulk density materials decrease in volume during shear, it is reasonable to expect that at some intermediate bulk density the material will exhibit no volume change. In soil mechanics it is more common to characterize the degree of packing of a soil by void ratio rather than bulk density. Void ratio is the ratio of the volume of voids to the volume of the solids and should not be confused with porosity, or voidage, which is the ratio of the volume of voids to the total volume of the bulk material. In studies of the mechanics of particulate materials, the void ratio is more useful than bulk density to characterize the packing of the system. Bulk density is a function of particulate density as well as packing of the particles; therefore if the packing of different materials is to be compared, void ratio provides a less ambiguous number. Also, in most granular materials the major portion of the strain is the result of interparticle movement and deformation results from change in the void volume and not in solid volume change. A high void ratio particulate system would be a low bulk density system, whereas low void ratio corresponds to high bulk density. A material at the intermediate bulk density at which no volume change occurs during shear is at the critical void ratio.

The lowest graph in Figure 2 shows how the void ratio varies with axial strain. Both materials, regardless of their initial void ratio, tend toward the same void ratio at large strains. Thus, at large strains both the strength and final void ratio of the sand are independent of the initial void ratio. The strength of a dense sand at large strains is often referred to as the constant volume or ultimate strength, as contrasted with the peak shear strength which is the yield stress which occurs at small strains.

From the preceding discussion it can be seen that if a dense granular material is held at constant volume while it is being loaded axially, there will be a tendency for an increase in lateral stress, whereas with a loose material there will be a tendency for a lateral stress decrease during axial loading.

If a bulk solid is loaded and unloaded there will be some rebound, but very little of the strain produced upon the loading will be recovered. Once again, this is because the majority of the strain in a particulate system is due to the rearrangement of the particles. Figure 3 shows an idealized stress-strain curve for a particulate system which has been subjected to repeated loading and unloading cycles. The effect of this repeated loading is to densify the granular material and decrease the void ratio. For a more complete discussion of the stress-strain behavior of soils see Lambe and Whitman.[31]

### B. Shear Strength

The shear resistance of bulk solids is frictional in nature, that is, the shear stress required to cause sliding along a plane within the mass of the material increases with increasing normal stress on the failure plane. Before discussing the factors which affect the shear strength of the bulk material, it is useful to examine the nature of frictional resistance between particles.

Figure 4a shows a block of solid material resting on a surface of the same material. The shear force, F, is increased until the block begins to slide. If the normal force N′ is increased,

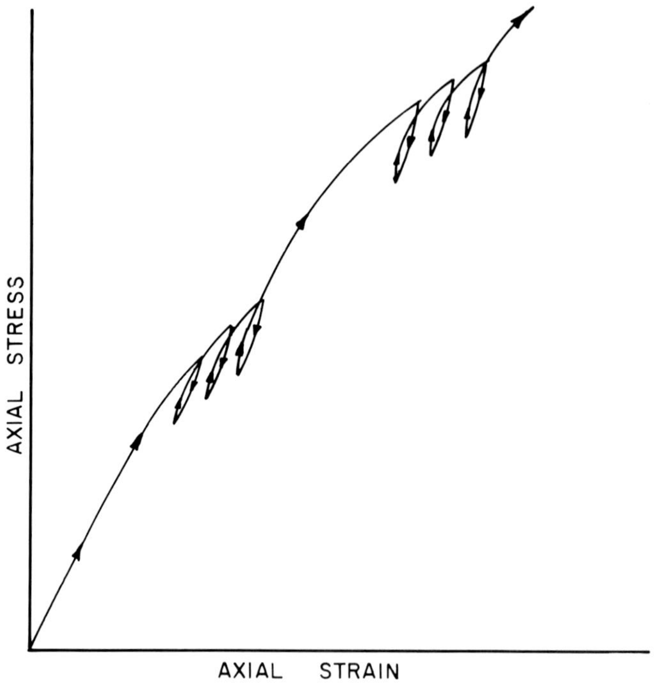


FIGURE 3. Stress-strain curve for bulk solid under cyclic loading.

it will take a greater shear force to cause the slippage. The resultant force, R, makes an angle $\phi_\mu$ with the normal, referred to as the friction angle. If the forces are converted to stresses by dividing by the cross-sectional area of the contact surface, and the shear stress, $\tau$, is plotted vs. the normal stress, $\sigma'$, a linear relationship exists (as shown in Figure 5) which is described by the equation:

$$\tau = \sigma' \tan \phi_\mu \tag{1}$$

Thus $\tan \phi_\mu$ is the coefficient of sliding friction, for which the symbol $\mu$ is often used, and characterizes the shear resistance of the material. The friction angle has been interpreted as a material property caused by chemical bonds at submicroscopic contact points.[32-34] This property will vary somewhat depending upon the microscopic roughness of the surface of the solid, the presence of contaminants on the surface, and the temperature. Values of the coefficient of sliding friction and the friction angles are shown in Table 1.

Although the coefficient of sliding friction is a material property, the shear resistance of the same material in bulk as an aggregate of particles depends upon several other factors. A simple model that can be used to describe the shear behavior of a particulate system as contrasted with the shear resistance of a solid is shown in Figure 4b, which shows two blocks of the same material in contact on a saw-tooth surface which has the teeth inclined at an angle, i, with the horizontal. The normal force on the block is N and the shear force which would cause the upper block to slide relative to the lower block is T. If the force normal to the inclined plane is N′ and the resisting shear force which is tangential to the inclined plane is F, a resolution of forces and solution of the static equilibrium shows that the shearing resistance of the two blocks can be described as

$$T = N \tan (\phi_\mu + i) \tag{2}$$

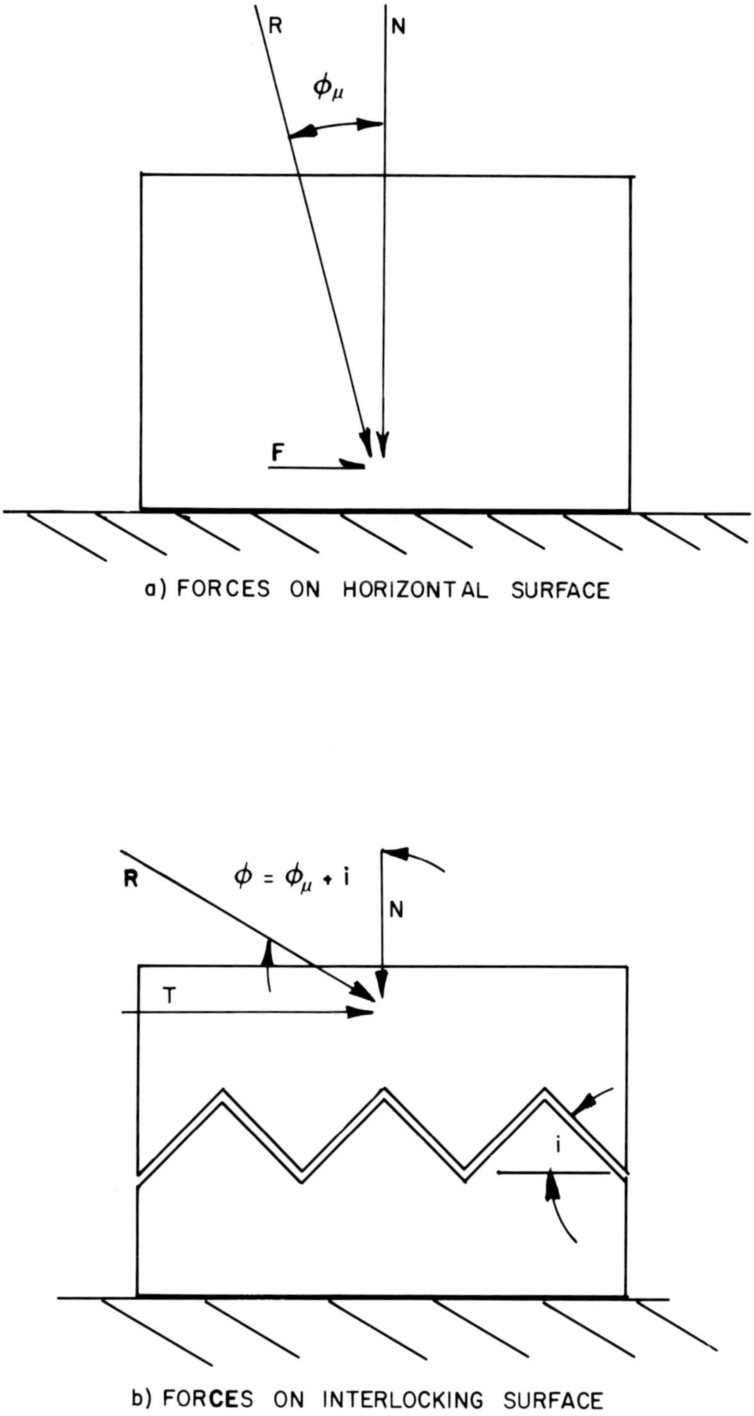


FIGURE 4. Frictional resistance models.

or if the forces N and T are expressed as stresses by dividing by the cross-sectional area, the relationship between shear stress, $\tau$, and normal stress, $\sigma$, is

$$\tau = \sigma \tan (\phi_\mu + i) \tag{3}$$

**Table 1**
**FRICTIONAL CHARACTERISTICS OF SOLIDS**

| | $\mu$ | $\phi_\mu$ (deg) |
|---|---|---|
| Aluminum[a] | 1.9 | 62 |
| Calcite[b] | 0.14 | 8 |
| Feldspar[b] | 0.12 | 7 |
| Glass[a] | 0.9—1.0 | 42—45 |
| Mica[a] | 1.0 | 45 |
| Polyethylene[a] | 0.2 | 11 |
| Polytetrafluoroethylene[a] | 0.04 | 2 |
| Quartz[b] | 0.13 | 7 |
| Steel[b] | 0.58 | 30 |

[a] Data for clean, dry surfaces taken from *Handbook of Chemistry and Physics*.[35]
[b] Data from Horn and Deere.[36]

It can be seen that as the angle, i, is increased the shearing resistance between the blocks increases. It is convenient to characterize the shearing resistance of the particulate system as

$$\tau = \sigma \tan \phi \tag{4}$$

where $\phi = \phi_\mu + i$. The system of the two blocks is analogous to a particulate system in which the angle i describes the degree of interlocking between individual particles, $\phi_\mu$, is the sliding friction angle between individual particles, and $\phi$ is the angle of internal friction of the particulate system. Although the nature of the contacts in a bulk solid is too complex to be described by a single angle, the model does show that $\phi$ is a parameter rather than a property because $\phi$ depends upon the interlocking in addition to $\phi_\mu$, the frictional resistance between individual particles.

The previous section described the effect of interlocking or density upon the stress-strain and strength behavior of a sand at a given confining stress. High bulk density materials have higher values of $\phi$ than low bulk density materials because dilation must occur in dense materials to overcome the effect of interlocking.

It was also pointed out that at high strains the strength of the particulate system is independent of the initial bulk density. This indicates that for a given bulk solid it is possible to have a third value of internal friction angle called the ultimate friction angle, $\phi_{cv}$. This value is used in soil mechanics applications where large strains occur. It might be expected that if the difference between the peak and ultimate friction angle is the result of the interlocking of grains, then the ultimate friction angle and the friction angle for the sliding friction should be the same. However, $\phi_{cv}$ is always greater than $\phi_\mu$. This is explained as the result of local volume increases in some zones and local volume decreases in other zones within the bulk solid which counteract to produce no overall volume change. Because $\phi_{cv}$ is independent of the void ratio it can be considered a material property of the bulk material, which is the result of $\phi_\mu$ plus some interlocking which occurs with no volume change at large strains. If a bulk solid is allowed to free fall and accumulate on a flat surface, a cone of the material will form, and if the experiment is repeated several times it will be observed that the angle that the surface of the cone makes with the horizontal is the same. This angle is referred to as the angle of repose, and an analysis of the statics of the stability of the side slope of the cone shows that the maximum angle that can be obtained is the friction angle of the bulk material. Thus, the maximum angle of repose is often assumed to be equal to the angle of internal friction, $\phi$. It has been shown that for a given particulate system, $\phi$ increases as the bulk density increases, with the lowest value being $\phi_{cv}$. If materials are dumped they are in a loose state, thus $\phi_{cv}$ is the friction angle which is equal to the maximum angle of repose.

### C. Factors Influencing the Strength of a Bulk Solid

The foregoing discussion has emphasized the importance of bulk density in controlling the strength of a bulk solid; however, there are several other factors which influence the magnitude of $\phi$ including composition, particle size distribution, and the angularity of individual grains. Some of these characteristics of the granular material affect bulk density which, in turn, affects shear strength.

The most obvious effect of composition upon the shear strength of a particulate system is that $\phi_\mu$ will depend upon the composition of the individual grains. It can be seen in Table 1 that a wide variation exists in $\phi_\mu$, with hydrocarbons and common rock-forming minerals having low values whereas materials such as glass and masonry are three to five times higher. Another effect that composition has upon the shear strength of an aggregate is the crushing strength or deformation characteristics of the individual grains. Although the simple frictional model implies a linear relationship between normal and shear stress, it has been shown that some bulk solids have strength envelopes which are concave downward. A quartz sand may exhibit a good linear relationship over a range of normal stresses between 500 and 1500 kPa, but a carbonate sand may have lower shearing strengths at the higher normal stresses. This is the result of crushing of the grains at the contact points, which has the effect of decreasing the interlocking angle, i. This aspect of composition suggests that aggregates of more deformable materials, such as agricultural grain, would exhibit curved strength envelopes whereas harder materials, such as ores, would exhibit a more nearly linear relationship between $\sigma$ and $\tau$.

A particulate system which is composed of grains of a given density, but of widely varying sizes, can attain higher bulk densities because the smaller particles can fill the voids between the larger particles. Conversely, a bulk solid composed of particles of more uniform sizes will tend to have a lower bulk density. The result is that less-uniformly graded bulk solids will have greater interlocking of particles and therefore higher values of $\phi$. A bulk solid composed of particles of widely varying sizes will tend to experience less crushing or deformation of grains when loaded because the load is distributed over many more contact points. Thus, a particulate system composed of grains of uniform size would have a greater tendency to exhibit a curved strength envelope.

Data on the strength characteristics of soils give insight to the effect of an average grain size of a bulk solid upon $\phi$. Sands of identical uniformity of gradation but with different average particle sizes were tested, and although a larger particle size results in a lower void ratio, the internal friction angle is affected very little.[38] The greater degree of interlocking produced by the larger particles is compensated by greater crushing and deformation at grain contacts because there is greater load per particle. It is to be expected that particulate systems of larger particle size will have more curved strength envelopes.

Angularity of particles will also influence $\phi$. Bulk solids composed of more rounded grains will have lower internal friction angles than those composed of angular particles because the more angular particles produce greater interlocking.

From this discussion it can be seen that the angle of internal friction depends upon many variables, but the most obvious, apart from the nature of the material itself, is its bulk density. As a result, most values of internal friction angle are tabulated as a range of values associated with a range of bulk densities. Table 2 shows some selected values of $\phi$ and bulk densities for various particulate materials. Although few data are available which can be used to compare $\phi$ with $\phi_\mu$ for a given material, it is possible to compare the frictional

**Table 2**
**PROPERTIES OF GRANULAR MATERIALS[a]**

| Material | Bulk density (kg/m³) | $\phi$ (deg) | $\phi'$ (deg) Against concrete | $\phi'$ (deg) Against steel |
|---|---|---|---|---|
| Portland cement | 1344—1600 | 24—30 | 20—24 | 17 |
| Coal | 800—1120 | 24—44 | 24—31 | 17 |
| Grain (small) | 736—990 | 23—37 | 16—25 | 15—23 |
| Barley[b] | 734—691 | 29—33 | — | — |
| Maize[b] | 764—821 | 23—27 | — | — |
| Wheat[b] | 836—898 | 24—27 | — | — |
| Iron ore | 2640 | 40 | 27 | 20 |
| Lime powder | 700 | 35 | 26 | 17 |
| Sand | 1600—2000 | 25—40 | 22—35 | 19—27 |
| Soybeans | 800—960 | 23 | 14 | 11 |

[a] Based on data from American Concrete Institute,[7] except as noted.
[b] Based on data from Smith.[37]

resistance of solid quartz with the internal friction angle of sand. A comparison of data in Tables 1 and 2 shows that the value of $\phi_\mu$ for quartz is 7°, whereas for sand $\phi$ ranges from 25 to 40° depending upon the degree of interlocking.

In addition to the internal shear resistance of a granular system, the designer of storage and conveyance facilities is concerned with the frictional resistance of the bulk solid on a surface such as concrete or steel. It would be expected that the angle of internal friction is the upper bound for this type of shearing resistance, and that the true value depends upon the roughness of the surface as well as the properties of the granular material. Data for the friction angle, $\phi'$, of the bulk solids against concrete and steel are also shown in Table 2.

In soil mechanics problems, it is often the case that experimental values of $\phi'$ are not available. The ultimate friction angle $\phi_{cv}$ is often used to calculate the frictional resistance of the sand on rough surfaces and $\phi_\mu$ used to calculate the frictional resistance on smooth surfaces.

### D. Mohr Theory as Applied to Bulk Solids

The previous discussion has considered the shear strength of bulk solids on an empirical basis without regard to theory. The strength behaviors of various materials have been explained by several theories, such as those of von Mises, Tresca, and Mohr, however, no single theory is adequate for all materials. Mohr's theory is most applicable to granular materials and states that on any plane the shearing stress is

$$\tau = \sigma \tan \psi \tag{5}$$

where $\psi$, the obliquity angle, is the angle between the resultant of the shearing and normal stresses and the normal stress. When the obliquity angle on the failure plane reaches its maximum value, failure occurs. Thus, if the maximum obliquity angle is $\psi_m$ the stresses on the failure plane at failure are described by the equation:

$$\tau = \sigma \tan \psi_m \tag{6}$$

For a particulate material at failure $\psi_m$ is equal to $\phi$.

Mohr also demonstrated that within any stressed material there are three orthogonal planes upon which the shear stresses are zero. These planes are called principal planes and the stresses which act on them are principal stresses. The largest stress is the major principal stress, $\sigma_1$, the smallest the minor principal stress, $\sigma_3$, and the one in between is the intermediate principal stress, $\sigma_2$. Most applications in soil mechanics are reduced to two dimensions and because the limiting stress conditions are defined by the major and minor principal stresses, the principal stresses of most interest are $\sigma_1$ and $\sigma_3$. Given the magnitude and orientation of the principal stresses, it is possible to compute the normal and shear stresses on any other plane from a consideration of static equilibrium according to the equations:

$$\sigma = \sigma_1 \cos^2\theta + \sigma_3 \sin^2\theta = \frac{\sigma_1 + \sigma_3}{2} + \frac{\sigma_1 - \sigma_3}{2} \cos 2\theta \tag{7}$$

$$\tau = (\sigma_1 - \sigma_3) \sin\theta \cos\theta = \frac{\sigma_1 - \sigma_3}{2} \sin 2\theta \tag{8}$$

where $\sigma$ and $\tau$ are the normal and shear stresses acting on a plane making an angle $\theta$, measured counterclockwise, with the major principal plane. It is possible to eliminate $\theta$ from these two equations, thus developing a single equation relating the major and minor principal stresses and the stresses on the plane of interest:

$$\tau^2 + \left[\sigma - \left(\frac{\sigma_1 + \sigma_3}{2}\right)\right]^2 = \left(\frac{\sigma_1 - \sigma_3}{2}\right)^2 \tag{9}$$

It can be seen that this equation is that of a circle with variables $\sigma$ and $\tau$, radius $(\sigma_1 - \sigma_3)/2$ and center $(\sigma_1 + \sigma_3)/2$ as shown in Figure 5. This is referred to as the Mohr circle and is used in soil mechanics applications with the sign convention that compressive stresses and counterclockwise shear stresses are positive. The Mohr circle, used to describe graphically the state of stress in any body is tangential to the failure envelope when the body is stressed to failure. These graphical relationships are shown in Figure 5. A description of Mohr theory as applied in soil mechanics can be found in Holtz and Kovacs.[39]

Another unique aspect of the application of Mohr theory to soil mechanics is the concept of the origin of planes. The origin of planes, $O_p$, is a point on the circumference of the Mohr circle which has the property that a line through the origin of planes, and any other point designating stresses, will be parallel to the plane upon which those stresses are acting. This relationship is shown in Figure 6.

### E. Stress Paths

If it is necessary to describe and analyze many states of stress, a graphical representation of a large number of Mohr circles becomes confusing and an alternative, such as "stress path analysis", is desirable. This method of analysis is discussed in two papers by Lambe.[40,41] A stress path analysis plots p vs. q where

$$p = \frac{\sigma_1 + \sigma_3}{2} \tag{10}$$

and

$$q = \pm \frac{\sigma_1 - \sigma_3}{2} \tag{11}$$

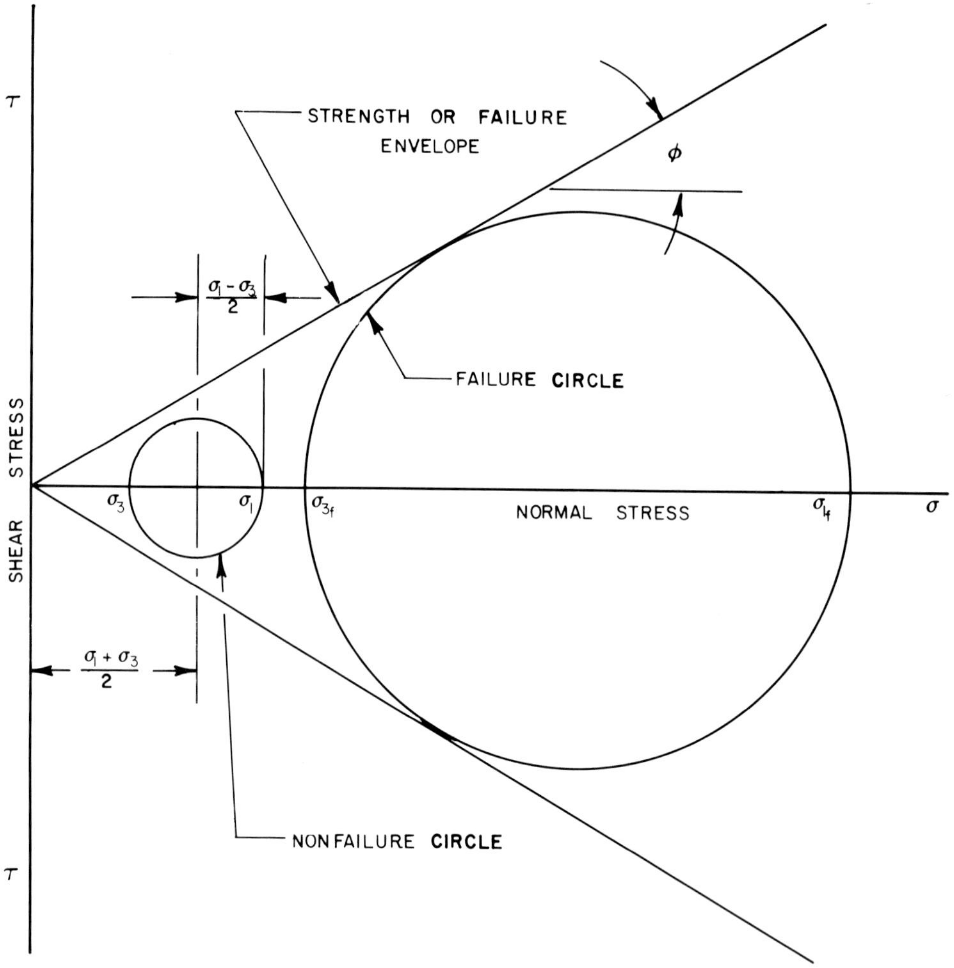


FIGURE 5. Mohr circles and strength envelope.

In other words, the top or bottom of the Mohr circle is plotted as a point. The coordinate q will be negative if $\sigma_1$ is inclined less than 45° to the horizontal. If p and q for a given stress state are known, it is possible to plot the corresponding Mohr circle.

Figure 7 shows the stress path on a p — q diagram for a body which has been isotropically consolidated and then loaded to failure by holding $\sigma_3$ constant while $\sigma_1$ is increased to failure. The isotropic consolidation is a line along the abscissa and the compression loading is a 45° line ascending to the right.

If each stress path terminates in failure, the locus of the failure points for various stress paths on a given material is called the $K_f$ lines, which pass through the uppermost and lowermost points of the limiting stress Mohr circles. The $K_f$ lines make an angle $\alpha$ with the horizontal as shown in Figure 7. Because the failure envelope is tangential to the same limiting stress circles, it can be shown from the geometry of the Mohr circle that sin $\phi$ is equal to tan $\alpha$. Thus, the $K_f$ lines are an alternative to the failure envelope; however, a point on the $K_f$ line represents the stresses on the plane of maximum shear stress, whereas a point on the failure envelope describes the stresses on the failure plane. Obviously, stress paths can be plotted for nonfailure loading conditions as well as for loading that results in shear failure.

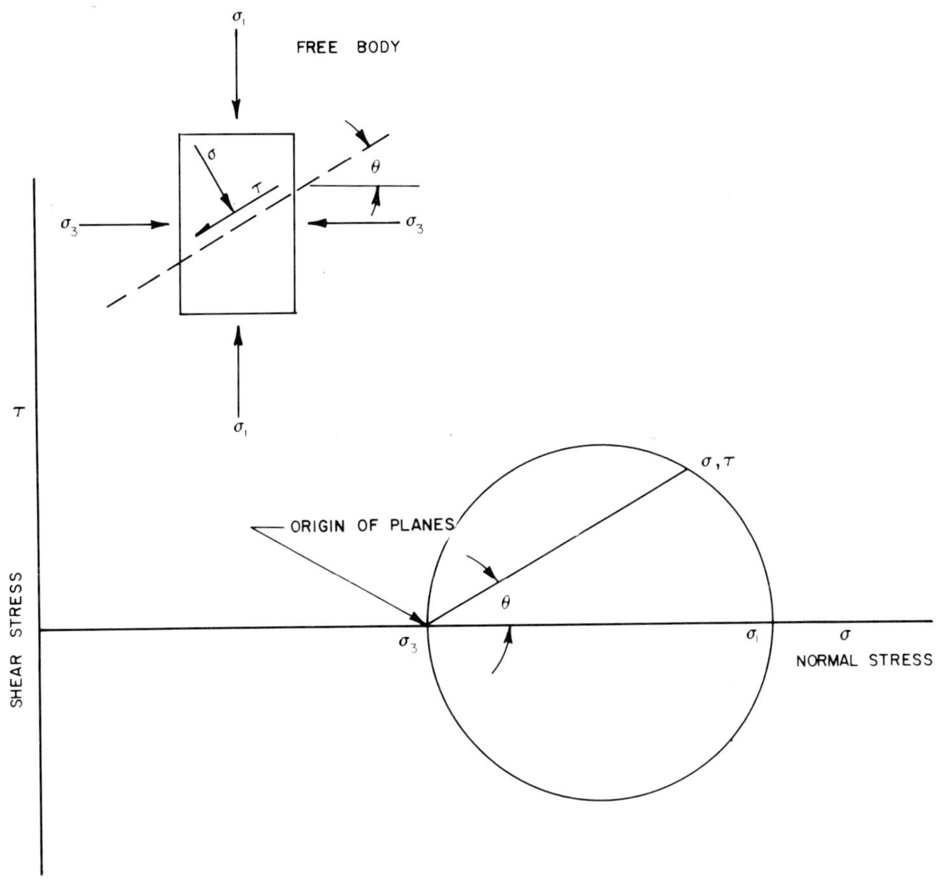


FIGURE 6. Mohr circle and origin of planes.

## F. Stress Paths for Various Loading Conditions

Granular materials have no tensile strength, therefore compression evaluation is most relevant. Two failure stress paths of interest to designers of storage facilities for bulk solids are compression unloading, in which the material is brought to failure by decreasing the minor stress while holding the major stress constant, and extension loading where the minor stress is increased while holding the major stress constant. In a bulk storage structure the major principal stress acts vertically and the minor principal stress acts horizontally. When a granular material is sheared by compression unloading the failure is referred to as an active case failure, whereas the extension loading is called passive case failure. These stress paths and the $K_f$ lines are shown in Figure 8. From a consideration of the geometry of the Mohr circle, if can be shown that the active case horizontal stress ratio is

$$K_a = \frac{\sigma_{hf}}{\sigma_{vf}} = \frac{1 - \sin\phi}{1 + \sin\phi} \tag{12}$$

where $K_a$ is the active case stress ratio, $\sigma_{h_f}$ is the horizontal stress at failure and $\sigma_{v_f}$ is the vertical stress at failure. The passive case stress ratio, $K_p$, is the reciprocal of $K_a$:

$$K_p = \frac{1}{K_a} = \frac{1 + \sin\phi}{1 - \sin\phi} \tag{13}$$

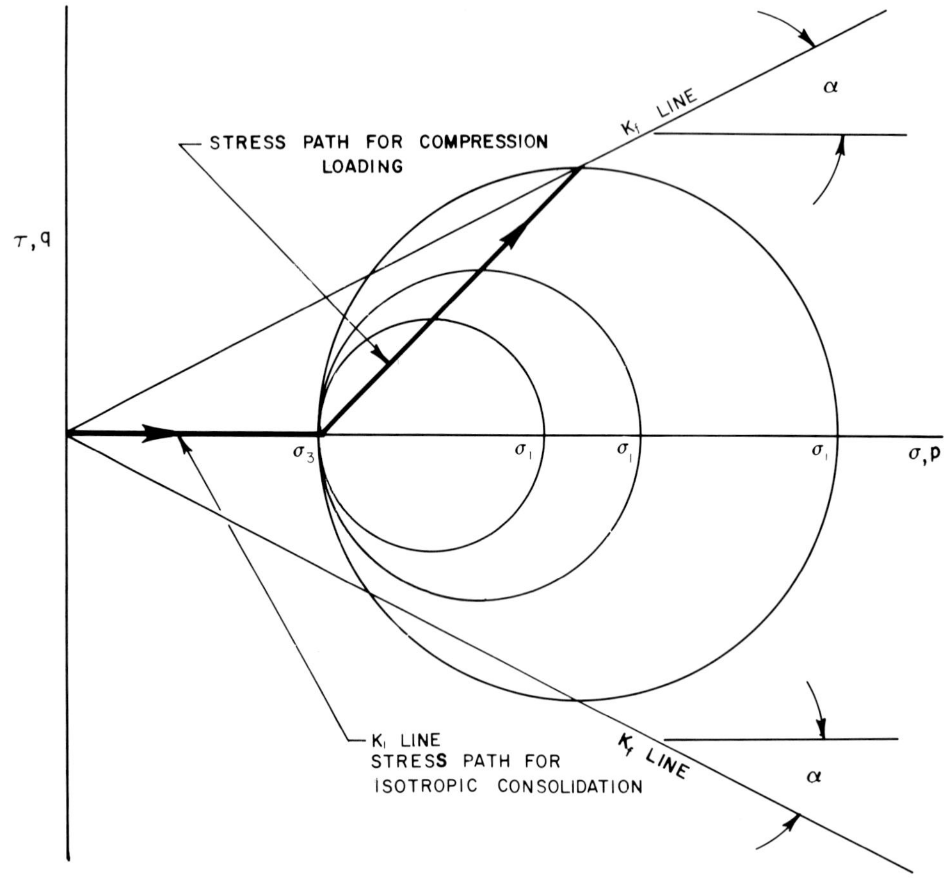


FIGURE 7. Stress path and Mohr circles for a specimen loaded to failure under compression loading.

Both the active and passive case stress ratios have been used to estimate lateral stresses in silos.

The confined compression condition is of interest in silo design because a test of this nature represents the stress condition of an element of granular material within a rigid silo which is being subjected to increasing vertical stress as more material is being added from the top. The ratio of lateral to vertical stress for the condition of no lateral strain is referred to as the "at rest" lateral stress ratio and given the symbol $K_o$. If $\sigma_h$ is the horizontal stress and $\sigma_v$ is the vertical stress then

$$K_o = \frac{\sigma_h}{\sigma_v} \tag{14}$$

A relationship between $\phi$ and $K_o$ was developed by Jàky:[42]

$$K_o = \frac{(1 - \sin \phi)(1 + 2/3 \sin \phi)}{(1 + \sin \phi)} \tag{15}$$

which, for friction angles between 20 and 45°, may be approximated by $K_o = 0.9\ (1 - \sin \phi)$. Jàky[43] later simplified this to $K_o = (1 - \sin \phi)$, in which form it continues to be used. Thus, in soil mechanics applications $K_o$ is related to $\phi$ by the empirical expression

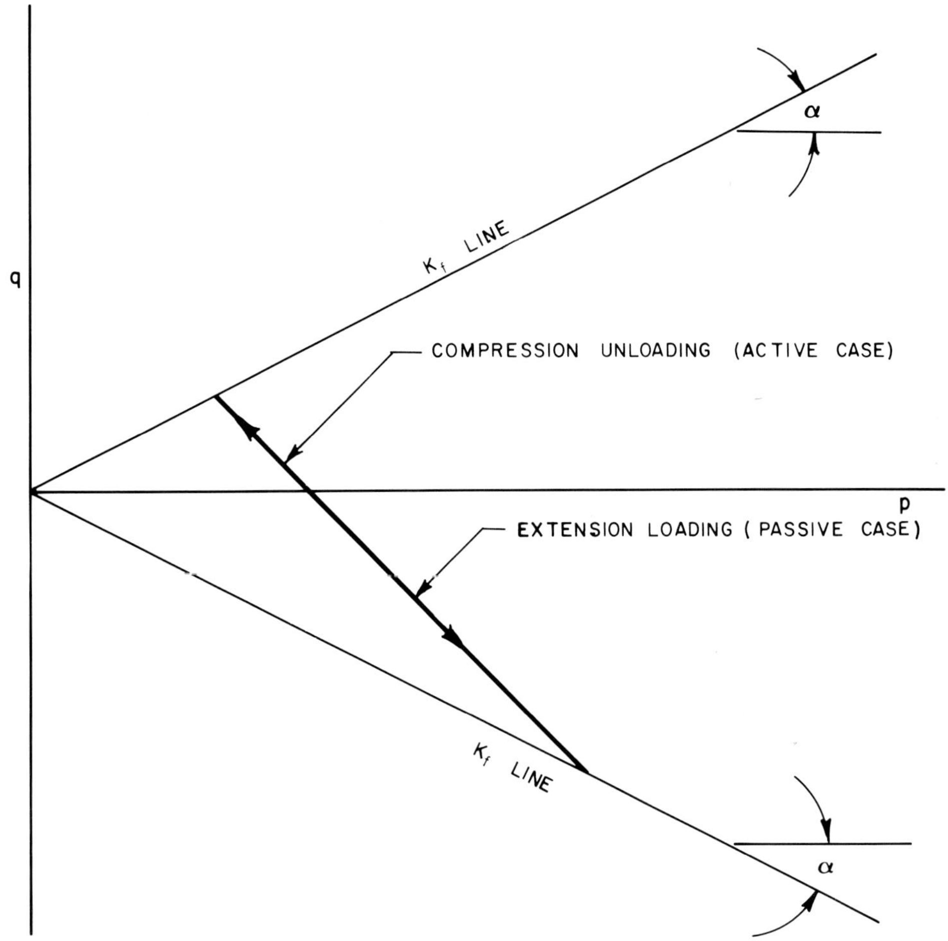


FIGURE 8. Stress paths for compression unloading and extension loading.

$$K_o = 1 - \sin \phi \tag{16}$$

The confined compression test can be shown on a p — q diagram as a line which makes an angle β with the abscissa. The slope of the $K_o$ line, tan β, is related to the internal friction angle, φ, by the relationship:

$$\tan \beta = \frac{\sin \phi}{2 - \sin \phi} \tag{17}$$

The $K_f$ lines and the $K_o$ lines for a granular material with $\phi = 30°$ are shown in Figure 9.

An important concept in the strength and stress-strain behavior of particulate systems is the effect of stress path upon the material behavior. The Mohr failure envelope and the $K_f$ line are independent of stress path. If a granular material is loaded to failure by compression loading, extension unloading, or any other application of stresses to induce failure, the strength parameters will be unaffected. However, the stress-strain and volume change characteristics are very much dependent upon stress path. In general, compression unloading will result in low values of principal stress difference for a given amount of strain, whereas compression loading will produce higher values of principal stress difference for the same strains. Extension loading, where the lateral stress is increased while the axial stress is held constant, will result in extremely high values of principal stress difference for a given strain. Although considerable data exist for the strength parameters of various bulk solids, with the exception of sands very little is known about the stress-strain behavior of other particulate systems when loaded by various stress paths.

Although current practice utilizes limiting stresses for design of bulk storage facilities, there is a growing interest in the use of finite element methods. This approach requires constitutive equations describing the stress-strain and volume change characteristics of particulate systems. The preceding discussion points out that because stress-strain characteristics of granular materials are dependent upon stress path, it is necessary that the laboratory tests simulate the stresses to be expected in the bin or silo. The following section describes test methods and an approach to the development of stress-strain parameters which may be useful in finite element analysis.

## III. TEST METHODS AND ANALYSES OF DATA

### A. Test Apparatus

From the previous section it can be seen that bulk materials are not linearly elastic and cannot, therefore, be characterized simply by evaluating Young's modulus and Poisson's ratio. In the field of soil mechanics, a number of tests have been developed to evaluate the stress-strain behavior of soils and other particulate materials subjected to various stress conditions. Apparatus used for many of these tests have been described in some detail by Bowles,[44] but the more common will be briefly described here.

In general, there are apparatus used primarily for measuring stress-strain characteristics, shear strength parameters, or both strength and stress-strain behavior. The one-dimensional consolidation test describes only stress-strain, whereas the direct shear test and Jenike shear test are used for strength determination. The triaxial test can be used for both strength and stress-strain measurements.

### B. One-Dimensional Compression

In the one-dimensional compression test, which is also known as the confined compression and consolidation test, a squat, cylindrical specimen is compressed axially while lateral strains are prevented. The ratio of the lateral stress to the axial stress is assumed to be constant and equal to $K_o$, the at-rest lateral stress ratio. The stress path for this test, the $K_o$ line, raises from the origin but always remains below the $K_f$ line and thus, the specimen never reaches shear failure as shown in Figure 9. The volumetric strain is the result of axial strain alone, and because axial strains are small due to the lateral constraint, the shear strain is small (see Figure 10a).

The apparatus, known as a consolidometer or an oedometer, is shown schematically in Figure 11. The axial stress is applied incrementally via porous stones which allow the egress of pore fluid. Lateral strain is prevented by a close-fitting confining ring against which the lateral stresses are developed. The main experimental objection to the apparatus is the side friction developed between the material and the confining ring, which violates the desired one-dimensional strain condition. This effect may be minimized by using a "floating" confining ring with a low-friction liner.

As each increment of axial stress is added, the axial deformation, which may be time dependent, is measured. The results may be presented graphically in terms of axial stress vs. axial strain or in terms of void ratio vs. axial stress (see Figure 12). The deformation is primarily plastic, and reduction of the axial stress results in the recovery of only the elastic deformation which is small by comparison.

From curves such as those in Figure 12, the loading and unloading tangent moduli (the slopes of the loading and unloading stress-strain curves) may be evaluated for given stresses

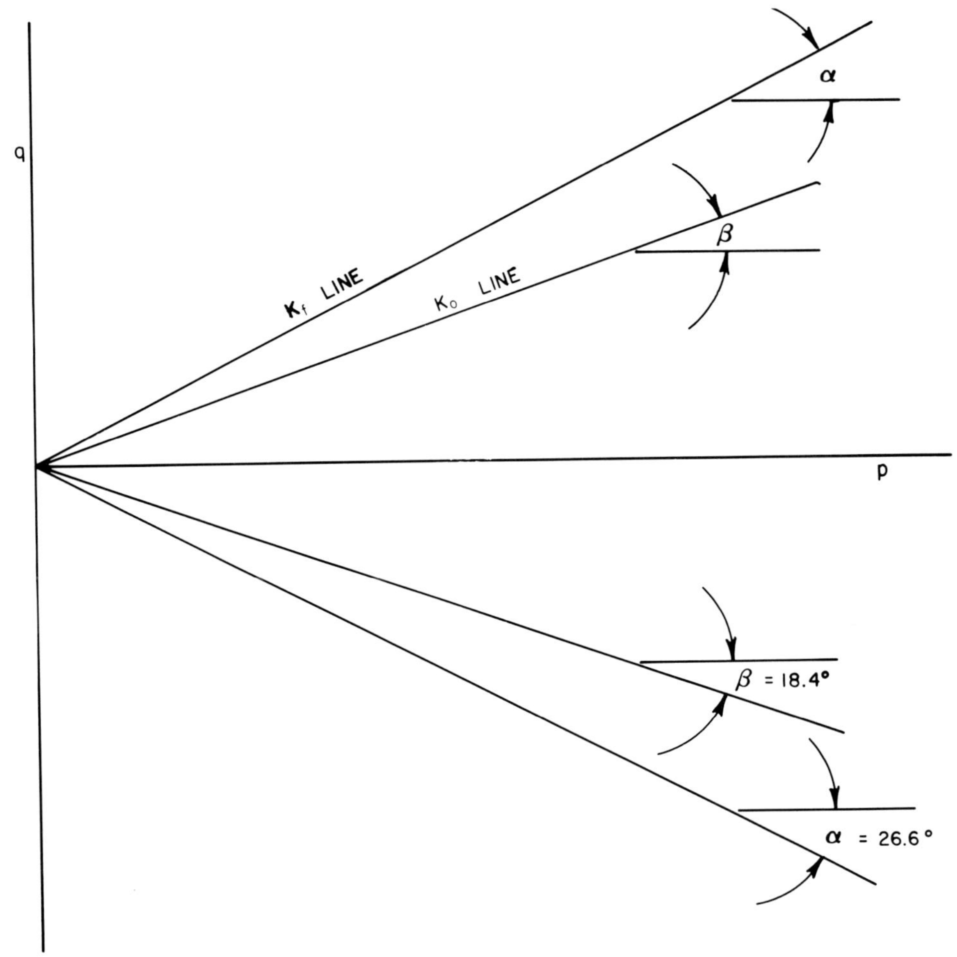


FIGURE 9. Stress path for anisotropic consolidation ($K_o$ line) relative to $K_f$ line.

or strains. For unconsolidated materials, the tangent modulus increases approximately exponentially with strain. The one-dimensional consolidation test is normally used for estimating the axial strain resulting from a change in axial stress, thus, the stress-strain relationship may be expressed by:

$$\epsilon_1 = A \log \left(\frac{\sigma_1}{\sigma_{1.o}}\right) \tag{18}$$

where $\epsilon_1$ = axial strain; $\sigma_1$ = axial stress; $\sigma_{1.o}$ = initial axial stress, ($\epsilon_1 = 0$); and A = an empirical constant.

Equation 18 is linear with variables $\epsilon_1$ and log $(\sigma_1/\sigma_{1.o})$ and with slope A which may be evaluated by the least squares procedure with no intercept. Rearranging Equation 18, differentiating and substituting, the constrained tangent modulus, $M_t$, is given by:

$$M_t = \frac{\sigma_1}{A} \tag{19}$$

Drnevich[45] has suggested an alternative hyperbolic stress-strain relationship:

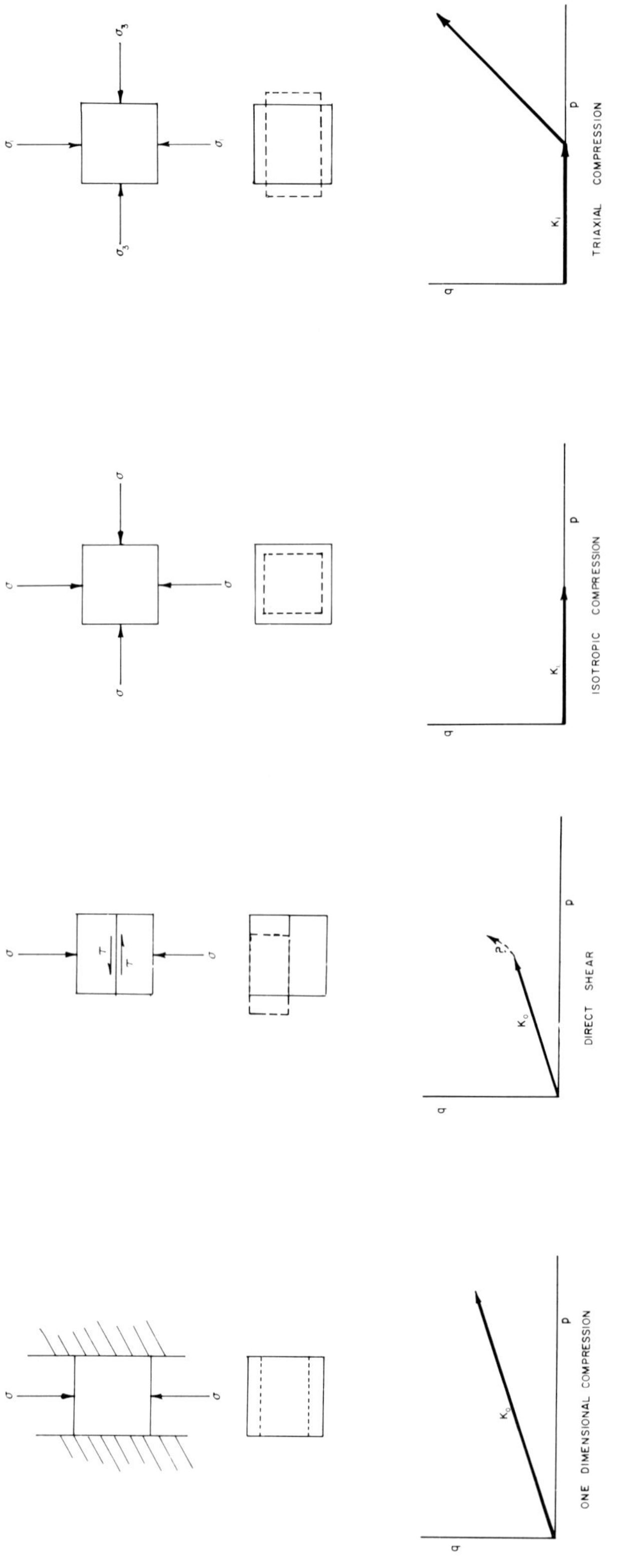


FIGURE 10. Stress, deformation, and stress paths for various test conditions.

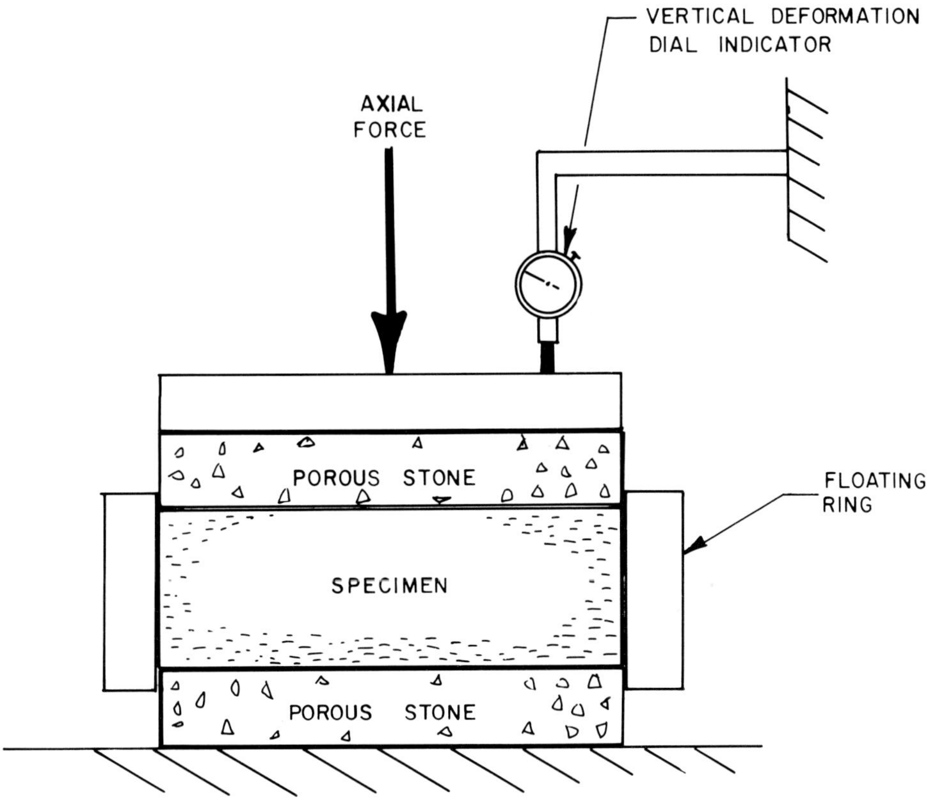


FIGURE 11. Floating ring consolidometer.

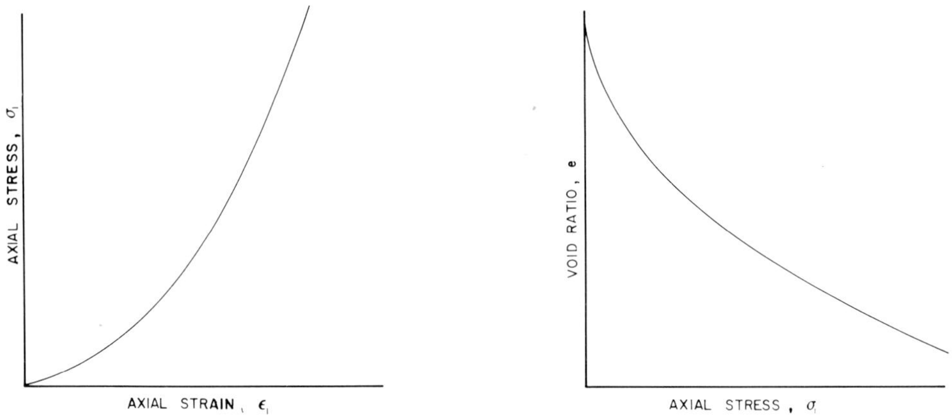


FIGURE 12. Stress-strain and void ratio-stress relationships for one-dimensional compression test.

$$\sigma_1 = \frac{M_i \epsilon_1}{1 - \dfrac{\epsilon_1}{\epsilon_{1.ult}}} \tag{20}$$

where $M_i$ = initial tangent modulus ($\epsilon_1 = 0$) and $\epsilon_{1.ult}$ = the asymptotic maximum value of $\epsilon_1$ (see Figure 13).

The constants $M_i$ and $\epsilon_{1.ult}$ may be evaluated by rearranging Equation 20, which results in

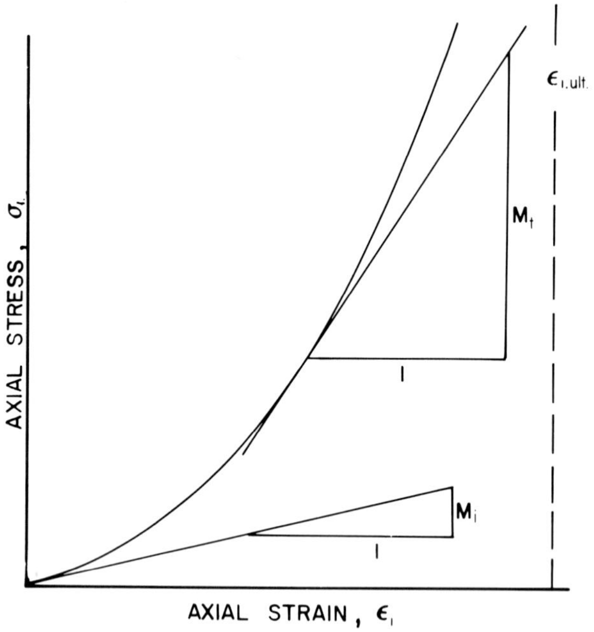


FIGURE 13. Hyperbolic one-dimensional compression stress-strain relationship.

$$\frac{\sigma_1}{\epsilon_1} = M_i + \frac{\sigma_1}{\epsilon_{1.ult}} \tag{21}$$

which is linear with variables $\sigma_1/\epsilon_1$ and $\sigma_1$. The least-squares procedure may be used to evaluate $M_i$ and $\epsilon_{1.ult}$ which are the intercept and the inverse slope, respectively.

Rearranging Equation 20 and differentiating, the constrained tangent modulus is given by

$$M_t = M_i \frac{\epsilon_{1.ult}^2}{(\epsilon_{1.ult} - \epsilon_1)^2} \tag{22}$$

or, if the use of the axial stress is more convenient than the use of the axial strain, by

$$M_t = M_i \left[\frac{\sigma_1 + 1}{M_i\ \epsilon_{1.ult}}\right]^2 \tag{23}$$

## C. Direct Shear

In the direct shear test, the specimen is compressed or consolidated one-dimensionally and then sheared along a horizontal plane. During the consolidation phase the stress path is assumed to follow the $K_o$ line, but during the shearing phase the principal stresses rotate unpredictably and the stress path to failure is unknown. The applied shear stress produces a large shear strain at failure, but volumetric strain is also possible (see Figure 10b).

Two basic types of direct shear apparatus exist, the translational shear box and the rotational shear box. The translational shear box, shown schematically in Figure 14, confines the specimen in a circular or square box which is split horizontally. The axial stress, which is applied via ridged plates, consolidates the specimen as in the one-dimensional compression test. The horizontal shear force is applied to the free portion of the box, which slides over the fixed portion as the specimen is sheared. The main experimental objections to the

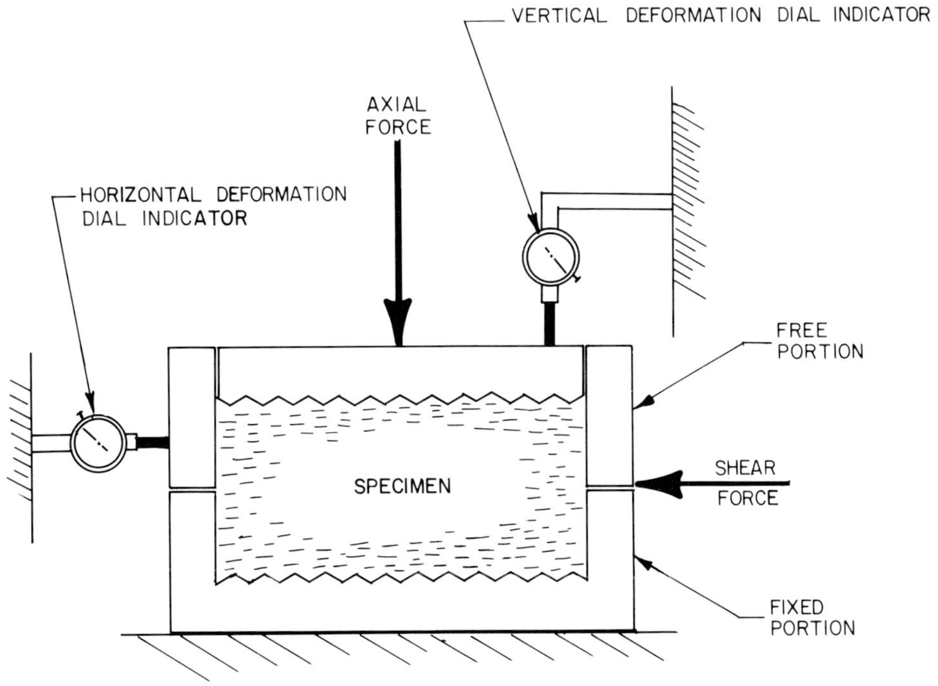


FIGURE 14. Translational direct shear test apparatus.

apparatus are that the failure plane is forced to be horizontal and that failure occurs only in a central lenticular zone rather than throughout the whole sample (see Figure 15), so that shear strains cannot be evaluated. Despite these objections, the direct shear apparatus is useful for determining shear strength parameters.

The simple shear apparatus overcomes the experimental objections somewhat by confining the specimen in a rubber membrane reinforced with wire rings or by confining the specimen within a hinged box.[46] Although these apparatus allow shear strains to be evaluated, since the whole sample distorts, the shearing stresses are not uniformly distributed (see Figure 15).

On completion of the consolidation phase, the specimen is normally subjected to a uniform shear deformation rate although controlled deformation and stress application rates are possible. The shear force is recorded as shear deformation proceeds allowing shear stress-shear deformation curves to be plotted (see Figure 16). Peak or ultimate values of shear stress may be taken as the shear strengths and plotted against normal stress to define the failure envelope of the material.

The initial position of the shear/stress-shear deformation relationship for cohesionless materials may be represented by an hyperbola, but because strains cannot be evaluated an expression for the tangent shear modulus cannot be derived. An expression for the tangent stiffness modulus (the slope of the shear/stress-shear deformation curve) can be found,[47] although this is of little value in describing the behavior of bulk material. It may be used, however, to describe bulk material-solid surface interaction if the bulk material in one portion of the shear box is replaced with a piece of the solid material under investigation, so that shearing takes place between the bulk and solid materials. The shear parameters evaluated, referred to as adhesion and wall friction, and the tangent stiffness modulus are commonly used in the finite element analysis of bulk storage structures and retaining walls.

The rotational shear box is similar to the circular translational shear box except that the shear stress is applied by rotating the top portion of the box. The experimental objections

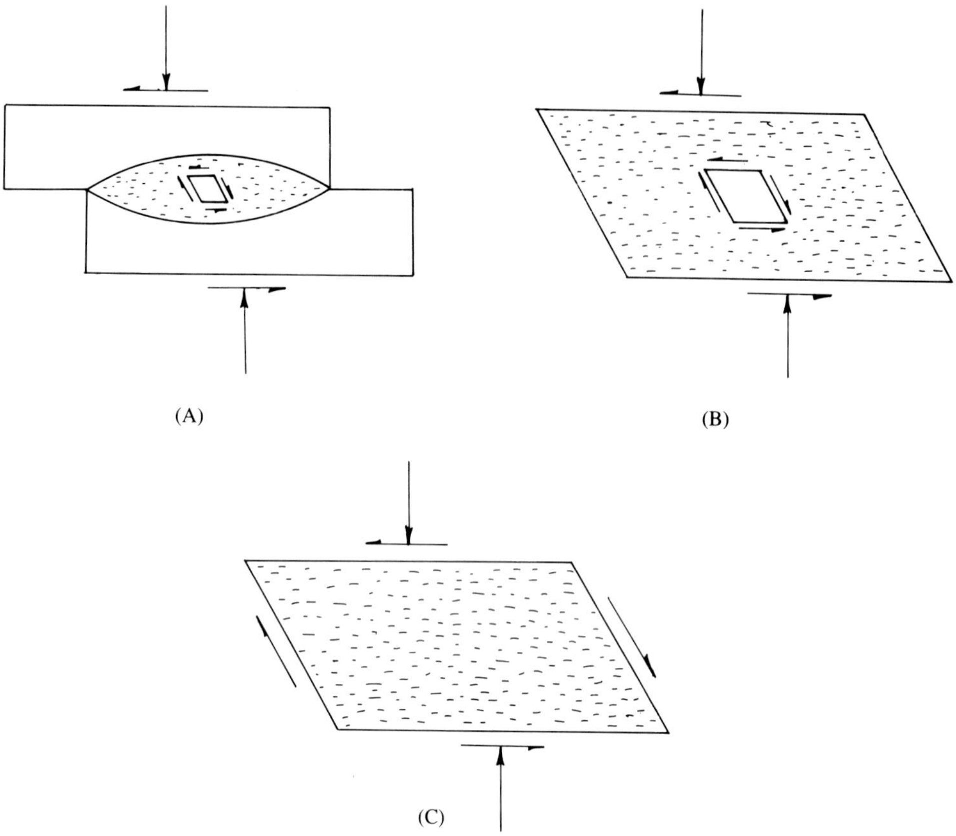


FIGURE 15. Stresses and deformations in: (A) direct shear, (B) simple shear, and (C) pure shear.

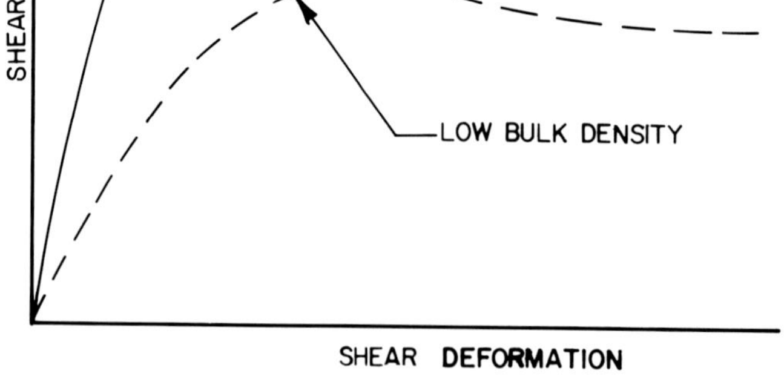


FIGURE 16. Shear stress-deformation relationships for direct shear test.

to the translational shear box also apply to the rotational shear box; in addition, the shear deformation in the failure plane is not constant but proportional to the distance from the axis of rotation. The advantage of the rotational shear box is that large shear deformations

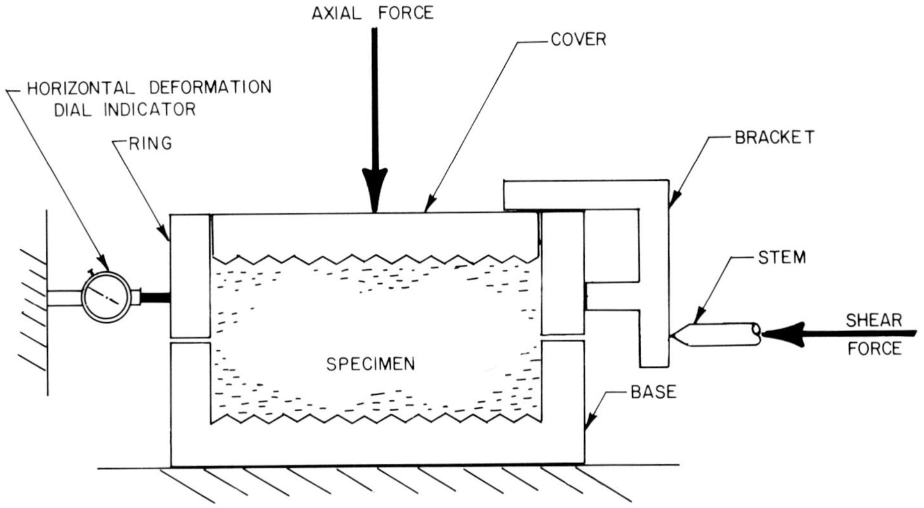


FIGURE 17. Jenike direct shear test apparatus.

are possible. Large shear deformations facilitate specialized tests such as those where the sample is used for several levels of axial stress.[48] To minimize the problem of variable shear strain the circular shear plane may be replaced by a thin, annular shear plane, but this in turn, increases the significance of the friction between the bulk material and the walls of the annular box. To overcome these difficulties, Scarlett and Todd[49] designed an annular direct shear box on the guard ring principle in which the normal stress is applied independently to three concentric annuli. The shear strength of the material is calculated from the torque necessary to constrain the center annulus when the lower portion of the box is rotated.

### D. Jenike Shear

In order to characterize the flowability of cohesive bulk solids, Jenike[20] developed a specialized translational direct shear test apparatus. The conventional direct shear apparatus used for soil testing was unsuitable, primarily because the range of cohesion values for which it is designed is greater than the useful range for bulk solids.

In the Jenike shear test, the specimen is consolidated plastically under normal and shear stresses and is then sheared under reduced normal stresses. The initial consolidation takes place one-dimensionally, as in the direct shear test, and the stress path is assumed to follow the $K_o$ line. During the shearing portion of the consolidation phase the principal stresses rotate unpredictably and the stress path to the plastic flow condition is unknown, as is the stress path during the later shearing phase of the test. The applied shear stresses produce large shear strains at failure, but volumetric strains are also possible.

The apparatus, shown schematically in Figure 17, consists of a cover, a ring, and a base. The base and the ring confine the bulk solid to be tested and the axial normal forces are applied to the cover. The horizontal shear forces are applied, via a stem and bracket, to the cover and the ring which slide over the base. The experimental objections to the translational shear box also apply to the Jenike shear apparatus, although the method of application of the shear force to the ring and cover is claimed to improve the distribution of the stresses across the specimen.

The specimen is preconsolidated with a constant axial normal stress and an oscillating horizontal shear stress and is then consolidated by its being sheared plastically in a horizontal plane. On completion of the consolidation phase, the axial normal stress is reduced and the specimen again sheared plastically. The maximum shear stress developed is taken as the shear strength of the material. A series of such tests, at different reduced axial stresses, allows definition of the failure envelope of the material for the particular normal consolidating stress. The failure envelope of a noncohesive material is independent of the consolidation stress, approximately linear, and passes through the origin of the Mohr diagram. The failure envelope of a cohesive material is dependent upon the magnitude and duration of the consolidation stress with the intercept on the shear axis of the Mohr diagram (i.e., the cohesion) approximately directly related to the magnitude of the consolidation stress. The failure envelope generated from Jenike shear tests on a cohesive material is similar to that of an overconsolidated clay. More details of the apparatus and its use may be found in papers by Jenike[20,21] and Schwedes.[50]

### E. Triaxial Shear

In the triaxial shear test the specimen may be subjected to controlled stresses in three orthogonal directions. Although the name of the apparatus suggests that the stresses may be controlled independently, in all but highly complex research apparatus two of the orthogonal stresses are equal. Typically, the specimen is consolidated isotropically and then compressed axially under constant lateral confining stresses. The stress path for such a test is coincident with the p axis during the consolidation phase and raises at 45° to the right to meet the $K_f$ line during the axial compression phase (Figure 7). With suitably instrumented triaxial shear apparatus, any stress path or series of stress paths may be followed to emulate the prototypical stress conditions. Isotropic consolidation of an isotropic specimen produces axial and lateral normal strains which lead to volumetric strain but little shear strain (Figure 10c). Anisotropic stress conditions normally produce volumetric and shear strains (Figure 10d).

The basic triaxial apparatus, shown in Figure 18, consists of a pressure chamber in which a cylindrical specimen is subjected to controlled axial and lateral stresses. The specimen is enclosed by a flexible rubber membrane and by rigid end caps. Isotropic stresses are imparted to the specimen by pressurizing the confining fluid in the cell. Anisotropic stress conditions are achieved by mechanically applying axial stresses via the end caps. It is assumed that the principal stresses act on vertical and horizontal planes throughout the specimen, however they rotate unpredictably at the ends due to the friction developed between the specimen and the rigid end caps.

On completion of the consolidation phase, the specimen is normally subjected to uniform axial deformation rate although controlled deformation and stress application rates are possible. The mechanically applied axial force is recorded as axial compression proceeds, allowing principal stress difference-axial strain curves to be plotted, as in Figure 19. The peak, or ultimate, values of principal stress difference may be taken as the diameters of the Mohr failure circles which define the failure envelope of the material.

The principal stress difference-axial strain relationship may be approximated by a hyperbola[51] of the form:

$$\Delta\sigma = \frac{\epsilon_a}{\dfrac{1}{E_i} + \dfrac{\epsilon_a}{\Delta\sigma_{ult}}} \tag{24}$$

where $\Delta\sigma = \sigma_1 - \sigma_3$ (where $\sigma_1$ and $\sigma_3$ are the major and minor principal stresses, respectively); $\epsilon_a$ = axial strain; $E_i$ = initial tangent modulus of elasticity; and $\Delta\sigma_{ult}$ = the asymptotical maximum value of $\Delta\sigma$, termed the asymptotic compressive strength (see Figure 19).

$E_i$ and $\Delta\sigma_{ult}$ may be evaluated by rearranging Equation 24, which results in

$$\frac{\epsilon_a}{\Delta\sigma} = \frac{1}{E_i} + \frac{\epsilon_a}{\Delta\sigma_{ult}} \tag{25}$$

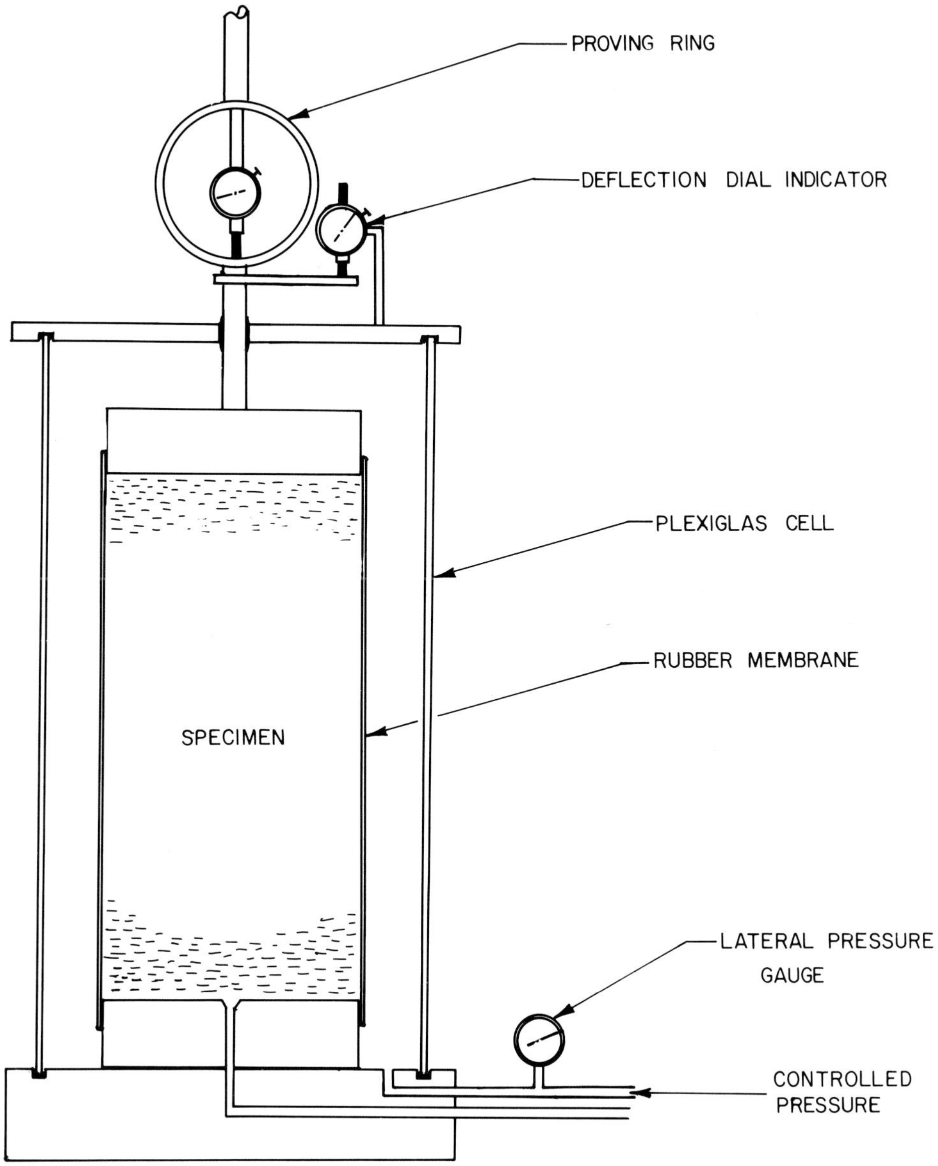


FIGURE 18. Triaxial shear apparatus.

which is linear with variables $\epsilon_a/\Delta\sigma$ and $\epsilon_a$. The least-squares procedure may be used to evaluate $E_i$ and $\Delta\sigma_{ult}$, which are the inverse of the intercept and slope, respectively.

Janbu[52] has shown that the relationship between the initial tangent modulus and the minor principal stress may be expressed by

$$E_i = k_h P_a \left(\frac{\sigma_3}{P_a}\right)^n \tag{26}$$

where $k_h$ and n are empirical, dimensionless constants and $P_a$ = atmospheric pressure.

The constants may be evaluated by the least-squares procedure in logarithmic space.

Substituting for $E_i$ in Equation 24, rearranging and differentiating with respect to axial strain, the tangent modulus is given by

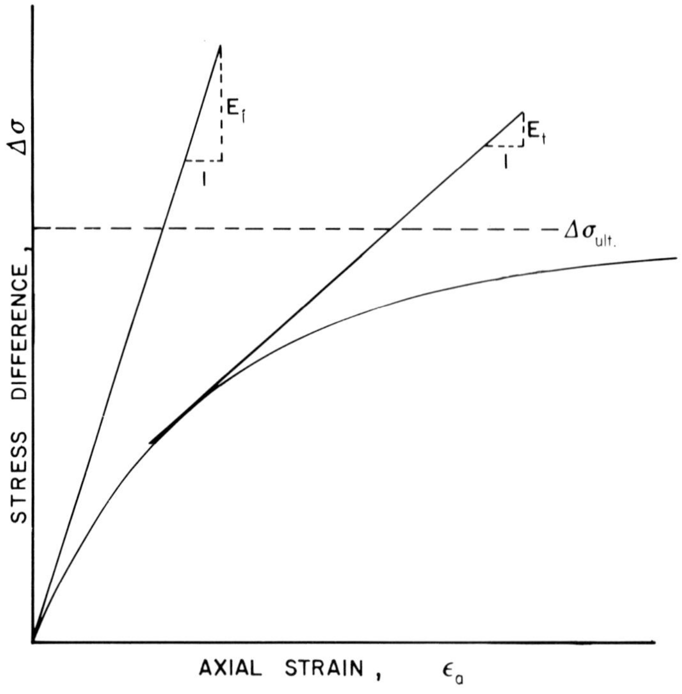


FIGURE 19. Hyperbolic, triaxial compression stress-strain relationship.

$$E_t = k_h P_a \left(\frac{\sigma_3}{P_a}\right)^n \left[1 - \frac{\Delta\sigma}{\Delta\sigma_{ult}}\right]^2 \tag{27}$$

Further details of this analysis and its application to agricultural grains are reported by Smith and Lohnes.[30]

The conventional triaxial shear test apparatus, which was developed for testing soils, is unsuitable for testing bulk solids primarily because of the high stress range for which it is designed. Marchant and Schofield[53] and Smith and Lohnes[54] developed a low-stress triaxial shear test apparatus suitable for testing agricultural grains. A unique feature of the latter apparatus (see Figure 20), which makes the one-dimensional consolidation test (also known as a zero lateral strain test) possible, is that the ram which enters the cell to apply the additional axial stress has the same cross-sectional areas as the specimen. The lateral stress is applied by partially evacuating the specimen pore space, while maintaining the cell air space at atmospheric pressure. The volume change of the specimen is monitored indirectly by coupling the cell air space to a volume monitor which consists of a bead of mercury in a horizontal capillary tube. Any change in the volume of the cell air space during axial loading will cause air to enter or leave the capillary tube, thereby displacing the mercury bead. Since the cross-sectional area of the entering ram and shortening specimen are equal, a constant cell air volume creates a condition of zero lateral strain for the specimen. Thus, it is possible to achieve zero lateral strain by adjusting the specimen pore space vacuum such that the mercury bead remains stationary in the volume monitor. Tests may also be performed such that the specimen is maintained at constant volume during the axial compression phase. The lateral stress is applied by increasing the air pressure in the cell and adjusting it throughout the test such that the volume of the specimen pore space, and thus the volume of the specimen, is held constant. The latter is achieved by coupling the specimen pore space to the volume monitor and adjusting the cell air pressure such that the mercury bead remains stationary in the capillary tube.

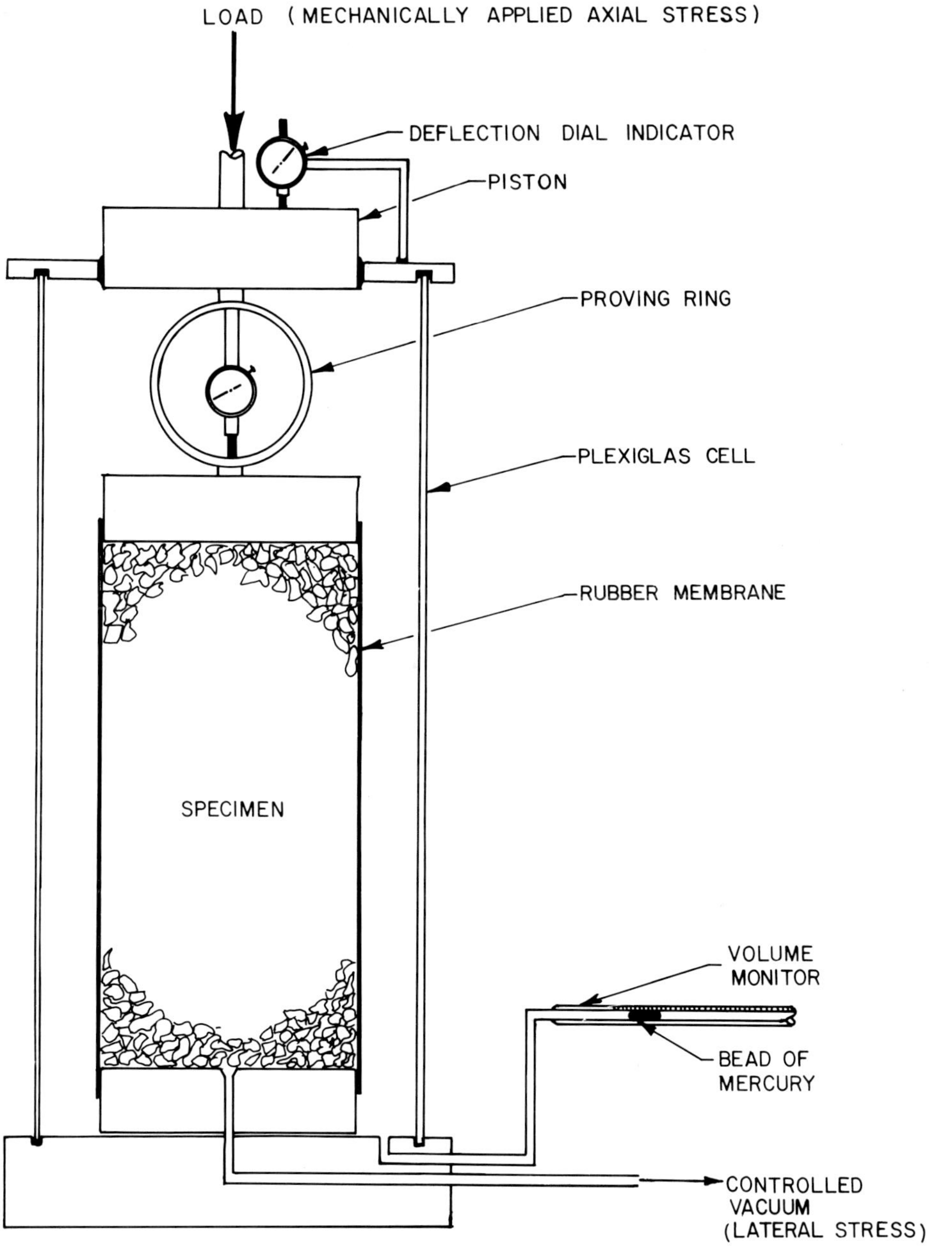


FIGURE 20. Low stress, triaxial shear test apparatus as set up for zero lateral strain test.

## IV. DESIGN STRESSES

Early designers assumed that vertical and lateral stresses were equal and designed storage structures for hydrostatic pressures based on the bulk densities of the granular materials. This procedure, which ignores the shear strength of the granular material, overestimates lateral stresses if the confining structure remains stationary or is allowed to move away from the granular material and may over- or underestimate lateral stresses if the structure is moved toward the granular material, depending upon the magnitude of the movement. Rankine[2] recognized the importance of the frictional shear strength mobilized within the granular material behind a retaining structure and developed expressions for the ratio of the lateral stress to the vertical stress assuming that all the internal shear strength of the granular material is developed (i.e., the granular material is in a state of plastic failure). If the structure is considered to move away from the granular material, the ratio, known as the active case lateral stress ratio, is given by Equation 12. If the structure is considered to be moved toward the granular material the ratio, known as the passive case lateral stress ratio, is given by Equation 13. If the full shear strength of the material is not mobilized, the lateral stress ratio lies between the active and passive case lateral stress ratios which represent the lower and upper bounds, respectively.

Jàky[42,43] developed an expression for the lateral stress ratio for material retained by a stationary structure in which a portion of the material shear strength is mobilized. For this zero lateral strain condition, the ratio, known as the at-rest lateral stress ratio, is usually assumed to be as described in Equation 16.

Rankine's and Jàky's analyses assume frictionless retaining structures, therefore the major and minor principal stresses act vertically and horizontally, respectively, in the at-rest and active case and vice versa in the passive case. It follows that the lower bound for the ratio of the minor to the major principal stress is $(1 - \sin \phi)/(1 + \sin \phi)$, the upper bound being unity.

Rankine's solution adequately describes the lateral stresses developed in shallow bins and is still used today for such structures. In tall structures, however, the assumption of zero friction between the granular material and the wall leads to overestimates of both lateral and vertical stresses. Janssen[5,6] recognized the importance of this friction and, by equating the vertical forces acting on a horizontal element within the granular mass, derived an exponential expression for the vertical stress at any depth. The lateral stress was determined by multiplying the vertical stress by a lateral stress ratio which Janssen suggested should be found by experiment. Koenen[55] suggested using the active case lateral stress ratio and, in this form, Janssen's solution continues to be used in the design of tall structures.

Reimbert[56] developed a theory, which was based on extensive model tests, in which the lateral stress distribution is hyperbolic, but Isaacson and Boyd[57] showed that this semiempirical solution and Janssen's theoretical solution are both special cases of a more general solution.

Janssen's and Reimbert's solutions adequately describe the lateral stress during the loading and storage of granular materials, but in some instances underestimates the lateral stresses developed during the unloading of the material. Takhtamishev[16] noted that the unloading stresses depend upon the flow patterns, of which there are two major types. The first type, called funnel-flow, is characterized by material proceeding to the discharge opening via a central funnel which is supplied only by material near the surface. The second type, called mass-flow, is characterized by the entire granular mass moving simultaneously toward the discharge opening. Takhtamishev concluded that mass-flow, which occurred in materials of low bulk density, was responsible for increased lateral stresses, whereas funnel-flow, which occurred in materials of high bulk density, produced no lateral stress increases. Although the increased stresses were attributed to bulk density, no mechanistic explanation was presented.

Kötter, as cited by Jenike and Johanson,[22] suggested that, whereas active stress conditions prevail during loading and storage requiring use of the active case lateral stress ratio, passive stress conditions prevail during unloading which require use of the passive case lateral stress ratio. This concept was incorporated in the stress prediction equations developed by a number of workers in the 1950s. None of these equations has been widely accepted, although, more recently, Gaylord and Gaylord[58] and Vivancos[59] have reintroduced the idea of using the passive case lateral stress ratio with the Janssen and Reimbert solutions for the prediction of unloading lateral stresses. Reimbert and Reimbert,[60,61] however, suggest using their solutions with the active case lateral stress ratio increased by a dynamic coefficient to allow for the increased lateral stresses associated with discharge from a mass-flow bin. Modern codes tend to recommend this approach using Janssen's or Reimbert's solutions.

Jenike[20,21] developed equations to describe the stresses induced in a mass-flow hopper during discharge. He assumed a radial stress field, i.e., a field in which stresses increase linearly from the vertex of the hopper. He also assumed that the wall friction was fully developed and that the bulk material flowed plastically in a passive state. The passive case lateral stress ratio is given by

$$K = \frac{(1 + \sin \delta)}{(1 - \sin \delta)} \tag{28}$$

where $\delta$ is the effective angle of friction which is equal to $\phi$ for cohesionless materials. Jenike and Johanson[22] modified Jenike's equations to describe the static stresses induced in a mass-flow hopper by assuming an elastic-active stress state. They pointed out that it was not sufficient to design only for the larger of the stress distributions, but that the stresses produced by the switch from the active to the passive state must be accommodated. When flow commences, the switch travels up the hopper causing high stresses or overpressures at its contact with the hopper wall due to the unsupported weight of the material in transition.

The solutions of Jenike and Johanson are rapidly gaining acceptance for the design of mass- and funnel-flow storage structures. A complete overview of the design processes may be found in the papers by Jenike et al.[62]

## V. CONCLUDING REMARKS

Limiting stress analysis has been the traditional basis for the design of bulk solid storage and handling facilities. There has been a tendency to assign constant values to $\phi$ and $\phi'$ for a given granular material, thereby treating its strength as an invariable material property. The problems associated with many of these facilities[63] suggest that the failure to recognize the strength of a granular material as a parameter may contribute to design inadequacies. Designers should be aware that, for a given material, $\phi$ will vary with stress level, void ratio, moisture content, temperature, etc.

Recent interest in finite element analysis has created the need for a better understanding of the stress-strain characteristics of bulk solids. At present, analytical techniques are far in advance of knowledge on material properties, necessary for the formulation of constitutive equations. Tests on soils and agricultural grain have shown that stress-strain behavior of these materials is dependent upon stress path and it is likely that other particulate systems will show similar sensitivity to stress history.

The triaxial test is probably the most reliable means of evaluating strength and stress-strain characteristics because of its ability to measure strains and volume changes and to apply stresses along any stress path. Although the triaxial test may be the best conventional method, innovative tests are needed to more accurately simulate the stress conditions in storage and handling facilities.

Tests on agricultural grains have provided qualitative evidence that their bulk frictional characteristics may be more dependent upon individual particle morphology than upon interlocking and stress history.[29] The characterization of individual particle morphology has advanced greatly as illustrated by Beddow and Meloy[64] and Kaye;[65] however, data on bulk stress-strain properties for granular material other than soils is virtually nonexistent. It may be possible to relate bulk stress-strain relationships to particle stress-strain relationships, shape, surface roughness, and simple bulk properties such as void ratio. This has not, so far, been possible for natural soil due to the complex nature of its particle minerology, morphology, and packing and to the presence of pore water and cohesion. Conversely, agricultural products, coal, mineral ores, and processed bulk solids such as polyethylene pellets have complex particle morphology but are more uniform in composition, shape, and gradation than soils. It may be possible to predict the bulk behavior of these bulk solids from their particulate characteristics because of their less complex nature.

## ACKNOWLEDGMENT

The authors gratefully acknowledge the support given by the Engineering Research Institute of Iowa State University.

## ALPHABETICAL SYMBOLS

| | |
|---|---|
| A | Empirical constant (one-dimensional compression) |
| $E_i$ | Initial tangent modulus |
| $E_t$ | Tangent modulus |
| F | Shear force |
| i | Incliniation of teeth |
| K | Lateral stress ratio |
| $K_a$ | Active case lateral stress ratio |
| $k_h$ | Empirical constant (triaxial shear) |
| $K_\circ$ | At rest lateral stress ratio |
| $K_p$ | Passive case lateral stress ratio |
| $M_i$ | Initial constrained tangent modulus |
| $M_t$ | Constrained tangent modulus |
| n | Empirical constant (triaxial shear) |
| N | Normal force |
| N′ | Normal force on plane |
| $O_p$ | Origin of planes |
| (p,q) | Coordinates of stress path |
| Pa | Atmospheric pressure |
| R | Resultant force on plane |
| T | Shear force |

## GREEK SYMBOLS

| | |
|---|---|
| $\alpha$ | Angle between $K_f$ line and the abscissa |
| $\beta$ | Angle between $K_\circ$ line and the abscissa |
| $\delta$ | Effective angle of internal friction |
| $\epsilon_1$ | Axial strain (one-dimensional compression) |
| $\epsilon_{1.ult}$ | Asymptotic maximum value of $\epsilon_1$ |
| $\epsilon_a$ | Axial strain (triaxial shear) |
| $\theta$ | Angle between planes on which $\sigma$ and $\sigma_1$ act |
| $\mu$ | Coefficient of sliding friction |
| $\sigma$ | Normal stress |
| $\sigma'$ | Normal stress on plane |

| | |
|---|---|
| $\sigma_1$ | Axial stress, major principal stress |
| $\sigma_{1.0}$ | Initial axial stress (one-dimensional compression) |
| $\sigma_2$ | Intermediate principal stress |
| $\sigma_3$ | Lateral stress, minor principal stress |
| $\sigma_h$ | Horizontal normal stress |
| $\sigma_{hf}$ | Horizontal normal stress at failure |
| $\sigma_v$ | Vertical normal stress |
| $\sigma_{vf}$ | Vertical normal stress at failure |
| $\Delta\sigma$ | Principal stress difference |
| $\Delta\sigma_{ult}$ | Asymptotic maximum value of $\Delta\sigma$ |
| $\tau$ | Shear stress |
| $\phi$ | Internal friction angle |
| $\phi'$ | Bulk solid-solid friction angle |
| $\phi_{cv}$ | Ultimate internal friction angle (at critical void ratio) |

## REFERENCES

1. **Coulomb, C. A.,** Essai sur une application des règles de maximis et minimis à quelques problèmes de statique à l'architecture, *R. Acad. Sci. Paris,* 5, 7, 1776.
2. **Rankine, W. J. M.,** On the stability of loose earth, *Philos. Trans. R. Soc. London,* 1857.
3. **Roberts, I.,** The pressure of stored grain, *Engineering (London),* 34, 399, 1882.
4. **Roberts, I.,** Determination of the vertical and lateral pressures of granular material substances, *Proc. R. Soc. London,* 1884, 225.
5. **Janssen, H. A.,** Versuche über Getreidedruck in Silozellen, *Z. Ver. Dtsch. Ing.,* 39, 1045, 1895.
6. **Janssen, H. A.,** On the pressure of grain in silos, in *Proc. Inst. Civ. Eng., London,* 1896, 553.
7. American Concrete Institute, *Recommended Practice for Design and Construction of Concrete Bins, Silos and Bunkers for Storing Granular Materials (ACI 313-77) and Commentary,* American Concrete Institute, Detroit, 1977.
8. **Marston, A.,** *The Theory of Loads of Closed Conduits In The Light Of The Latest Experiments,* Bull. 96, Iowa Engineering Experimental Station, Ames, 1930.
9. **Prante, M.,** Messungen des Getreidedruckes gegen Silowandungen, *Z. Ver. Dtsch. Ing.,* 40, 1122, 1896.
10. **Jamieson, J. A.,** Grain pressures in deep bins, *Trans. Can. Soc. Civ. Eng.,* 17, 554, 1903.
11. **Bovey, H. T.,** Experiments on grain pressures in deep bins and the strength of wooden bins, *Eng. News,* 52, 32, 1904.
12. **Pleissner, J.,** Versuche zur Ermittlung der Boden-und Seitenwanddrucke in Getreidesilos, *Z. Ver. Dtsch. Ing.,* 50, 976, 1906.
13. **Toltz, M.,** Discussion on grain pressures in deep bins, *Trans. Can. Soc. Civ. Eng.,* 17, 641, 1903.
14. **Lufft, E.,** Tests of grain pressure in deep bins at Buenos Aires, Argentina, *Eng. News,* 52, 531, 1904.
15. **Ketchum, M. S.,** *The Design of Walls, Bins and Grain Elevators,* 1st ed., McGraw-Hill, New York, 1907, 307.
16. **Turitzin, A. M.,** Dynamic pressure of granular material in deep bins, *Proc. Am. Soc. Civ. Eng.,* ST2, 49, 1963.
17. **Reimbert, M.,** Surpression dans les silos lors de la vidange, *Travaux,* 38, 780, 1954.
18. **Caquot, A. and Kérisel, J.,** *Traité De Mécanique Des Sols,* 2nd ed., Gauthier-Villars, Paris, 1949, 350.
19. **Caquot, A.,** La pression dans les silos, in *Proc. 4th Int. Conf. Soil Mech. Found. Eng.,* Butterworths, London, 1957, 2, 191.
20. **Jenike, A. W.,** *Gravity Flow Of Bulk Solids,* Bull. 108, Utah Engineering Experimental Station, Salt Lake City, 1961.
21. **Jenike, A. W.,** *Storage And Flow Of Solids,* Bull. 123, Utah Engineering Experimental Station, Salt Lake City, 1964.
22. **Jenike, A. W. and Johanson, J. R.,** Bin loads, *Proc. Am. Soc. Civ. Eng.,* ST4, 1011, 1968.

23. **Stewart, B. R.,** Effect of moisture and specific weight on internal properties of sorghum grain, *Trans. Am. Soc. Agric. Eng.*, 11, 260, 1968.
24. **Mohsenin, N. N.,** *Physical Properties of Plant and Animal Materials,* Vol. 1, Gordon & Breach, New York, 1970.
25. **Munroe, J. A. and Moysey, E. B.,** Friction characteristics of rapeseed and flaxseed, *Can. Agric. Eng.*, 16, 38, 1974.
26. **Moysey, E. B., Gagnon, R. H. and Brown, T.,** *Factors Affecting Internal Friction in Grain,* Pap. No. NCR79-303, American Society of Agricultural Engineers, St. Joseph, Mo., 1979.
27. **Lawton, P. J. and Marchant, J. A.,** Direct shear testing of seeds in bulk, *J. Agric. Eng. Res.*, 25, 189, 1980.
28. **Marchant, J. A.,** An incremental stress/strain law for cohesionless granular materials, *J. Agric. Eng. Res.*, 25, 421, 1980.
29. **Smith, D. L. O. and Lohnes, R. A.,** Frictional and stress-strain characteristics of selected agricultural grains as indicated by triaxial testing, in *Proc. Tech. Prog. Int. Powder Bulk Solids Handling Processing,* Cahners Exposition Group, Chicago, 1981, 103.
30. **Smith, D. L. O. and Lohnes, R. A.,** Frictional properties and stress-strain relationships for use in the finite element analysis of grain silos, *J. Powder Bulk Solids Technol.*, 6(3), 4, 1982.
31. **Lambe, T. W. and Whitman, R. V.,** *Soil Mechanics,* John Wiley & Sons, New York, 1979.
32. **Terzaghi, K.,** *Erdbaumechanik Auf Bodenphysikalischer Grundlage,* Deuticke, Vienna, 1925.
33. **Bowden, F. P. and Tabor, D.,** *The Friction And Lubrication of Solids, Part 1,* Clarendon Press, Oxford, 1950.
34. **Bowden, F. P. and Tabor, D.,** *The Friction And Lubrication of Solids, Part 2,* Clarendon Press, Oxford, 1964.
35. **Weast, R. C., Ed.,** *Handbook of Chemistry and Physics,* 60th ed., CRC Press, Boca Raton, Fla., 1979, F21.
36. **Horn, H. M. and Deere, D. U.,** Frictional characteristics of minerals, *Geotechnique,* 12, 319, 1962.
37. **Smith, D. L. O.,** The Triaxial Load Response of Grain, Ph.D. Dissertation, Iowa State University, Ames, 1981.
38. **Leslie, D. D.,** Large-scale triaxial tests on gravelly soils, *Proc. 2nd Pan-Am. Conf. Soil Mech. Found. Eng.*, Associacao Brasileira de Mecanicia dos Solos, Sao Paulo, 1963, 181.
39. **Holtz, R. D. and Kovacs, W. D.,** *An Introduction to Geotechnical Engineering,* Prentice-Hall, Englewood Cliffs, N.J., 1981, 432.
40. **Lambe, T. W.,** Stress path method, *Proc. Am. Soc. Civ. Eng.*, 93(SM6), 309, 1967.
41. **Lambe, T. W.,** Stress path method: second ed., *Proc. Am. Soc. Civ. Eng.*, 105(GT6), 727, 1979.
42. **Jàky, J.,** A nyugalmi nyomás tenyezöje (The coefficient of earth pressure at rest), *Magy. Mérn. Épitész Egylet Köz.*, October, 355, 1944.
43. **Jàky, J.,** Pressure in silos, Proc. 2nd Int. Conf. Soil Mech. Found. Eng., Rotterdam, 1948, 103.
44. **Bowles, J. E.,** *Engineering Properties of Soils and Their Measurement,* McGraw-Hill, New York, 1978.
45. **Drnevich, V. P.,** Constrained and shear moduli for finite elements, *Proc. Am. Soc. Civ. Eng.*, 101(GT5), 459, 1975.
46. **Roscoe, K. H.,** An apparatus for the application of simple shear to soil samples, *Proc. 3rd Int. Conf. Soil Mech. Found. Eng.*, 1, 186, 1953.
47. **Clough, G. W. and Duncan, J. M.,** Finite element analyses of retaining wall behavior, *Proc. Am. Soc. Civ. Eng.*, 97(SM12), 1657, 1971.
48. **Nicholson, D. E.,** Gravity flow of powder in a lunar environment. I. Testing of simulated lunar basalt powder for gravity flow, in Proc. 1st Int. Powder Bulk Granular Solids Technol. Conf., Harrogate, England, 1971, 163.
49. **Scarlett, B. and Todd, A. C.,** The critical porosity of free flowing solids, *Trans. Am. Soc. Mech. Eng.*, 91(Ser. B), 478, 1969.
50. **Schwedes, J.,** Measurement of powder properties for hopper design, *Trans. Am. Soc. Mech. Eng.*, 95(Ser. B), 55, 1973.
51. **Konder, R. L. and Zelasko, J. S.,** A hyperbolic stress-strain formulation for sands, *Proc. 2nd Pan-Am. Conf. Soil Mech. Found. Eng.*, Associacao Brasileira de Mecanicia dos Solos, Sao Paulo, 1963, 289.
52. **Janbu, N.,** Soil compressibility as determined by oedometer and triaxial tests, *Proc. Eur. Conf. Soil Mech. Found. Eng.*, Deutsche Gesellschaft fur Erd-und Grundbau, Essen, 1963, 19.
53. **Marchant, J. A. and Schofield, C. P.,** A combined constant pressure and volume change apparatus for triaxial tests at low pressures, *Geotechnique,* 28, 351, 1978.
54. **Smith, D. L. O. and Lohnes, R. A.,** Dilatancy as the cause of overpressures: experimental evidence, in *Proc. Int. Conf. Design of Silos for Strength and Flow,* Discussion Vol., Powder Advisory Centre, London, 1980.
55. **Koenen, M.,** Berechnung des Seiten und Bodendruckes in Silozellen, *Zentralbl. Baurerwaltung,* 5, 446, 1896.

56. **Reimbert, M.,** Recherches nouvelles sur les efforts exercés par les matières pulvérulentes ensilées sur les parois des silos, *Circulaire Série,* 1, 1, 1943.
57. **Isaacson, J. D. and Boyd, J. S.,** Mathematical analysis of lateral pressures in flat-bottomed, deep grain bins, *Trans. Am. Soc. Agric. Eng.,* 8, 358, 1965.
58. **Gaylord, E. H. and Gaylord, C. N.,** *Granular Material Pressure in Bins,* Pap. 77-4503, American Society of Agricultural Engineers, St. Joseph, Mo., 1977.
59. **Vivancos, J.,** Calculation of the loads produced in a high silo during the emptying process, *Am. Concrete Inst. J.,* 75, 13, 1978.
60. **Reimbert, M. and Reimbert, A.,** *Silos Traité Theoriqué Et Pratique,* 1st ed., Editions Eyrolles, Paris, 1956.
61. **Reimbert, M. and Reimbert, A.,** *Silos — Theory and Practice,* Trans. Tech Publ., Cleveland, 1976.
62. **Jenike, A. W., Johanson, J. R., and Carson, J. W.,** Bin loads, II, III, and IV. Concepts, Mass-flow bins, and Funnel-flow bins, *Trans. Am. Soc. Mech. Eng.,* 95(Ser. B), 1, 1973.
63. **Johnston, F. T.,** Silo problems, in *Proc. Tech. Prog. Int. Powder Bulk Solids Handling Processing,* Cahners Exposition Group, Chicago, 1981, 97.
64. **Beddow, J. K. and Meloy, T. P., Eds.,** *Advanced Particulate Morphology,* CRC Press, Boca Raton, Fla., 1980.
65. **Kaye, B. H.,** *Direct Characterization of Fineparticles,* John Wiley & Sons, New York, 1981.

Chapter 7

# EVALUATION OF EXPLOSION AND FIRE PROPERTIES OF POWDERS AND DUSTS*

**K. N. Palmer**

## TABLE OF CONTENTS

## I. INTRODUCTION

Dust explosions are caused by combustible dusts dispersed in a gas, usually air, and ignited. Many combustible dusts, ranging from very common to highly specialized materials, are capable of causing explosions. Typical categories of dust include food and agricultural dusts such as grain, sugar, and starch; synthetic materials such as plastics, chemicals, and pharmaceuticals; metals such as aluminum, magnesium, and titanium; and traditional fuels such as coal and wood.

The appearance of a dust explosion can be broadly compared with that of a gas or vapor explosion occurring under similar circumstances. For a cloud dispersed in the open air the result of the ignition is a flash of flame, usually developing little hazardous pressure. If the dust cloud is confined, as in a plant or a room, then pressure effects would be expected depending on the size of the cloud, the rate of burning of the dust, and the ease of discharge of pressure to the atmosphere. In many industrial operations dust is dispersed as a cloud, usually in air, and consideration must therefore be given to the possibility of a dust explosion should a source of ignition be present.

Fires can be caused by combustible dusts, provided sufficient oxygen and dust to support the combustion are present. The dusts may be as layers, heaps, or other deposits rather than dispersed as a cloud in air as is necessary for a dust explosion. Although relatively fine particles are needed for a dust explosion, fires can occur in a much wider range of particle sizes — up to individual lumps of the order of centimeters in diameter. The effect of particle size is much more severe with dust explosions than with fires. As will be seen, the ease of ignition is also dependent on particle size, both for explosions and fires.

There are two types of burning in dust fires: flaming and smoldering (or glowing). The type of burning obtained depends upon the ignition source and the characteristics of the dust including particle size, the geometry of the deposit, and the ease of access of air. In a dust explosion, flame must have been present to cause the rapid propagation and pressure development; a suspension of glowing or incandescent particles would not propagate into unburned dust and could not give an explosion. But if glowing particles were dispersed in suspension, and thereby generated flame, an explosion could be produced.

In the present state of knowledge it is not possible to predict whether a combustible dust will be capable of causing a dust explosion or the mode of combustion in a fire involving the material. In the case of dust explosions, this has led to the need for the development of tests to first determine whether or not a powder or dust can give rise to an explosion and then, if the explosion is possible, for the measurement of certain parameters. In the case of dust fires, corresponding tests have not been developed and the question of whether or not fires can be initiated depends upon ad hoc investigations.

The design of the test or investigation is affected by the type of ignition source to which the dust may be exposed in practice. For dust explosions, there are tests available to determine the ease of ignition from various practical sources, and further tests for assistance in the design of explosion protection measures. As would be expected, the various practical sources of ignition which are likely to cause dust explosions are also likely to cause dust fires, and although test methods for measuring the ignitability of dust fires are less highly developed than for dust explosions, it is of interest to consider the various practical sources of ignition and their relevance to the initiation of explosions and fires.

## II. SOURCES OF IGNITION

### A. General

For an explosion or fire to occur the dust cloud must be ignited. Apart from a few special instances where the act of dispersion of dust may ignite it, a separate source of ignition must be present and the concentration of dust in the cloud must be favorable.

In the safe design for industrial plants handling dusts, sources of ignition must be excluded as far as possible. Although stringent precautions may be taken in this respect, the protection obtained is not complete. Further methods of protection must therefore be applied, but the likelihood of a source of ignition being present can be much reduced by good design and management and these should be regarded as essential requirements for plant safety.

In some industries, e.g., foodstuffs, considerable protection against sources of ignition is obtained during stages of manufacture where the product is wet with water. It is at subsequent stages, where a product may be dried, milled, transported, screened, and packaged that the dust fire and explosion problems become serious. If the process should involve the use of flammable solvents, then the flammability of the solvent is important and may indeed be the governing factor in the explosion hazard. Experience in a wide range of industries has shown that where flammable solvents and flammable powders are handled together the hazard is considerable.

Although there is a vast number of different sources of ignition having various temperatures, energies, durations, etc., they can be conveniently grouped as follows:

Flames
Smoldering
Hot surfaces
Welding and cutting
Friction and impact
Electric sparks
Spontaneous heating

### B. Flames

The flames produced from burning of gases, liquids, or solids are potent sources of ignition for dusts. Ignition of a dust can be made by a permanently established flame as on a gas burner or torch, or from a pilot light, or from transient flames such as could be produced by ignition of a cloud of vapor.

Because of the effectiveness of flames for igniting dusts, the general safeguard is to prevent their coming together. This may mean excluding flames from within and from the vicinity of the plant handling the dust. In some industries, the presence of a flame and a flammable dust may be an inherent part of the process, e.g., the burning of fuel, but in many industries this should not be the case and prevention of flames coming into contact with dusts should be attainable.

### C. Smoldering

Smoldering or glowing is a form of combustion without the presence of flame. Powders which are able to smolder may generally be ignited by relatively low-energy ignition sources such as cigarettes, hot surfaces, and electric sparks. To enable the smoldering to continue it is necessary that the burning material should not wholly melt or liquefy, since smoldering requires a solid fuel in which to propagate. The solid material may, however, be a partial product of combustion, e.g., a carbonized residue, so that where smoldering is initiated in the center of a dust deposit and liquefied material cannot move away it may be broken down into solid residues which then continue to smolder.

If smoldering can stabilize in the material then it may easily be converted into flames, following which a fire can develop. Smoldering is, therefore, a source of ignition comparable in effectiveness to flames.

### D. Hot Surfaces

These may be present as part of the operating process or as a result of some fault, e.g., friction. Typical situations which could affect the dust are

1. Presence of a dust suspension within a heated enclosure, such as an oven or dryer
2. Deposition of dusts on a surface within a heated enclosure, for example, on the internal wall of a dryer
3. Dust dispersed around or deposited on the heated surface with cooling to the atmosphere, such as dust on an overheated bearing or an electric lamp

Since the dust may be in the vicinity of the heated surface for a relatively long time, lower temperatures than those present in flames can lead to ignition.

**E. Welding and Cutting**

Many dust explosions and fires have been caused by welding or hot cutting operations carried out while explosible dust suspensions or layers were present. Welding or cutting operations should always be rigorously controlled, preferably by means of a work permit system, since the operator may not be familiar with the hazards involved in applying heat in a plant containing flammable dust. The work permit system should be used whether the plant is shut down temporarily for a minor repair or whether it has been shut down more permanently for a substantial overhaul. In either case the plant should be cleaned free of powder, at least in those places where the welding and cutting operations are to take place, but experience has shown that even with these precautions flammable residues may remain which can lead to ignition and explosion. Generally, it is bad practice to permit welding or cutting operations in a plant while it is operating. Permission to undertake cutting or welding operations must always be coordinated with knowledge of the conditions inside the plant and in the area in which the operation is to be carried out. Because of the high temperatures involved in the operations, whether using gas or electricity, welding or cutting should always be regarded as potentially hazardous.

**F. Friction and Impact**

There is strong evidence from industrial experience that friction and impact sparks produced by powered sources are able to ignite dusts, although relatively little laboratory evidence is available to provide a detailed technical background. Consequently, information is lacking on the levels of mechanical energy needed to ignite various dusts, but it is generally recognized that powered impacts are more hazardous than manual ones. Frictional ignition can arise in various ways:

1. The material being processed, itself, generates excessive friction in the plant, for example by overloading valves, screws, etc.
2. The material being processed may contain a foreign object such as metal or stone.
3. Frictional impact can be caused by breakdown within a plant, since portions of metal may become detached.
4. The use of tools, whether powered or manual, can give rise to intense localized heating.
5. A dust layer, and possibly a suspension, can be ignited by sparks caused by the impact of steel on rusty steel carrying a smear of aluminum.

Although the occurrence of friction and impact sparks can sometimes be foreseen and the plant and its operation designed to avoid the risk, there will be cases where, due to failure of plant or personnel, ignition can occur. Attempting to prevent ignition is likely to be only partly successful, and alternative means of protection against a consequent fire or explosion must be taken.

**G. Electric Sparks**

Dust explosions and fires can arise from both electric power and electrostatic charging.

Electric power from the mains or batteries is a powerful source of ignition because the energies available are usually greatly in excess of those required to ignite dusts. The sparks produced by the normal operation of switches, contact breakers, can ignite dust. They need to be protected. The principal precautions are first, install the sparking equipment in clean air and second, when equipment must be used in the presence of dust it should preferably be of a dust-tight design. The design of dust-tight equipment should also be such that even if covered with a layer of dust, hazardous surface temperatures do not develop.

The production of hazardous sparks from electrostatic charging of dusts has been studied for many years. It is common knowledge that in the processing, transportation, and general handling of dusts electrical charges are generated. Usually the generation of charge cannot be avoided in the manipulation of powders and therefore attention must be paid to the safe removal of charges and the prevention of their accumulation. In industry, a considerable danger from static electricity arises because of the transfer of charge to isolated metallic conductors that are not grounded. Where discharge to earth does occur from conducting objects the charge flows readily from the whole of the object and, therefore, the energy in the spark is maximized. There is considerable information available on the methods of grounding and of permissible resistances to earth for removal of electrostatic charges.[1]

### H. Spontaneous Heating and Ignition

The initiation of dust fires by hot surfaces or dust deposits in hot environments is an important hazard and much attention has been devoted to it both by way of experiment and theory. Because of the practical importance of the hazard, routine tests for the various situations have been developed. These tests give minimum ignition temperatures which may then be applied directly to practical situations. Frequently, however, there are uncertainties arising because of the effects of increasing scale in industrial applications, since it is known that both the ignition temperature and the time required to produce ignition depend upon the dimensions of the dust layer or deposit.

A satisfactory approach is to use the thermal ignition theory, on the basis of which determinations carried out on a laboratory scale can be extrapolated with confidence to full-scale industrial applications. This is the case particularly where the small-scale tests are isothermal in type, i.e., the dust is tested at a fixed ambient temperature and the conditions for the sample to proceed to ignition are determined. The alternative approach involves either full adiabatic conditions or heating the dust in an environment whose temperature is raised at a controlled, but arbitrary, rate. The adiabatic type of experiment involves allowing the specimen to heat itself to ignition by preventing all heat losses, or by adding heat to compensate for losses which are unavoidable. There are a number of variations of technique and a good general picture has been provided by a recent symposium publication.[2]

There is evidence to suggest that the most satisfactory method, in theory, is that using isothermal techniques although there are disadvantages in the time required for the experimentation and, with particularly vigorous materials, the hazard which might develop should the relatively large mass react violently. For many common substances, however, the method is found to be straightforward and does have the advantage that it provides a basis for firm extrapolation to larger specimens than those tested.

Thermal ignition theory is based upon the fact that the ignition of a dust in a hot environment depends upon the balance between the heat generation produced by reaction of the dust with the air and the heat loss arising from cooling to its surroundings. Since heat generation takes place through the volume of the dust, but the heat losses are governed by its surface area, the conditions for ignition will depend upon the size of the dust layer or deposit. Further, as the rate of heat generation increases with temperature according to an exponential law, whereas the heat losses depend upon the difference between the temperature of the dust and its surroundings, a temperature can be defined at which the rate of generation

**Table 1**
**IGNITION TEMPERATURES (°C) OF CUBES OF POWDERS IN AN OVEN**

| | Cube size (mm) | | |
|---|---|---|---|
| **Material** | **50** | **75** | **150** |
| Activated carbon | 125 | 113 | 99 |
| Wood sawdust | 185 | 173 | 152 |
| Fishmeal | 173 | 158 | 135 |

of heat is just in balance with the rate of heat loss. At higher ambient temperature the rate of heat generation outstrips the cooling effect and the system can proceed to ignition. The temperature at which the heating and cooling are balanced is called the "critical ignition temperature". It may be regarded as the minimum ignition temperature for the particular size of specimen involved. Some typical values for small cubes heated in ovens are given in Table 1.

From the results, thermal ignition theory may be applied, the basic equation of which is

$$\log \frac{(\delta_c T_1^2)}{(r^2)} = M - \frac{N}{T_1} \tag{1}$$

where r is the characteristic length (radius of sphere, half side of cube, half thickness of layer, etc.), $T_1$ is the minimum ignition temperature, M and N are constants for a particular material, and $\delta_c$ is a dimensionless constant depending on geometrical properties ($\delta_c = 0.88$ for a layer, 2.52 for a cube, and 3.32 for a sphere). If results are available for two values of r, the constants M and N can be evaluated and, hence, ignition temperatures for other values of r and $\delta_c$ can be calculated. Such predictions are found to be accurate, in general.

Thermal ignition theory can also assist in the estimation of the time required to heat to ignition. The relationships are complex so high accuracy is not obtainable, but for temperatures near to the critical it may be shown that the time to ignition varies with $r^2$. However, in many cases, there is an induction period which arises from the presence of antioxidant in the dust being heated or to the completion of short-term preliminary chemical reactions, and the decay of antioxidant or the completion of the preliminary reactions may not follow the same law as the principal oxidation process. Thermal ignition theory does enable the presence of these precursor reactions to be identified and is more satisfactory in this respect than are methods involving rising ambient temperatures. Knowledge of the time required for ignition is important in industry because, on the larger scale, it may be possible to operate safely at a temperature above the minimum ignition temperature for the dimensions of the material being processed, safety being obtained by keeping the duration of the overheating well short of that required for ignition to occur. Thus transient excursions to higher temperatures, on the industrial scale in large bulks, may not lead to ignition.

The nonisothermal methods involving calorimetry or thermal analysis may be quicker to operate and involve smaller specimens than the isothermal technique. Application of the results to full-scale industrial situations is not so precise as with the isothermal method and misleading results can be obtained if induction periods are present. The misconceptions arise because the induction reactions prevent the evolution of heat until the temperature of the system as a whole has risen sufficiently for the preliminary reactions to be completed in the time available. This time would depend upon the rate of heating, which is arbitrarily chosen, and therefore the results are rendered less certain. The techniques, however, are quite suitable for comparative purposes or for quality control assessments, but their use with new materials

**Table 2**
**CLASSIFICATION TEST APPARATUS**

| Apparatus | Direction of dispersion of dust | Ignition source |
|---|---|---|
| Vertical tube | Upward | Spark or coil |
| Horizontal tube | Horizontal | Coil |
| Inflammator | Downward | Spark or coil |

must be made with reservation if extrapolation to full-scale industrial conditions is subsequently made.

For those with industrial problems there is a user's guide to fire and explosion hazards in the drying of dusts which covers the theory and its application to all the principal types of industrial dryers.[3]

## III. TEST METHODS FOR DUST EXPLOSIBILITY

### A. Classification Tests

Before embarking on tests to determine individual explosion properties, it may sometimes be helpful to obtain a decision on whether or not the dust is explosible, that is, capable of causing a dust explosion. This is particularly the case in the U.K. where the law requires that explosion prevention or protection shall be provided for dusts that are known to be explosible. In other countries a similar screening procedure may sometimes be of interest. The word "classification" in this context means that the dust is categorized into one of two groups, as follows: Group (a) — dusts which ignited and propagated flame in the test apparatus; Group (b) — dusts which did not propagate flame in the test apparatus. Dusts are classified as Group (a) if they ignited and propagated flame in any of three tests with a small source of ignition, either "as received" or after sieving and drying the sample. Group (a) dusts should be regarded as explosible, and liable to give rise to a dust explosion hazard, whereas Group (b) dusts are not explosible although they may present a fire risk. These classification tests give little indication of the severity of the dust explosion hazard, since only visual observations are made. The results should be regarded as a guide which may require further tests for a more detailed assessment of the hazard. Furthermore, the dust is tested in apparatus at ambient temperature and the results cannot readily be extrapolated to conditions where dust is dispersed in an industrial plant which is heated. However, subject to these limitations, the information is obtained easily and is likely to be helpful.

For the tests, three types of apparatus are used and their arrangement is summarized in Table 2. Basically, the dust is dispersed by a blast of air over an electrical ignition source such as a spark or wire coil, with either one or both ends of the tube containing the ignition source open to atmosphere. Observation is made as to whether or not flame propagates away from the ignition source.

The vertical tube assembly, sometimes known as the Hartmann apparatus, is illustrated in Figure 1 and consists of a transparent plastic combustion tube 30.5 cm long and with a 6.4-cm inside diameter mounted over a brass dispersion cup 6.4 cm in diameter and 1.6 cm deep in the center. The volume of the apparatus is 1.2 ℓ. (The vertical tube is also used in the measurement of certain dust explosibility parameters; see below.) The dust under test is placed in the cup and is dispersed by air passing upward and around a deflector mounted in the cup. The ignition source is a pair of brass electrodes mounted at various positions in the plastic tube, with a spark gap of approximately 0.6 cm. The spark is obtained from an induction coil. When a wire coil igniter is used the plastic tube is replaced by a glass tube, but the procedure is otherwise similar.

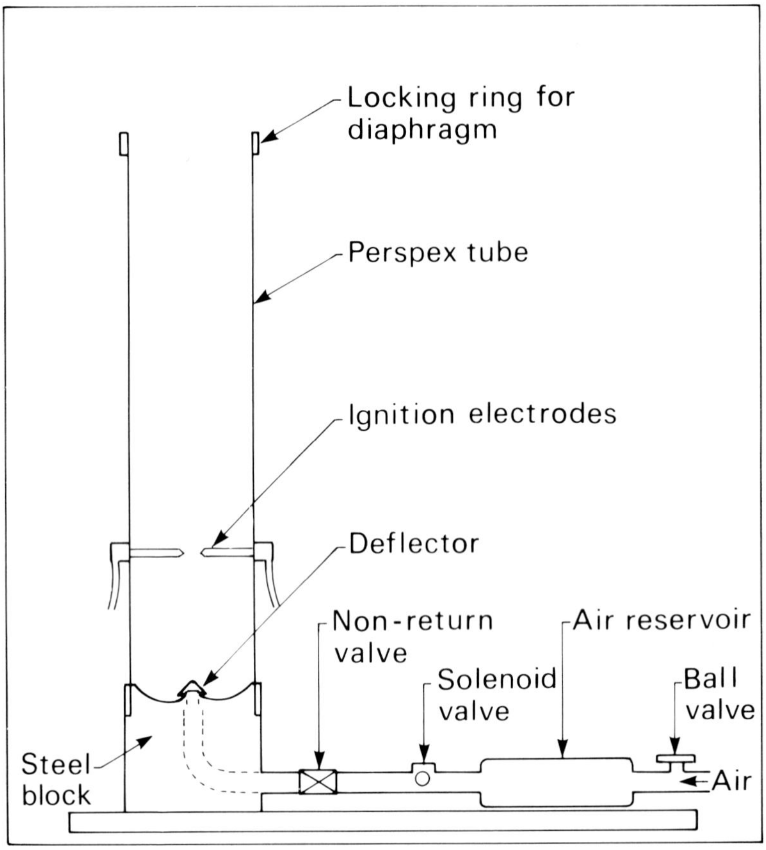


FIGURE 1. The vertical tube apparatus.

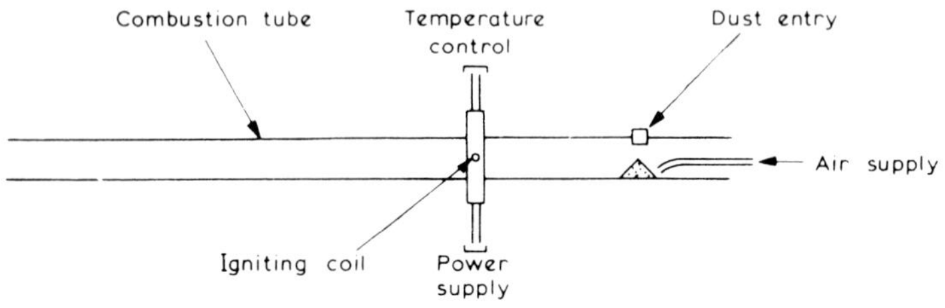


FIGURE 2. The horizontal tube apparatus.

The horizontal tube apparatus is constructed of glass and is 1.38 m long with an internal diameter of 7.6 cm (Figure 2). The tube is open at both ends and the dust is introduced through a hole in the upper surface of the tube 13 cm from one end, and deposited as a loose heap in the tube. The ignition coil is mounted 46 cm from the same end of the tube and consists of a platinum wire coil mounted on a ceramic former placed horizontally across the diameter of the tube. The dust is dispersed by a metered blast of air which carries the dust down the tube past the igniter. The temperature of the coil is maintained thermostatically at 1300°C.

The inflammator apparatus is shown in Figure 3. The dust falls vertically downward onto the igniting source through a vertical glass tube 1.02-m long and 7.6 cm in diameter. The tube is open at the top and closed at the bottom with a rubber bung. The dust holder is

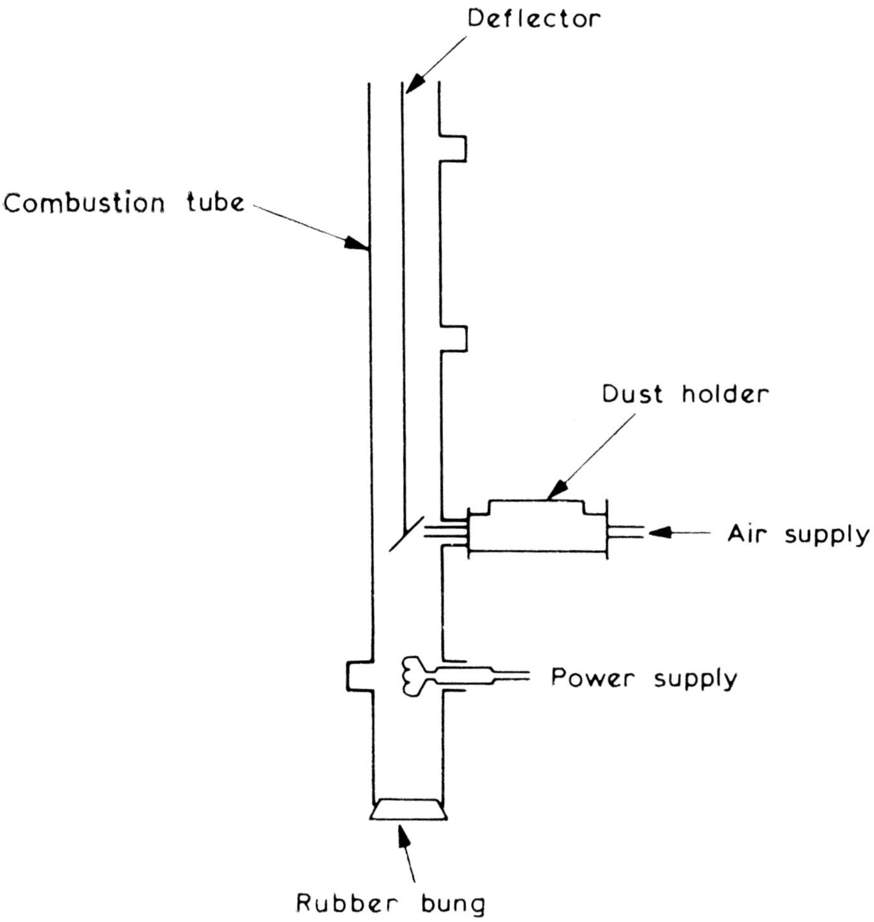


FIGURE 3. The inflammator.

attached by a side arm, and two further side arms 13 cm from the bottom of the tube enable a spark electrode or wire coil igniter to be inserted. When the dust is dispersed from the holder by the air it strikes a deflector plate mounted on the axis of the tube and then falls, under gravity, past the igniter.

With all of these apparatus, the dust concentration may be varied by altering the amount of dust and the pressure of the dispersion air, and the tests can be repeated an arbitrary number of times. Further details are described elsewhere.[4] A list of dusts which have been classified according to these tests is revised periodically[5] and should be consulted at an early stage if it is suspected that a material could be an explosible dust.

### B. Individual Tests for Explosibility

If a dust is known to be explosible, or has been classified as a Group (a) material, then quantitative information on various explosion parameters may be necessary for plant design or explosion protection. A series of tests has become established over the years, in which individual explosion parameters are measured. These parameters are

1. Minimum ignition temperature
2. Minimum explosible concentration
3. Minimum ignition energy

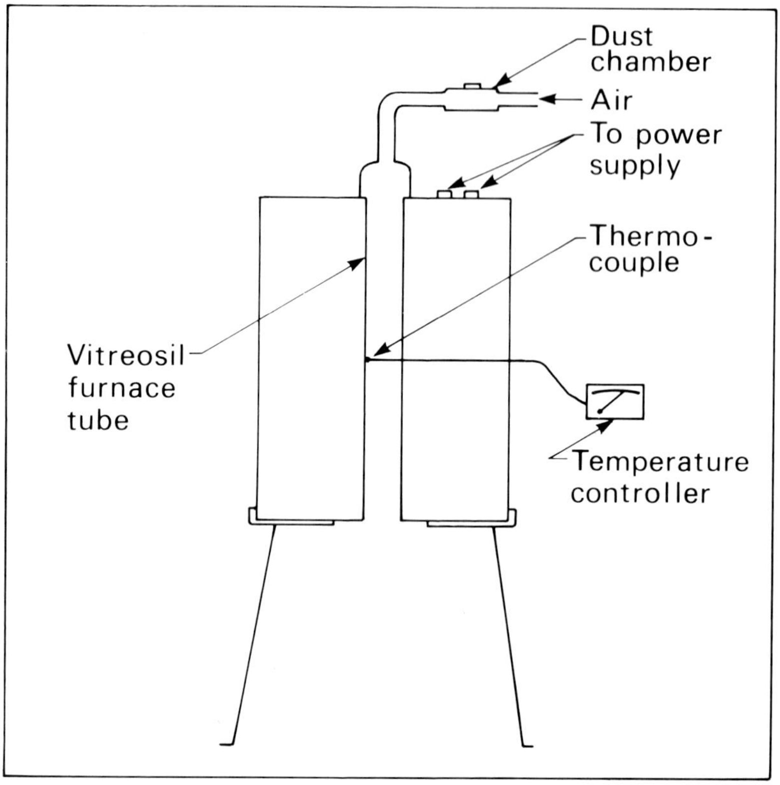


FIGURE 4. Furnace apparatus for the determination of the minimum ignition temperature of dust clouds.

4. Maximum permissible oxygen concentration to prevent ignition
5. Maximum explosion pressure and rate of pressure rise

The methods used in the U.S. are described in full in a report from the U.S. Bureau of Mines[6] and a similar series of test methods is used in the U.K.[4] One of the most important tests is that for explosion pressure measurement. Much work has been carried out in recent years in the Federal German Republic and in Switzerland, as a result of which an alternative test method has been promulgated.[7]

The results of the tests can be used in various ways. One is the use of explosibility data in plant design to avoid ignition from envisaged sources such as hot surfaces, electric sparks, etc. Alternatively, the data can be used in the design of the explosion protection since information on the severity of the explosion can be related to the design of appropriate precautions. While there are some differences in the test methods between countries, the aims of carrying out the tests are broadly similar. The test methods and their applications are now briefly summarized.

*1. Minimum Ignition Temperature*

Measurement of the minimum ignition temperature of a dust cloud in air is measured in a version of the Godbert-Greenwald furnace apparatus (Figure 4). It consists of a silica furnace tube 21.6-cm long and 3.7 cm internal diameter, heated externally by an electric winding. The tube is vertical and is open to atmosphere at its lower end. The top of the tube is closed by a glass adapter containing a 90° bend which connects to the dust holder. Dust is dispersed into the furnace by opening a solenoid valve, which releases compressed

**Table 3**
**EXPLOSION PARAMETERS OF SELECTED DUSTS**

| Dust | Min. ignition temp. (°C) | Min. explosible conc. ($kg/m^3$) | Min. ignition energy (mJ) | Max. explosion pressure (bar) | Max. rate of pressure rise (bar/sec) |
|---|---|---|---|---|---|
| Aluminum | 610 | 0.045 | 10 | 8.6 | 1350+ |
| Aspirin | 550 | 0.015 | 15 | 5.9 | 520 |
| Cellulose acetate | 340 | 0.035 | 20 | 7.9 | 450 |
| Coal, 37% volatiles | 610 | 0.055 | 60 | 6.1 | 160 |
| Cornstarch | 390 | 0.040 | 30 | 9.9 | 650 |
| Dextrin | 410 | 0.050 | 40 | 6.7 | 610 |
| Magnesium | 560 | 0.030 | 40 | 7.9 | 1020 |
| Nylon | 500 | 0.030 | 20 | 6.6 | 280 |
| Phenol formaldehyde | 450 | 0.015 | 10 | 7.4 | 450 |
| Polypropylene | 420 | 0.020 | 30 | 5.2 | 380 |
| Polystyrene | 500 | 0.020 | 15 | 6.9 | 480 |
| Rice | 440 | 0.050 | 50 | 7.1 | 180 |
| Sulfur | 190 | 0.035 | 15 | 5.3 | 320 |
| Wheat flour | 380 | 0.050 | 50 | 7.4 | 250 |
| Wood flour | 430 | 0.050 | 20 | 6.4 | 580 |

air from the reservoir. The furnace is mounted on a stand, enabling the lower end of the tube to be readily observed. Furnace temperatures up to 1000°C are maintained thermostatically by a temperature controller governed by a thermocouple attached to the furnace tube. The dust is blown from the holder into the furnace by releasing the air pressure from the reservoir and observations are made as to whether or not flame appears at the bottom open mouth of the tube. If ignition occurs the furnace temperature is lowered until two temperatures differing by 10 K are obtained, so that ignition takes place at the higher temperature only. The quantity of dust in the holder, and the dispersion pressure, are variable in order to obtain the minimum ignition temperature. The lowest furnace temperature at which ignition occurs is taken as the minimum ignition temperature. Values for some typical materials are included in Table 3.

Application of the results of the test should be made with due consideration of the inherent limitations of the method. For instance, there are theoretical reasons why the minimum ignition temperature should be dependent upon the dimensions of the apparatus in which it is measured. The test apparatus is small scale and, generally, ignition temperatures would be expected to fall as the scale of the hot enclosure was increased. Thus, on application to hot enclosures in an industrial plant, lower ignition temperatures than those measured by the test apparatus may be encountered. Clearly, for safety, the temperature of a larger enclosure should not be greater than the minimum ignition temperature as determined, but may well need to be less if ignition is to be avoided. A typical approach is that the temperature of the enclosure shall not exceed two thirds of the minimum ignition temperature of the dust (measured in degrees Celsius).

This test method may also be used for comparing the ignition temperatures of different dusts. This approach is sometimes very valuable since it enables the ignition temperature of a new or unknown material to be compared with others whose performance in industrial plants is familiar. In particular, it enables materials of low ignition temperatures, below 300°C, to be identified rapidly since such materials may well require special treatment in processing through plant units which are heated or have high mechanical energy inputs (such as grinders and micronizers).

*2. Minimum Explosible Concentration*

Customarily the minimum explosible concentration, or lower explosibility limit, is measured in the vertical tube apparatus (Figure 1). The source of ignition is a continuous electric spark. A weighed quantity of the dust is placed in the dispersion cup of the tube and a paper diaphragm with a small hole in the center is attached to the top of the tube by a ring. The dust is dispersed in the usual way and the criterion for indicating an explosion is the propagation of flame away from the ignition source; propagation is usually accompanied by bursting of the diaphragm. When propagation occurs, the amount of dust is reduced and testing continued until no explosion is observed in four successive trials. Tests are made with the electrodes 5 and 10 cm from the bottom of the tube and the mean of the two minimum masses of dust is divided by the volume of the tube for calculation of the minimum explosible concentration.

The method is convenient, but it cannot be expected to give high accuracy because of nonuniform distribution of the dust when dispersed. There is evidence that greater uniformity of dispersion can be obtained by modifying the cup in the vertical tube apparatus, but this approach has not yet become standardized. Also, the dispersion in the tube is under turbulent conditions and weak flames are likely to propagate more readily through stationary rather than stirred suspensions. However, information from larger scale and more accurately controlled experiments, and from the application of the tests to industrial processes, indicates that the values obtained in the test are sufficiently accurate for satisfactory application.

*3. Minimum Ignition Energy*

The vertical tube apparatus (Figure 1) is also used for measurement of minimum ignition energy. Electrodes with fine points are set with spark gaps of about 5 mm and are connected either directly to a bank of condensers or to the secondary winding of a step-up transformer, through the primary circuit of which condensers at known voltages are discharged. The time interval between dispersing the dust and passing the spark is varied and the criterion for ignition is the propagation of flame at least 10 cm into the tube. Up to four trials at each ignition energy are made. From the capacity of the condensers (C), and the voltage to which they are charged (V), the energy is calculated as $^1/_2\ CV^2$ J assuming no energy loss in the circuits.

Results from the tests may be applied to practical situations where ignition by static electricity sparks is a possible hazard. There are limitations on the accuracy of the measured values due to fluctuations in the dust dispersion during test and also to the circuit characteristics. Particularly where a transformer is used, there are possible energy losses in the circuit. Also, the simple calculation of spark energy assumes that the condensers discharge fully, which may not be the case. Furthermore, research has shown that the ability of a spark to ignite a dust suspension is a function of the duration of the spark as well as its energy, and therefore the electrical resistance of the circuit generating the spark is also relevant. By careful adjustment of circuit resistances, considerably lower ignition energies may be obtained than those from the usual test, but it may be argued that such special circumstances are unlikely to arise by chance in practical situations. However, as a broad guide, a dust which has a minimum ignition energy in the test of less than 25 mJ should be regarded as prone to ignition by static electricity even though the plant in which it is handled is grounded, since energies up to this level may be obtained under favorable circumstances from charged personnel, who may cause a spark in the vicinity of the dust cloud.

*4. Maximum Permissible Oxygen Concentration to Prevent Ignition*

Two methods are currently used for measuring the extent to which an inert gas, such as nitrogen or carbon dioxide, needs to be added to air to prevent ignition of a dust cloud. The first, and long established, method involves use of a modification of the furnace apparatus

(Figure 4). The adapter at the top of the furnace is replaced by another which is fitted with a side arm through which the air/inert gas mixture enters. The mixture then flows through the furnace and emerges from the base into a glass tube extension which prevents the backward flow of air into the apparatus. The air/inert gas mixture is also used to disperse the dust into the furnace. The temperature of the furnace is maintained at 850°C. By varying the composition of the air/inert gas mixture the maximum permissible oxygen concentration is found.

An alternative method has recently been developed in the U.K. involving the use of a modified version of the vertical tube apparatus (Figure 1). The top of the tube is closed with a paper diaphragm to prevent ingress of air and the tube is provided with a short side arm near its top from which the air/inert gas mixture issues and where it can be checked for composition. Ingress of the air/inert gas mixture provides for the purging of air from the vertical tube and also from the reservoir which is pressurized to disperse the dust. The ignition source is either an electric spark or a heated wire coil. The tube is operated in the usual way and the criterion for ignition is that the flames must be seen to move away from the ignition source. The test is repeated with reduced oxygen levels until no flame propagation is observed in at least ten consecutive tests in which the gas concentration is unchanged.

Generally, the two tests do not give equivalent results. The modified furnace apparatus tends to demand lower oxygen concentrations to prevent ignition, principally because the dust is dispersed in a hot environment although no additional ignition source is present. It may be argued that the principal application of this test should be for cases where dusts are present in a hot environment and where protection against explosion by the use of inert gas is required. It will be expected that lower oxygen concentrations would be necessary in heated enclosure and the values given by the modified furnace test are generally lower than by the alternative method. Before the modified vertical tube apparatus was introduced a comprehensive research program was carried out at the Fire Research Station, Borehamwood, Hertfordshire, U.K., to establish the maximum oxygen concentrations for various dusts in an experimental rig of full industrial size. The test using the modified vertical tube apparatus was then designed to give results consistent with those obtained on the larger scale, and this test has now gained acceptance in the U.K. for applications where the dust is dispersed in plant at ambient temperatures.

The choice of the dust apparatus thus depends upon the type of industrial process to which the results are to be applied.

*5. Maximum Explosion Pressure and Rate of Pressure Rise*

For many years in the U.S. and the U.K. the maximum pressure characteristics of a dust have been measured in a closed version of the vertical tube apparatus (see Figure 5). The apparatus consists of a vertical steel tube of the same dimensions as that in Figure 1, with fittings on either side for the ignition source and two small observation windows. The top of the tube is closed, with a fitting for a pressure transducer. A weighed quantity of the dust under test is placed in the dispersion cupboard at the base of the tube, and its concentration is calculated assuming uniform dispersion. The dust is dispersed by the release of air from a reservoir through a solenoid valve. A nonreturn valve, fitted between the solenoid valve and the tube, prevents the combustion gases from escaping during an explosion. The output from the pressure transducer is amplified and recorded. The peak value of pressure is the maximum explosion pressure, and the maximum gradient of the pressure-time curve is the maximum rate of pressure rise. The dust cloud is usually ignited by an electric spark source, but heated wire coil and guncotton flame sources are also available.

More recently, in Europe, the use of a larger vessel for measurement of explosion parameters has been introduced since it enables the scaling up of values for maximum rate of pressure rise to full-scale industrial equipment to be made more readily.[7] The basic standard

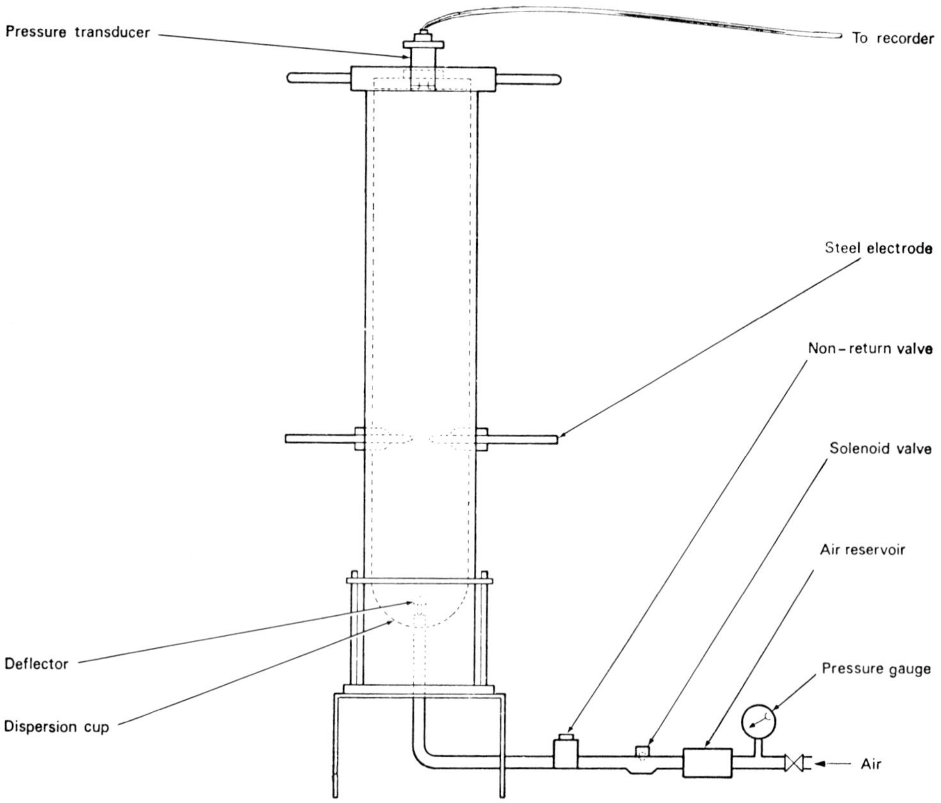


FIGURE 5. Measurement of the explosion pressure in a vertical tube apparatus of 1.2 ℓ volume.

is a vessel of volume 1 $m^3$, and much of the systematic investigation of explosion pressure characteristics and their relationship to explosion relief venting was carried out on this and larger vessels.

However, for routine use the 1 $m^3$ vessel is inconveniently large and a spherical vessel with a volume of 20 ℓ is now becoming standardized (Figure 6). The sphere is constructed of stainless steel and surrounded by a water jacket that permits test temperatures to be closely controlled. The required quantity of dust is dispersed into the chamber from a pressurized cylinder which forces the dust through a perforated ring within the sphere. Prior to dispersion of the dust the sphere is evacuated to 0.4 bar absolute pressure and the cylinder containing the dust is pressurized to 21 bar absolute pressure, so that after the air has entered the sphere the pressure at the time of ignition is 1 bar absolute, i.e., atmospheric pressure. In order to obtain results that are commensurate with those for larger vessels a delay of 60 msec between the initiation of dispersion and the activation of the ignition source is imposed. The usual ignition source is two 5 kJ chemical igniters.

The vertical tube and the spherical apparatus usually give reasonably similar values for the maximum explosion pressure, but the maximum rates of pressure rise are often different and the law which enables the results from the 20-ℓ sphere to be extrapolated to larger volume appears not to hold systematically for volumes down to that of the vertical tube apparatus. Research is currently being carried out to explore this difference, in order that the considerable bank of results available from tests with the vertical tube apparatus can be correlated with those from the sphere. In the meantime an empirical approach has been advanced.[8]

The principal use of measurement of maximum rate of pressure rise is in connection with

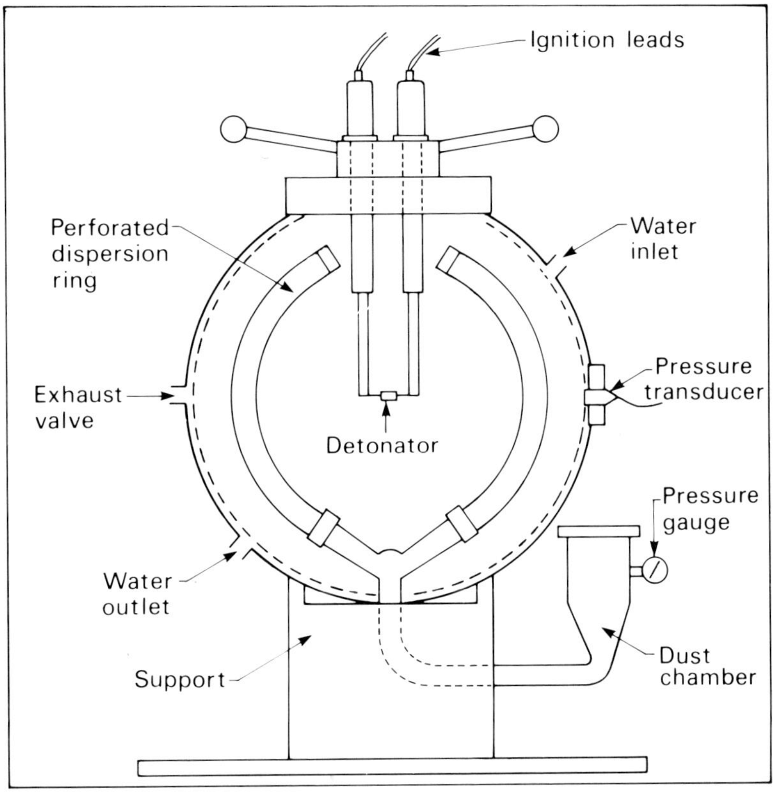


FIGURE 6. Measurement of explosion pressure in the 20-ℓ sphere.

**Table 4**
**VENT RATIO REQUIREMENTS**

| Max. rate of pressure rise (bar/sec$^{-1}$) | Vent ratio/m$^{-1}$ |
|---|---|
| –350 | 1/6 |
| 350—700 | 1/5 |
| 700+ | 1/3 |

the design of explosion relief vents in industrial plants. Several approaches to this problem are currently in existence. The simplest and most long-standing is the use of the vent ratio. This is defined as the area of vent per unit volume of enclosure being vented, and is not dimensionless. The vent ratio can be related to the maximum rate of pressure rise as measured in the vertical tube apparatus,[4] as shown in Table 4. This approach is convenient to use in volumes up to 30 m$^3$, but for larger volumes the area of vent may become inconvenient and some reduction in the vent ratio may be justifiable.[4] Because of the lack of firm information on this extrapolation, a substantial investigation was made in European laboratories which culminated in the proposal that 1-m$^3$ and 20-ℓ spheres be used as standard test equipment. Experiments were carried out in volumes up to 30 m$^3$. The basis of the method is that the maximum rate of pressure rise, $(dp/dt)_{max}$, is related to the volume of the vessel (V) and to a constant for the particular dust ($K_{st}$) by the equation

$$(dp/dt)_{max} \cdot V^{1/3} = K_{st} \qquad (2)$$

This is the so-called "cubic law" and by using it in conjunction with results from experiments with relief vents a series of nomographs have been produced which relate the volume of vessel, area vent, pressure at which the vent opens, and characteristics of the dust to the reduced explosion pressure (that is, the maximum pressure in a vented explosion).[7] In recent years, this approach has been adopted in the U.S., but with the use of the vertical tube apparatus for measurement of the maximum rate of pressure rise for the dust.[8] There is, then, an empirical correlation between the maximum rate of pressure rise measured by this method, and the nomographs. Although there is still some technical uncertainty, particularly as regards the correlation, the difficulties of extrapolating the vent ratio approach to large volumes is avoided. When the uncertainty involving the correlation of results from the vertical tube and the 20-$\ell$ sphere is resolved a major step forward will have been completed.

### C. Data on Dust Explosibility Parameters

In Table 3 it was possible only to give a very limited selection of materials for which data have been published and much fuller compilations are available elsewhere.[4,6,9,11,12] Many of these compilations are derived from the same sources, but data are now available for practically all the common dusts and powders which can cause a dust explosion hazard. In using data from compilations it is essential that the limitations of the data are fully kept in mind. The methods used in determining the parameters are usually as stated and their limitations have already been indicated. The test laboratories also may have used different methods in preparing the dust samples for test. In particular, in some laboratories the samples are reduced by grinding to a standard particle diameter, whereas in other laboratories the samples are tested as received. The effect of particle size on explosion parameters can be important and should be remembered when designing explosion precautions. For instance, the particle size of material obtained from a bag filter unit may be much smaller than that of the main product in a hopper, so that the explosion hazard is correspondingly increased. With a particular sample, where there is doubt as to the applicability of published values for the explosion parameters, it is always advisable to carry out tests specifically on the sample. Many of the substances listed in the compilations are single materials, whether natural or synthetic in origin, and the explosibility of mixtures or formulations cannot readily be predicted from that of the individual constituents. Direct measurement of the parameters may again be necessary.

The moisture content of the dust also affects the explosibility parameters, both in terms of the heat required to evaporate moisture and also the effect on dispersability of the powder by an increase in its moisture content. In some laboratories the problem is avoided by drying samples prior to testing, whereas in other cases it is argued that the dried material may not be typical of the dust actually present in industry. Particularly with natural materials, where moisture contents of the order of 10% by weight can be expected under equilibrium conditions of the atmosphere, the importance of moisture content must be considered. Unfortunately, in most cases the proportion of moisture needed to render a dust nonexplosible is too high for this to be a practical method of explosion protection. However, it is legitimate to take account of the moisture content of the dust in the overall assessment of the explosion hazard.

### D. Hazard Index

From the various tests for dust explosibility, a group of explosibility parameters can be obtained directly. These are essentially independent measurements and it would be convenient if they could be grouped together to give an overall assessment of explosibility in numerical terms. An accurate assessment of this type is currently not possible, but an empirical approach has been put forward by the U.S. Bureau of Mines.[4,11] Ratios called the ignition sensitivity and the explosion severity are calculated for the dust, and an index of explosibility can be obtained as a product of the two ratios. The ignition sensitivity and explosion severity for

a dust are obtained by comparing various explosibility parameters with the values obtained in the same tests using a standard dust, for which a Pittsburgh coal dust is used. The ratios are calculated as follows:
Ignition sensitivity

$$\frac{(\text{Min.ign.temp.} \times \text{Min.ign.energy} \times \text{Min.expl.conc.})\ \text{Standard dust}}{(\text{Min.ign.temp.} \times \text{Min.ign.energy} \times \text{Min.expl.conc.})\ \text{Sample dust}}$$

Explosion severity

$$\frac{(\text{Max.expl.press.} \times \text{Max rate press.rise})\ \text{Sample dust}}{(\text{Max.expl.press.} \times \text{Max rate press.rise})\ \text{Standard dust}}$$

Index of explosibility = ignition sensitivity × explosion severity

The index of explosibility has been correlated with the explosion hazard rating on an empirical basis, with a range of ratings from weak to severe. The index of explosibility for Pittsburgh coal dust is, of course, 1, which is regarded as a boundary between moderate and strong explosion hazard ratings. A dust which yields a rating well above that of coal dust is clearly likely to be a major dust explosion hazard.

The index is empirical and equal weight is given in the ratios to each of the parameters, but this may not be completely justifiable. For instance, minimum ignition energies and minimum explosible concentrations can vary over relatively wide ranges, whereas minimum ignition temperatures are well above room temperature and, due to limitations of the apparatus, cannot exceed 1000°C. The ratio is more heavily dependent on minimum ignition energy and minimum explosible concentration than on minimum ignition temperature.

The index of explosibility is an attempt to put on a numerical basis the explosion hazard rating of the dust, and as such it is useful, although it should be used with caution because of the limitations of an empirical relationship. The hazard rating of course, applies only to the dust itself and in any practical situation consideration would also have to be given to the processing the dust was likely to receive and the mass involved. The hazard rating relates to the dust suspended as a cloud in air and ignited, and does not take account of the burning properties of layers of dust.

## IV. IGNITION OF DUST LAYERS

### A. Introduction

Combustible dusts can cause fires, providing sufficient dust and oxygen to support the combustion are present. The dusts may be as layers, heaps, or other deposits rather than dispersed as a cloud in air as is necessary for a dust explosion. There are obvious differences in the appearances of dust fires and explosions. In fires, the rate of burning is much slower than in an explosion, and because the oxygen supply to particles of a layer is reduced, the burning rate of individual particles in a fire is correspondingly slower. The obvious characteristics of a dust explosion — the rapid rate of flame propagation and the generation of damaging pressure if the burning material is confined — are generally absent from dust fires. The quantity of dust in a unit volume of a layer is far greater than in the same volume of an explosible suspension, and the total amount of heat released when the layer is completely burned is greater than with the suspension, although the duration of the fire is much longer. In addition, certain dusts which are not explosible, such as anthracite and coke, can be ignited and cause fires.

There are two types of burning in dust fires, flaming and smoldering (or glowing). The type of burning obtained depends upon ignition source, the characteristics of the dust, the geometry of the deposit, and the ease of access of air. In dust fires, smoldering or glowing can propagate into unburned material, or flame may propagate over or through it.

In broad terms, the principal sources of ignition of dust fires are the same as for dust explosions. Fires can be initiated by flames, glowing particles, hot surfaces, welding and cutting, friction, electric sparks, and spontaneous heating. Although research has been carried out on the interaction of these ignition sources with various powders, the amount of work done has been relatively limited and recognized test methods have not yet evolved, except in the case of ignition by hot surfaces and spontaneous heating. Attention will therefore be restricted to these two sources which, in fact, can be dealt with by the same experimental technique. The action of hot surfaces may be upon a layer of dust with its upper surface exposed to ambient temperature or within a heated enclosure. Spontaneous heating is usually recognized as the heating of a mass of material whose surface is at ambient temperature. In both cases, the process involved is reaction of the dust with oxygen of the air to generate heat within the specimen. If the rate of generation of heat is greater than rate of heat loss, then the system may proceed to ignition and a fire results. The process has already been referred to in Table 1 and Equation 1. The test method involves the use of a thermostatically controlled hotplate or an oven at constant temperature. With this equipment, the minimum ignition temperature of a dust deposit of known geometry can be determined.

### B. The Ignition Temperature of a Dust Layer

The test apparatus consists of a circular stainless steel hotplate 200 mm in diameter and at least 20-mm thick. The plate is electrically heated and its temperature controlled by a thermocouple mounted in the plate at the center. The apparatus is shown in Figure 7 and further details are available in Reference 10. Dust layers are prepared by filling the cavity formed when a stainless steel ring is placed on the surface of a hotplate and by leveling the layer to the top of the ring. The ring is 100 mm in diameter and usually 5- or 12.5-mm deep. A fine thermocouple is stretched across the heated plate panel to the surface, and located so that the junction is at the geometric center of the dust sample.

The temperature of the plate is adjusted to the desired value and allowed to become steady. The ring is then placed centrally on the surface and filled with the dust, and leveled off. The temperatures within the dust layer and at the hotplate surface are measured by automatic recorders.

Typical temperature-time curves are shown in Figure 8. The ignition temperature is the lowest temperature of the hot surface, rounded down to the nearest multiple of 10 K at which ignition occurs in a dust layer of given thickness on the hot surface.

If the dust is seen to flame or glow, this is sufficient evidence of ignition. If flaming or glowing is not seen, the temperature-time record of the thermocouple within the dust layer must be examined. It will usually be found that, provided the temperature of the hotplate is high enough, the temperature of the layer will increase slowly to a maximum value that may be in excess of the temperature of the hotplate and then fall slowly to a steady value below the temperature of the hotplate. This behavior is evidence of self heating in the dust layer and may often be accompanied by discoloration of the dust but not by progressive combustion of the layer. Discoloration is not considered to be ignition. If the temperature of the hotplate is slightly higher it will be found that the temperature measured in a dust layer will continue to rise instead of passing through a maximum, and lead to ignition. In the determination of ignition temperature, repeated trials are carried out using a fresh layer of dust each time and adjusting the temperature of the hotplate until a temperature is found that is just high enough to cause ignition in a layer.

If ignition temperatures are measured from at least two depths of layer, then Equation 1 may be used to calculate the constants M and N for the material under test. From the equation, minimum ignition temperatures of other depths or other geometries can be calculated.

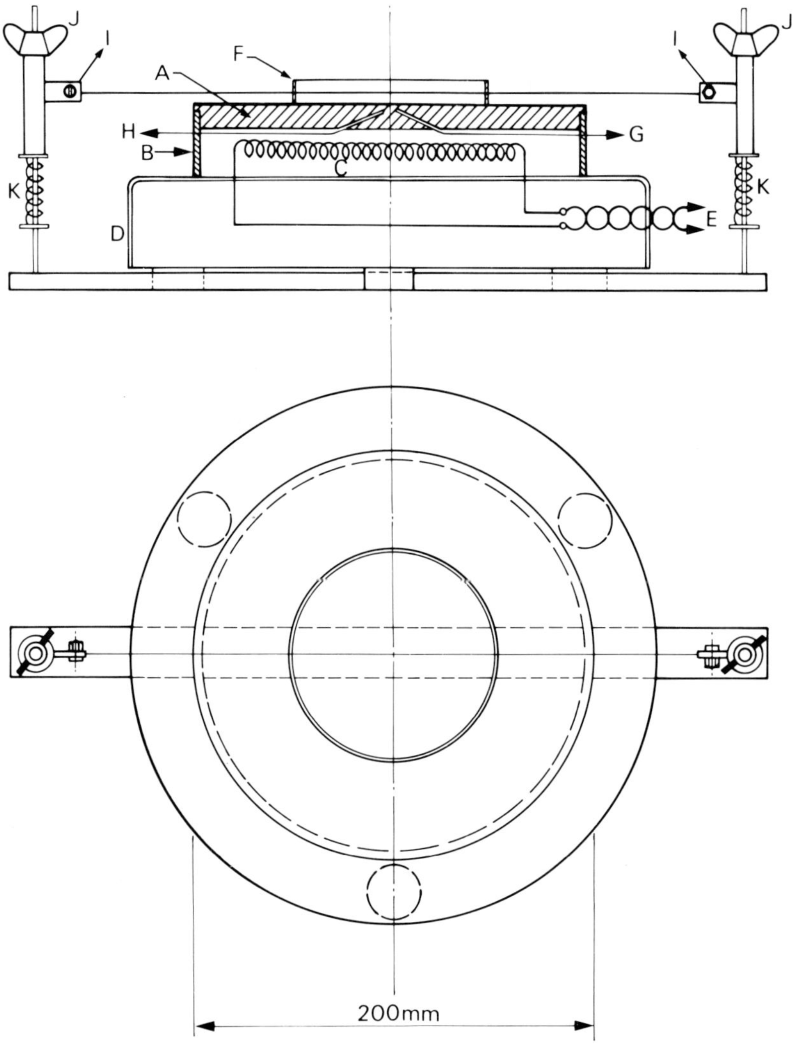


FIGURE 7. Hot plate apparatus for determination of the minimum ignition temperature of dust layers. (A) Heated plate; (B) skirt; (C) heater; (D) heater base; (E) heater connection to power supply and controller; (F) ring for dust layer; (G) plate thermocouple to controller; (H) plate thermocouple to recorder; (I) dust layer thermocouple to recorder; (J) screw adjustment for thermocouple height; (K) coil spring.

## C. Ignition of Dust Deposits

Where it is desired to obtain results directly for dust deposits in a hot environment, the dust may be heated in a laboratory oven. This should be fan-stirred and run at a constant temperature. The dust, either in the form of a layer of known thickness or as a cube in a wire gauze basket, is placed within the oven with a thermocouple at the center of the dust. The interior temperature of the oven is also monitored and the thermocouple connected to a recorder. Temperature-time curves similar to those in Figure 8 are again obtained and the criterion for ignition is the same. By means of Equation 1, the ignition temperatures of other geometries and sizes of specimen can be calculated.

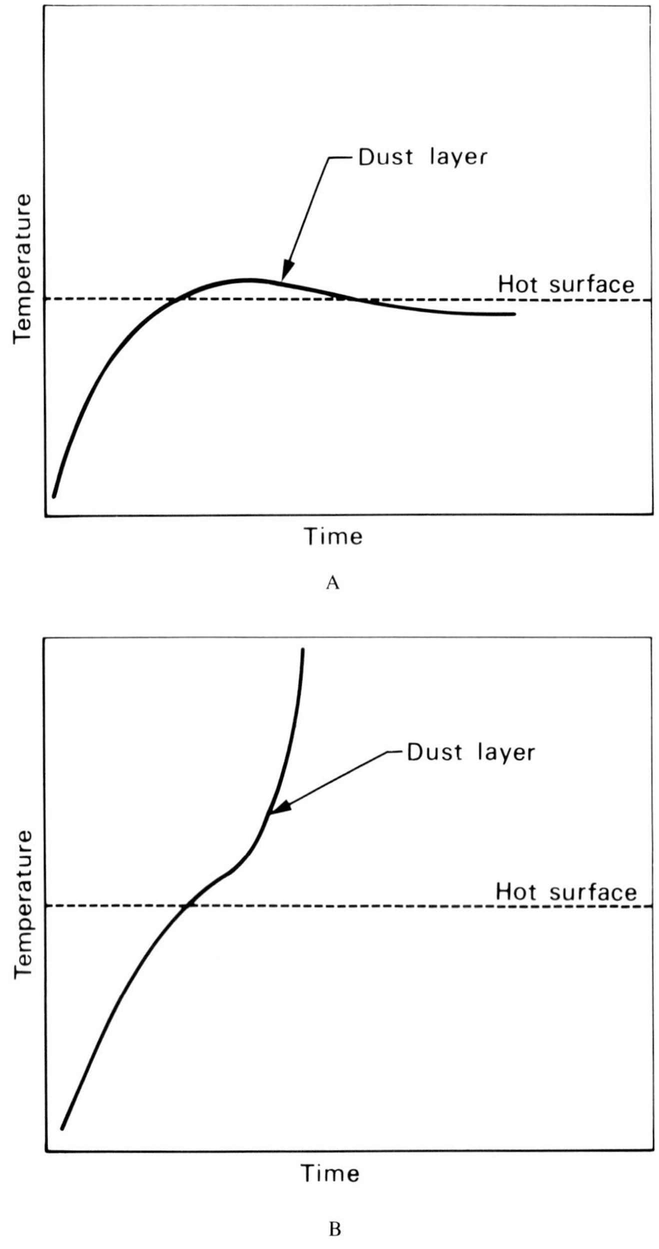


FIGURE 8. Typical temperature-time curves for ignition of dust layer on hot surface. (A) Self-heating without ignition; (B) ignition.

## V. CONCLUSIONS

The tests currently available for the measurement of explosibility parameters of dust clouds and the ignition of dust layers cover the majority of those of interest. Although technical advances in the existing methods are to be expected it is likely that in the foreseeable future these will be in the nature of improvements to existing techniques rather than complete innovations. The test methods themselves are of limited interest; it is their application to practical problems and their use for design data which are of greater importance. At present, much of this application is only semiquantitative and it is to be hoped that resources will be made available in the future for advances in this area.

There are some outstanding cases of commonplace sources of ignition not having a test method available for their assessment. Particularly, the processes of ignition by flames and glowing material, friction, and intensive mechanical energy such as in grinders, do not have tests available which will give quantitative assessments of explosibility and ignition hazards. Even modest advances would be particularly welcome because currently there are no means whereby design data can be obtained in a recognized manner. The problem is of importance since many serious dust explosions have recently been attributed to ignition by mechanical failure or the application of flames in industrial plants and installations.

## ALPHABETICAL SYMBOLS

$K_{st}$ Dust explosibility constant
M Material constant
N Material constant
p Explosion pressure
r Characteristic length
t Time
$T_l$ Minimum ignition temperature of dust layer
V Volume

## GREEK SYMBOLS

$\delta_c$ Geometrical constant

## REFERENCES

1. British Standard 5958, Part 1, British Standards Institution, London, 1980.
2. **Anon.,** *Runaway Reactions, Unstable Products and Combustible Powders,* Symp. Ser. No. 68, Institution of Chemical Engineers, Rugby, U.K., 1981.
3. **Anon.,** *User Guide to Fire and Explosion Hazards in the Drying of Particulate Materials,* Institution of Chemical Engineers, Rugby, U.K., 1977.
4. **Palmer K.N.,** *Dust Explosions and Fires,* Chapman & Hall, London, 1973.
5. **Anon.,** Classified lists of dusts that have been tested for explosibility in the form of a dust cloud. Dust Explosions in Factories, SHW 830, Her Majesty's Stationery Office, London, 1975.
6. **Dorsett, H. G., Jacobson, M., Nagy, J., and Williams, R.P.,** U.S. Bureau of Mines, RI 5624, U.S. Goverment Printing Office, Washington, D.C., 1960.
7. **Anon.,** *Druckentlastung von Staubexplosionen,* VDI 3673, Verein Deutscher Ingenieure, Dusseldorf, 1979.
8. **Anon.,** *Explosion Venting, 1978,* Guide 68, National Fire Proctection Association, Boston, 1978.
9. National Materials Advisory Board, *Matrix of Combustion-Relavent Properties and Classifications of Gases, Vapors, and Selected Solids,* NMAB 353-1, National Academy of Sciences, Washington, D.C., 1979.
10. National Materials Advisory Board, *Test Equipment For Use in Determining Classifications of Combustible Dusts,* NMAB 353-2, National Academy of Sciences, Wasington, D.C., 1979.
11. National Materials Advisory Board,*Classification of Combustible Dusts in Accordance with the National Electrical Code,* NMAB353-3, National Academy of Sciences, Washington, D.C., 1980.
12. Brenn-und Explosions-Kenngrossen von Stauben, *Forschungsbericht Staubexplosionen,* Berufsgenossenschaften e.V., Bonn, German Federal Republic, 1980.

Chapter 8

# PRACTICAL ASPECTS OF ELECTROZONE* SIZE ANALYSIS

**R. F. Karuhn and R. H. Berg**

## TABLE OF CONTENTS



* The word "electrozone", a registered trade name of Particle Data, Inc., has been released for use as language. Its etymology is self-evident and it fills the need for a succinct, generic term for this major sensing zone method of particle measurement. Parallel terms "photozone" and "sonozone" are also appropriate, as the authors have discussed in recent years with others eminent in the field of fineparticle measurement. Of these, "photozone" has the most pronounced subtypes, e.g., beam scattering, beam blockage, and beam diffraction.

# I. INTRODUCTION

The modern particle measurement laboratory should be equipped for various methods of particle size analysis including electrozone, photozone, elutriation, sedimentation, sieving, and microscopy. Experience at Particle Data Laboratories Ltd.* has proved the electrozone method to be dominant, although one or more of the other methods may be deemed ancillary, or sometimes better, for a given sample.

The electrozone method has the best combination of high resolution, size range, sensitivity, and mildness of secondary effects, thus justifying this thorough presentation of its usage. This well-established method was first detailed extensively by Berg,[1] and is based upon (1) causing a uniform suspension of particles in an electrolyte to flow, substantially singly, through a small orifice simultaneously with an electric current, and (2) electronically processing the resulting sequence of particle-pulses, the amplitude of each pulse being proportional to the volume of its particle.

# II. ANALYTICAL CONSIDERATIONS

In the operation of a particle measurement laboratory, there are initial points to consider for incoming samples for size analysis:

1. Safety (explosive, radioactive, poisonous, septic, etc.)
2. Protection of sample integrity (atmosphere, thermal limits)
3. Condition as received (wet/dry, packaging, frozen/caked)

Next for consideration are points affecting the choice of analysis method to be used, the particulate sample properties:

4. How it behaves in fluid media (air, aqueous, nonaqueous): dissolves, agglomerates, reacts, or otherwise changes from the state desired for analysis
5. Physical aspects: friability, porosity, shape, density
6. Probable size distribution span extremes

Then, assuming the electrozone method is to be used, consider the points for measurement. Our service laboratory uses the ELZONE®** computerized, multichannel, analyzing system. This differs from other electrozone measurement systems, but procedures are parallel:

7. Electrolyte selection and preparation
8. Sample preparation and presentation to electrozone sensor
9. Operational steps for data acquisition and processing
10. Selection of parameters and forms for data presentation

These and many other facets of particle size analysis as conducted on a computerized ELZONE® analyzer are detailed below.

## A. Laboratory Safety

Incoming samples can pose a health hazard or other danger. Powder samples often may generate airborne dust from aliquot sampling, sieving, air classifying, etc. Both wet and dry samples can have poisonous, flammable, septic, radioactive, or other dangerous properties.

* Particle Data Laboratories, Ltd. offers particle size analysis and other pertinent particle measurement services.

** ELZONE is a registered trade name of Particle Data, Inc.

Suitable precautions are taken against such dangers, based on data and recommendations from the sample supplier plus the experience and resources of the laboratory. Samples of unknown composition and/or properties are held aside until such information has been obtained from the sample source. On occasion, samples with danger potentials that may exceed the defenses of the laboratory are not accepted for analysis.

Storage, handling, and disposal techniques are adopted accordingly, including the presence of the sample supplier for on-the-spot guidance when deemed advisable. Disposal is strictly in accordance with the rules of public health and safety.

**B. Protection of Sample Integrity**

In addition to being held in a closed container, samples may have special environmental requirements as specified by the supplier. These include holding at deep-freeze, icewater, or room temperatures prior to analysis, or protecting from exposure to air, bacteria, etc. by appropriate isolation or inert atmosphere purging. Samples that are extremely prone to deterioration, often being delivered to the laboratory by overnight courier, are given first priority for analysis.

Stable samples are simply saved in sealed containers for several months, in case of the need to reanalyze. Those that may grow bacteria are stored cold for shorter periods before disposal. Unstable or dangerous samples are held in a safe manner for periods of some days, at most.

**C. Sample Condition as Received**

Careful examination of each sample upon receipt is essential for identification and assurance of undamaged condition. Without these, analysis should not be undertaken. As a result of mishandling in packaging and transport, identity labels may be loose or contents may have deteriorated.

To avoid error and confusion in processing thousands of samples annually, each must have incoming attention involving, at least, laboratory identification and routing for storage, analysis, and disposal. Some may require repackaging where containers are damaged or improper (e.g., fiber-shedding paper bags). Other treatment may be appropriate in cases where unintended wetting, freezing, or the like have occurred.

**D. Choice of Size Analysis Method**

Although the electrozone is the most advantageous method for size analysis, there are often reasons for using other methods. A major reason is that existing data for similar samples are by a particular method and conformity is required for maximum validity of data comparisons. Other reasons arise generally from particulate sample properties and environments, as indicated in Table 1 and the discussions below.

**E. Behavior of Samples in Liquid Media**

Materials that are soluble in both aqueous and nonaqueous polar liquids (electrolyte bases) are usually measurable when a suitable electrolyte is presaturated with that material and then used as the measurement medium. Often, nonconductive liquid components of a sample can be solvated with an organic electrolyte such as ''KM''*. Thus, for materials having complex organic media such as ink, paint, liquid toner, crude/lube oils, etc. the electrozone method is superb.

Sometimes adding an electrolyte, or filtering and resuspending in an electrolyte is not feasible because the material agglomerates, dissolves, or otherwise changes. Then, if circumstances permit, a photozone or sedimentation measurement might be used.

* ''KM'' refers to ''Karuhn's Medium'', a product of Particle Data Laboratories, Ltd.

**Table 1**
**CORRELATION OF SAMPLE TYPES, ANALYSIS DESIRED, AND INSTRUMENTAL METHODS**

| Sample type/source | Analysis desired | Instrumental methods |
|---|---|---|
| Stack emissions, quarry dusts, etc. | Aerodynamic settling velocity sizes | Air elutriators: ''Bahco®'', ''Roller'' |
| Marine or process sediments | Liquid sedimentation Stokesian sizes | ''Andreasen Pipette'', ''Sedigraph®'', etc. |
| Coal, rock, pellets (coarse materials) | Mass fractions for size intervals | Sieves: shaken, air-lift or sonic |
| Aerosols, gas/air contaminants | Concentration counts or size intervals | Light scattering or blocking (''HIAC®'') |
| Contaminants in nonconductive liquid (petroleum products) | Concentration counts for size intervals | Light scattering or blocking (''HIAC®'') |
| General dusts, powders, emulsions and slurries of biological, natural and synthetic origins | Morphology | Microscope, SEM |
| | Size distribution by count, area, or volume | Electrozone: ELZONE® COULTER® Counter |

*Note:* Trade names: ''Bahco®'' — Harry W. Dietert Co., ''Sedigraph®'' — Micromeritics Instrument Corp., ''HIAC'' — Pacific Scientific, Inc., ''COULTER® Counter'' — Coulter Electronics, Inc., ''ELZONE®'' — Particle Data, Inc.

Some materials that are initially airborne or liquidborne, but not suitable for photozone or electrozone methods, may be caught on a filter. These can have size distribution measurement by microscopy, with other characterization, of course.

### F. Physical Attributes of Samples

If the reason for measurement relates to sedimentation, then that measurement basis should be used, especially if fragile or odd shapes and/or mixed particle densities exist. Materials for extreme fragility also might be best measured by microscopy.

Sieve analysis is usually most effective for particles with combined size and density too great to permit a uniform electrozone sample suspension. Even given efficient agitation and high viscosity for the suspension, electrozone data will not generally be acceptable under the sample size-density condition expressed by:

$$(\mu)|\rho' - \rho^0|^{1.5} > 500 \qquad (1)$$

where $\mu$ is microns diameter and $\rho'$ and $\rho^0$ are densities of particles and liquid, respectively. Below a value of 100 for this function of $\mu$ and $\rho' - \rho^0$, sieving becomes increasingly ineffective. Many samples have particle size-density ranges requiring both electrozone and sieve analyses, with blending of the two data sets in the overlap region of 100 to 500 for this size-density function. Where the size-density function is well under 100, allowing gentle sample agitation, the electrozone method may be used on quite fragile materials. For acceptable measurement in cases where particles are quite porous or aspherical, the calibration of the electrozone apparatus must be adjusted so that data on a reference sample are matched to its accepted analysis, usually microscopic.

Samples of magnetic or conductive material are usually quite readily and successfully analyzed, using appropriate techniques and/or precautions as described later.

### G. Size Distribution Span Extremes

Size span advice from the sample supplier is often checked at the laboratory by microscope, along with such factors as agglomeration and shape. The presence of high size-density particles calls for sieving in part or in whole, as discussed above. In cases where little or none of the particulate system mass lies above 0.25 μm, electron microscopy or centrifugally aided sedimentation will be chosen as the method of analysis.

## III. ELECTROZONE SYSTEM FUNCTIONING

In the electrozone measurement principle, particles suspended in an electrolyte are caused to flow through a small orifice in a nonconductive material, along with an electric current. Particles traverse the orifice essentially singly, causing electrical pulses at rates from a few thousand to a few hundred per second or lower, depending on flow velocity, orifice size, and particle concentration. The amplitude of each pulse is directly proportional to the volume of the particle as sensed by its "electrical envelope" displacement within the sensing orifice. When diameter is the expressed dimension for electrozone size, it means the diameter of a sphere of equal electrozone volume response; the term "equivalent spherical diameter" is often used, akin to the term "equivalent Stokesian diameter" for the sizes in sedimentation measurements. Numerous experiments and theoretical models have been reported in the literature describing the mechanism of the electrical response.[2-5]

The size span measurable by a given electrozone orifice can exceed 25:1 in diameter, i.e., 1 to 3% to 60 to 80% of orifice diameter. This represents a signal amplitude range in excess of 16,000:1, for which the ELZONE® system provides logarithmic conversion.[6] For general usage, orifice diameters range from 12 to 1200 μm, with a total measurable diameter span of 0.15 to 900 μm for practical laboratory work. Particulate systems having size spans greater than 25:1 in diameter are handled by a two- or three-orifice, data-blending technique and/or extrapolation via Gaussian function-fit in the ELZONE® computer programming. For narrower distributions, lesser portions of the logarithmic range are expanded to full scale via a selector switch.

Thus, logarithmic conversion is a dominant feature of the ELZONE® system, which allows selection of log span to suit the breadth of sample size distribution so that full-scale usage is achieved for all analyses. Further, the log scale is "natural" for presentation of size distributions and makes easy the blending of two or three ranges of size data for the wider distributions. Such performance cannot be achieved practically with linear scales.

The following conditions are typical for the ELZONE® system:

1. Flow velocity and current through the sensing orifice are held constant at selected values in the ranges of 1 to 5 m/sec and 30 to 3000 μA, respectively
2. Pulse durations are most often 20 to 30 μsec, with extremes of 5 to 200 μsec
3. Before logarithmic conversion, pulses are amplified by a selectable factor ranging from 20 to 1800
4. For conversion of pulse amplitudes to digital (integer) values, 128 channels full-scale is usual, with 256 or 64 channels selectable
5. Orifice diameter is selected to be 50 to 100% greater than that of the largest particles in the sample
6. Sample suspension volume ranges from 5 to 500 mℓ, correlating directly with the square of the orifice diameter
7. Concentration of particulate material ranges from 100 to 10 ppm by volume, or lower, which allows an acceptable maximum of 3 to 4% for particle coincidence in the sensing orifice
8. The amount of particulate material used thus falls in the range of 0.1 to 200 mg, usually

The acquisition of size distribution data by a computerized ELZONE® system normally

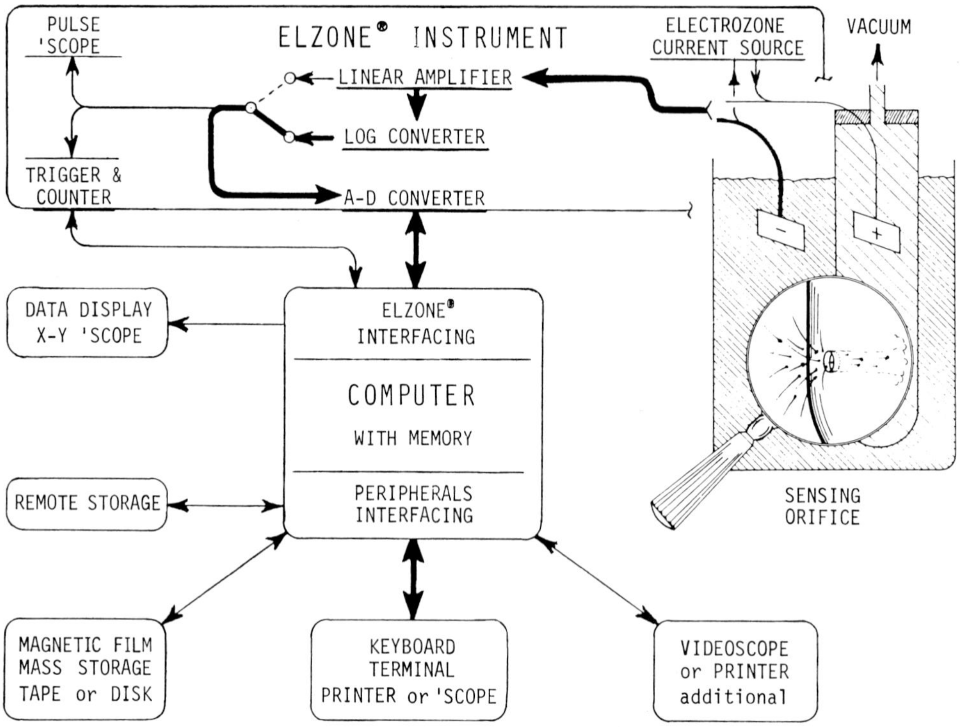


FIGURE 1. General schematic of complete ELZONE® system.

takes several minutes and consists of transfer of the digitized values for many thousands of particle-pulses to their proper locations in computer memory. The raw data format is thus a frequency histogram or differential distribution, normally having a logarithmic size scale.

Comeasurement of particle volume, length, and shape has been initiated at Particle Data,[7,8] and development is proceeding toward a mature capability. Comeasured volume and length data are accumulated in a three-dimensional frequency distribution for which volume and length are the coordinates for the horizontal plane and collected counts are the vertical axis. Length is sensed as a selected portion of the electrozone pulse duration. Shape is determined from pulse characteristics as the particle exits the electrozone orifice and may be shown as a codified sphericity index (prolate-ortho-oblate) in the data.

Throughout data acquisition and processing, a videoscope display provides interactive, operative monitoring of size distribution data and associated items.

The ELZONE® programming includes several data processing means including smoothing, scale normalizing or converting (both size and quantity axes), marking, editing, data-set blending, extrapolating, subtracting, and ratioing. The output means include graphing of histogram or cumulative data in frequency, area or volume (mass) form, alphanumeric printing of parameters and channel contents in complete or partial tabulations, transmission of data for storage or further processing in other devices, and overlay display and markers on the videoscope. The programming also provides means for forming customized series of processing and output steps, to be performed automatically in sequence, saving much operator time. Special data processing may be added, as requirements dictate.

## IV. COMPUTERIZED ELZONE® APPARATUS

The "Computerized ELZONE® System" used at Particle Data Laboratories is functionally organized as shown in Figure 1. Its physical array, for effective operational convenience,

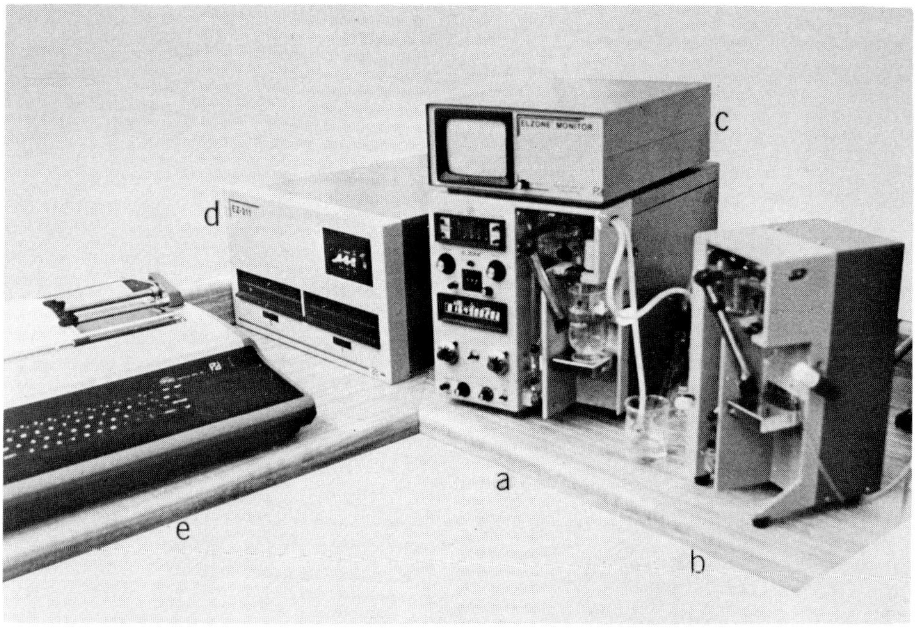


FIGURE 2. Practical bench-top array: (a) ELZONE® unit, (b) separate stand addition, (c) video monitor, (d) computer and floppy disc unit, (e) printing keyboard terminal.

is shown in Figure 2. Its performance efficiency is maximized by this careful convenience of the equipment combined with its speed and mature, advanced designs of both hardware and software.

The equipment location is, as much as possible, free of dust or electronic or sonic noise sources. Any of these can cause erroneous data and/or troublesome operation, especially when measurements are being done at high sensitivity in the submicron size range. Preventive measures may include air filtering of clean-room or clean-bench quality, vibration-isolating equipment mountings, avoidance of electrically noisy motors or lighting or SCR-regulated ovens and baths, and quiet power mains with proper grounding. During data acquisition at high sensitivity conditions, the sample stirrer is turned off and no loud talking is done.

## V. ELECTROLYTE SELECTION AND PREPARATION

Electrolytes for electrozone measurements must have effective properties; specifically, a suitable electrical conductivity and a physicochemical neutrality toward the particles. A listing of the most useful electrolytes and their applications is given in Table 2. Virtually all materials will have more than one suitable electrolyte, so extensive listings of materials vs. electrolytes need not be wholly relied upon, although such have been prepared and published.[9]

The practical range of conductivity is 2 to 200 mmho/cm, corresponding approximately to 0.1 to 10% NaCl in water as a reference. Higher conductivities are used for smaller orifices, i.e., over the orifice diameter range of 10 to 1000 μm, diameter relates inversely to conductivity as indicated in Table 2A. The concentration of the electrolytes suggested in Table 2 should be varied accordingly. This guide is not rigid, but general adherence to it aids achievement of full size span coverage and low noise effects from the sensing zone. For a given material, the electrolyte used must not adversely affect the particles, i.e., it

**Table 2**
**PRINCIPAL ELECTROLYTES AND THEIR MAIN APPLICATIONS**

| | |
|---|---|
| Physiological normal saline (aqueous) 0.9% NaCl, phos-buffered to pH 7.4 (mono- & dibasic sodium phosphates) | Biological cells of all kinds |
| Basic salt solutions (aqueous), e.g., sodium ortho- & pyrophosphate, sodium carbonate, sodium silicate, also parallel potassium salts<br>Note: basic conditions generally aid dispersion and wetting | Clays & ceramic/refractory raw materials<br>Minerals, earths & metals in general (high pH tolerant)<br>Dusts: stack, mine & mill<br>Powdered fuels & lubes<br>Plastics (Teflon®, PVC, etc.)<br>Inert materials in general |
| Neutral salt solutions (aqueous), e.g., NaCl, KCl, LiCl or sulfates, nitrates, as specific ions might be desired (see text) | Pharmaceuticals in general<br>Insect- pest- herb- fung- bacter-icides<br>Amphoteric metals (aluminum)<br>Minerals requiring neutral pH<br>Xerographic toners<br>Foodstuffs<br>Emulsions: oil, wax, latex, photographic, etc.<br>Inert materials generally (not difficult to disperse) |
| Self-electrolytes | Beer, soft drinks<br>Sea, river & lake waters<br>Urine<br>Parenteral fluids<br>Brines, acids, bases<br>Crystallizer liquor<br>Plating and process solutions |
| Salt solutions (polar organic liquids), e.g., ammonium thiocyanate (Na, K) lithium chloride in: alcohols, ketones, ethers, formamides, glacial acetic acid, with coupling solvents as occasion requires (tetrahydrofuran, cyclohexane, toluene, chlorinated hydrocarbons) | Foodstuffs that dissolve or swell in water<br>Deliquescent and anhydrous or semihydrated materials<br>Water-soluble salts, bases, acids, esters, etc.<br>Petroleum fractions (simple lube and hydraulic oils) |
| Solvating organic electrolytes, e.g., Karuhn's Medium ("KM") | Inks: printing, silkscreen, ball-point<br>Liquid xerographic toners<br>Paints, oil-based<br>Fillers and other particles in organic gels, pastes or nonpolar liquids<br>Crude oils<br>Liquid and gel lubricants |

**Table 2A**
**APPROXIMATE CONDUCTIVITY VS. ORIFICE DIAMETER**

| | | | | | |
|---|---|---|---|---|---|
| Diam. (μm) | 1000 | 320 | 100 | 32 | 10 |
| mmho/cm | 2 | 8 | 20 | 63 | 200 |
| % NaCl | 0.1 | 0.32 | 1.0 | 3.16 | 10 |

must not cause them to agglomerate, react, dissolve, disintegrate, shrink, or swell. Ideally, there should be the same degree of dispersion in the sample suspension as in the particulate system environment for which the measurement is to have meaning. To cover these requirements, electrolytes may be classed into three groups: aqueous, organic, and modified organic.

### A. Aqueous Media

These are generally safe, economical, and easy to handle, therefore used whenever possible. Given that a material is not soluble or slightly soluble in water, an aqueous electrolyte

**Table 3**
**USEFUL DISPERSANTS AND MAIN APPLICATIONS**

| | |
|---|---|
| Organic liquids, especially alcohols (1—4 carbons), glycerol, cyclohexanol | Carbon, stearates, pollens, chocolates |
| Aqueous solutions of phosphates (sodium): ortho- pyro- polymeta- (Calgon) | Inorganic materials not affected by high pH, water, or phosphate ions (frequently, just the stock electrolyte in daily use) |
| Triton® X-100 (Rohm & Hass, Philadelphia) | Nonionic, general purpose, often first to be tried for an unknown |
| Niaproof 4 (Niacet Corp., Niagara Falls) | Anionic, quite effective for majority of materials of (−) charge, perfect for polyfluorocarbons (Teflon®); also often first to be tried for an unknown |
| Lomar PW (Diamond Shamrock, Dallas) | Anionic — carbon, coal, coke, charcoal |
| Sulframin (Purex Co., Chicago) | Anionic — carbides, polymers, oil shale, stearates |
| Daxad 23 (W. R. Grace, Lexington, Maine) | Anionic — metals and metal oxides |
| Variquat 50-MC (Sherex Chemical, Dublin, Ohio) | Cationic — quite effective for materials of (+) charge; e.g., cationic biological exchange resin |
| Tween® 20 (ICI Americas, Inc., Wilmington, Del.) | Nonionic — latex stabilizing (polystyrene, polytoluene) |

*Note:* A modest selection such as the above covers virtually all materials. An unusually difficult sample may require some search for a suitable dispersion agent procedure. Dispersants commonly used for manufacturing and for cleaning number in the thousands and are often readily obtainable from manufacturers directly, in small sample quantities. Other names: Surfynol, Teepol, Nonidet, Darvan, Aerosol OT, Span, ad infinitum. (See *McCutcheon's Emulsifiers and Detergents,* McCutcheon Publishing Co., Glen Rock, N.J.)

may be used. However, certain properties of aqueous electrolytes must be considered, namely pH, wetting ability, and types of ions.

Moderately basic solutions, typically those of the alkali phosphates, generally favor suspension stability and good wetting action for most nonbiological materials, glass container surfaces, etc. Most often, a 4% solution of sodium pyrophosphate is used, since its conductivity is quite satisfactory in most cases and its general wetting and dispersing abilities are notable. Sometimes other phosphates such as sodium hexa-meta-phosphate provide superior wetting and dispersing action. However, should a material prove to require an acid condition for stability and/or good wetting, or should there be some dissolution at basic pH values well above 8, then an acid or neutral electrolyte is used.

The pH requirement may result in a poorly wetted/dispersed sample or the particle surfaces may have some property such as a hydrophobic oiliness, thus calling for the addition of an effective wetting agent or dispersant. There are undoubtedly many effective wetting agents for virtually every material somewhere among the myriad of such compounds on the market. Indeed, lacking an organized group of dispersants "on the shelf" in the laboratory, household detergents for cleaning dishes, walls, ovens, etc. can often be used effectively. As a practical matter, time is saved if all samples that can accept a dispersant at all are prepared using one. Table 3 lists the most popular compounds and their uses, as adopted over the past 25 years at Particle Data Laboratories.

Wetting/dispersing is a highly variable matter, associated with the physical chemistry of the particulate surfaces. Thus, only the loosest guidance can be provided — trial and error with microscopic examination being the usual method. Any negative effect of a given dispersant, e.g., solubilizing, can then be readily observed. Although surface areas range from tens of square centimeters per gram to hundreds of square meters per gram, a few drops of dispersant suffice for any given sample. The dispersant is generally added directly to the sample, the mix then being added to the electrolyte.

In some cases it is necessary to include or exclude specific ions in an electrolyte to obtain a desired result or to avoid an undesired one. When the particulate material is but slightly soluble in water, the electrolyte should first be saturated with that material and filtered for usage. This provides both ions, to suppress sample dissolution. Ions that will react unfavorably with the sample must be avoided, of course, as in the case of the chloride ion and photographic film particles. Whenever the possibility of specific ion constraint is suspected, the sample supplier is questioned on this point.

There are two other general cases of limited choice of aqueous electrolyte: (1) where the particles are already in an effective electrolyte such as sea water, plating solution, beer, crystallizer liquor; and (2) where the particles are biological cells requiring isotonic salinity (physiologically "normal" saline) to avoid swelling, shrinking, or other damage. In case (1) it may be necessary to dilute with particle-free water when electrolytic strength is too high for effective electrozone usage. In case (2), the wide usage of electrozone blood cell counting in hematology laboratories has resulted in the common availability of isotonic saline from lab supply houses. Formulations vary slightly, but all include mono- and dibasic phosphate buffering to pH 7.4, with an osmolality of 335 to 345 mosmos. Some are filtered for assurance of good data down to 3 μm (red cells), others down to 1 μm (platelets). These may be used for many applications, but cannot be regarded as usable and/or economical for all.

**B. Organic Media**

These are needed for many water soluble materials, including most salts and sugars, many esters, and foodstuffs. Other materials may be already suspended in media that are immiscible with water. For these types of samples, polar organic liquids having dielectric constants of 10 or more, with readily ionized materials such as lithium chloride, ammonium thiocyanate, or ammonium perchlorate, form a very useful group of electrolytes, as outlined in Table 2.

Alcohols and ketones having from one to four carbon atoms form the most useful organic media, but others such as amides and ethers are used per sample requirements. Often, when some other water-immiscible medium is involved, the polar liquid alone may not suffice to form a single liquid phase. In other cases, the material may be somewhat soluble in polar liquids. For both conditions, liquids having little or no polarity may be used as coupling solvents or for suppression of particle solubility. These liquids include chlorinated hydrocarbons, tetrahydrofuran, and aliphatic and aromatic hydrocarbons.

Since toxicity is almost always to be considered where organic media are used, adequate ventilation and handling precautions must be observed. Reduced measurement sensitivity (raising of the lowest measurable size) is a further consideration; i.e., organic media generally have lower conductivities and lower boiling points, both conducive to increased electronic source noise.

**C. Modified Organic Media**

In 1978, the proprietary electrolyte "Karuhn's Medium" was developed at Particle Data Laboratories to measure samples such as crude oil, complex hydraulic and lubricating oils, chocolates, paints, inks, and liquid toners.[10] Previously, these had been difficult or impossible to measure. When such materials are added to the "KM" electrolyte, the various complex organic liquid components are solvated, which allows ready measurement of the particulate components. "KM" will also solvate substantial quantities of useful organics, e.g., a solution having 20% toluene is readily formed. In sum, the capabilities of the electrozone method are greatly extended by "KM".

This solvating medium has good conductivity, low vapor pressure, miscibility with water, and excellent temperature and time stability. "KM" is a strong irritant, so precautions for usage are like those for many such materials, i.e., goggles, gloves, lab coats, and good ventilation.

**Table 4**
**ORIFICE FLOW AND COINCIDENCE RATES**

| Orifice diam. | Count/mℓ 1% coinc. | Flow (mℓ/min) at 16 cmHG ΔP | Count (1% coinc.) for 30 sec flow | Uncertainty ± 2σ (%) |
|---|---|---|---|---|
| 12 | 4,400,000 | 0.024 | 52,800 | 0.9 |
| 15 | 2,200,000 | 0.039 | 42,900 | 1.0 |
| 19 | 1,100,000 | 0.064 | 35,200 | 1.1 |
| 24 | 560,000 | 0.115 | 32,200 | 1.2 |
| 30 | 280,000 | 0.17 | 23,800 | 1.3 |
| 38 | 140,000 | 0.30 | 21,000 | 1.4 |
| 48 | 70,000 | 0.44 | 15,400 | 1.6 |
| 60 | 35,000 | 0.78 | 13,600 | 1.7 |
| 76 | 17,500 | 1.28 | 11,200 | 1.9 |
| 95 | 8,800 | 2.14 | 9,400 | 2.1 |
| 120 | 4,400 | 3.4 | 7,500 | 2.3 |
| 150 | 2,200 | 5.4 | 6,000 | 2.6 |
| 190 | 1,100 | 8.6 | 4,700 | 2.9 |
| 240 | 560 | 14.0 | 3,900 | 3.2 |
| 300 | 280 | 22.0 | 3,100 | 3.6 |
| 380 | 140 | 34.0 | 2,400 | 4.1 |
| 480 | 70 | 52.0 | 1,820 | 4.7 |
| 600 | 35 | 82.0 | 1,430 | 5.3 |
| 760 | 17.5 | 140.0 | 1,200 | 5.8 |
| 950 | 9.8 | 220.0 | 990 | 6.4 |
| 1200 | 4.4 | 360.0 | 790 | 7.1 |

### D. Electrolyte Preparation

Over the years, satisfactory electrolytes have been prepared using distilled water and "technical" grades of salts and anhydrous organic liquids. Electrolyte temperatures are kept the same as the analysis room, except when an analysis is run at some artificially maintained higher or lower temperature.

When an aqueous or an organic electrolyte must be presaturated with the particulate material of concern, it is added in substantial excess after the electrolyte salt has dissolved. In some cases, no salt other than this presaturating sample material is needed. For presaturation, the electrolyte with excess of sample is held at room temperature for many hours or even a few days, with a periodic or continuous agitation, and then filtered. Testing of the presaturated electrolyte is done on the ELZONE® system, using the sample material, to observe for stability of analysis with time. If dissolving or crystallizing is indicated by loss or gain of material with time, further saturating and/or filtering is done until the electrolyte is suitable for a satisfactory analysis.

For highly accurate particle size analyses, electrolytes must be virtually free of particles. This is defined at Particle Data Laboratories as a concentration that is below 0.5% of that for a coincidence rate of 1% in the orifice being used, counting all particles larger than the smallest size being measured. Table 4 lists 1% coincidence levels for many orifice sizes used by the ELZONE® system.

Filter cartridges of the high-area type provide the best quality of filtration and the easiest handling. As shown in Figure 3, no pumping or pressure other than a small liquid head is required. Filtering rates of 100 mℓ/min are readily achieved for 0.2 μm (sterile) filtering. A filter that will last at least 1 year, processing many hundreds of liters, costs but $30 to $50 (1983). Multicomponent filtration apparatus, by contrast, requires parts assemblies and filtrate transfers that can cause extra work and contaminated filtrate.

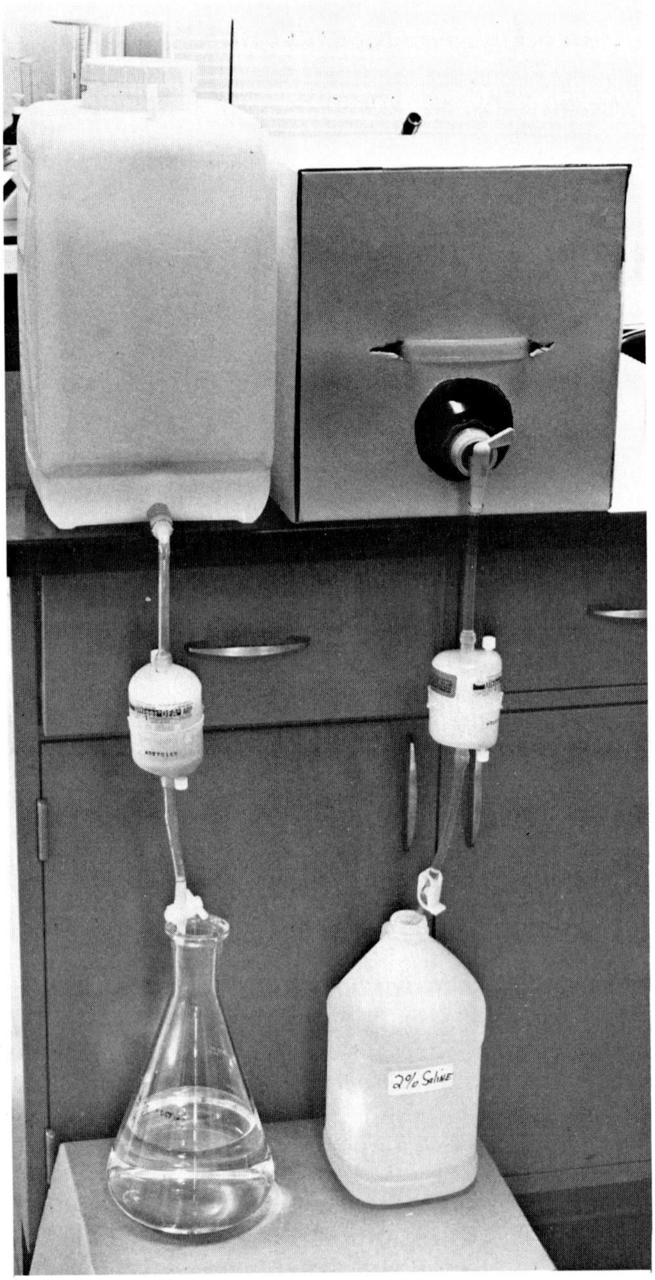


FIGURE 3. Simple system for electrolyte filtration in two configurations. (Left) rigid, vented, 20l supply container with end stopcock of polyethylene/ Teflon® (Nalgene 6460-0004). (Right) collapsible, cubic, 20l polyethylene container (Nalgene 2320), with end pinchvalve (locking) $9^{1/2}$ mm Tygon® tubing is used. Filter cartridges are 0.2 μm "Ultipor®" (Pall Filter Corp. #DFA3001ARP)

Initially, after the setup of the filtration system and filling of the supply container, the system should be purged by filtering and recycling about 4 ℓ of electrolyte. This cleans the lines and receiving flask surfaces and performs the "conditioning" of the filter membrane. The return of the flushing filtrate to the supply container is done as may be convenient. A few milliliters of "technical" formaldehyde solution per liter will prevent bacteria growth in aqueous electrolytes, thus protecting filter life.

In the numerous cases where sample materials do not affect the electrolyte, used amounts may be recycled through the filter for effective savings. In cases of volatile organics, periodic replenishing of the liquid component only is usually practical. However, those using ammonium thiocyanate are best made up as needed, as this salt decomposes upon exposure to light, causing operational problems as well as shortening filter life. The same applies to any electrolytes that may decompose with time.

Only enough electrolyte for one day of operation or less should be filtered one time. This avoids problems due to contamination of the electrolyte in active use. The container in active use should be kept covered with film or foil when nothing is being poured. These steps assure virtually particle-free electrolytes down to a size level of 0.2 to 0.3 μm.

## VI. SAMPLE PREPARATION AND PRESENTATION TO THE SENSOR

For preparation purposes, samples may be classified into five types:

Oil in water emulsions
Solids in aqueous liquids
Solids in organic liquids
Dry powders
Special problems

In all cases, there must be a proper choice of electrolyte, an effective method for achieving satisfactory dispersion of particles in it, and the correct presentation of the particles to the sensing orifice. When these preparation and operating procedures are not known, preliminary testing must be done to establish them.

Adequacy and stability of dispersion of the particles are primary concerns. These are best determined by observation of particle behavior via a 100X phase contrast microscope. In the absence of definite input to the contrary, the analyst must take the position that all samples must be completely deagglomerated and dispersed in suspension for proper measurement. A hurried inspection can easily prevent correct analysis, for instability may occur in many forms, including:

1. Coalescence of oil emulsion droplets
2. Agglomeration of surface-charged particles
3. Agglomeration of magnetized particles
4. Dissolving of particles
5. Crystal nucleation and growth
6. Deliquescence (in normal atmospheric conditions)

The preparatory elements, determined primarily with the aid of the microscope, include: electrolyte choice, proper wetting, degaussing, mixing (via ultrasonic bath and/or spatulation). At the same time, the size of the largest particles is observed in order to guide the selection of the orifice diameter to be used (about twice that of the largest particles).

Stability may be further confirmed by a series of data acquisitions with the ELZONE® system. In extreme cases, samples may require very quick analysis (in seconds) after being added to the electrolyte on the sample stand. Table 5 shows several sets of symptoms of instability for various particulate conditions.

A major aspect of proper presentation of the sample to the sensing orifice is the assurance of representative sampling. This often involves coning and quartering of dry powder sample

**Table 5**
**DISPERSION INSTABILITY GUIDE**

| Change in relative amount: At max size | Change in relative amount: At min size | Possible reason |
|---|---|---|
| + | + | Crystallization |
| − | − | Dissolution |
| − | + | Deagglomeration |
| + | − | Agglomeration |
| − | 0 | Settling of larger sizes |

or dynamic thieving from briskly agitated slurries. However, with particles of high size-density index, care must also be taken to be sure of a uniform suspension via raised viscosity of electrolyte and a high level of agitation of the sample suspension. Further, the orientation of the orifice should be toward the oncoming "ring flow" current in the suspension for isokinetic intake of particles. Again, a series of data acquisitions with the ELZONE® system may be used to establish the correctness of the operating method in presenting the particles to the orifice.

Special problems in sample preparation can take many forms. Some interesting solutions to these have been developed at Particle Data Laboratories and elsewhere. For example, the problems encountered when the particles in inks, paints, and other materials based on complex organic liquids gave rise to the development of the "KM" solvating organic electrolyte, as said above. Examples of other difficult sample preparations are discussed below, to illustrate this important aspect of fine particle laboratory practice.

Objectionable trace contamination is present in many liquids such as beverages, parenteral solutions, and oils and hydrocarbon fuels in amounts as low as 10 ppb or so. Measurement of such contaminants in a statistically satisfactory manner calls for the examination of thousands of times as much liquid as ordinarily scanned in a suspension that has been prepared for the electrozone measurement. This is even more of a problem if electrolyte must be added, as the added liquid should be much cleaner than the sample if background problems are to be avoided. This is especially so for nonaqueous media which typically require large portions of added liquid.

Highly effective measurement is readily achieved if such trace-level contaminants are concentrated by filtering some liters of the liquid in question and resuspension of the contaminants from the filter into a few milliliters of suitable electrolyte. In the case of oily liquids, it has been found that the pores of the filter portion that is clamped in the holder might not be washed free of oil upon solvent washing, resulting in error due to oil droplets being included with the resuspended material. This has been solved either by a special rewashing and refiltering prior to preparing for measurement, or by using a filter that does not hold oil in the clamp-covered pores.

Magnetic powders may be kept dispersed for a long enough time to permit electrozone measurement if they are degaussed first for 20 min at temperatures as high as 750°C (but short of sintering, of course) and/or spatulated with very viscous honey to inhibit reagglomeration. The honey suspension on a microscope slide is then dipped into the electrolyte in the stirred sample beaker. In this manner, data acquisition is initiated instantly while the suspension is being formed from the dissolving honey.

Semen, animal or human, poses two problems: elimination of interfering background particles in the semen and determining the percentage of live spermatozoa in the total. Triton® X100 for the dispersion of background particles has proven to be quite practical, even though it also removes the tails of the spermatozoa (for humans but not for cattle, oddly). Passing the diluted semen through a chromatographic column removes the dead spermatozoa, so the measurement of percent-live is simply a before-and-after data comparison.

**Table 6**
**COINCIDENCE EFFECTS ON ELECTROZONE DATA**

| Frequency distribution | | | | | Volume distribution | | | |
|---|---|---|---|---|---|---|---|---|
| **Median** | **Mean** | **S.D.** | **Skew** | **Coinc. (%)** | **Median** | **Mean** | **S.D.** | **Skew** |
| 5.09 | 5.19 | 2.52 | −0.006 | 1 | 7.72 | 7.77 | 3.85 | −0.28 |
| 5.21 | 5.22 | 2.58 | −0.09 | 5 | 8.66 | 8.51 | 4.04 | −0.04 |
| 5.21 | 5.23 | 2.71 | −0.18 | 10 | 8.86 | 8.78 | 4.28 | −0.02 |

## VII. COINCIDENCE EFFECTS

A further major aspect of presentation of the particles to the sensor is the avoidance of excessive concentration which causes unacceptable levels of coincident particle passages. Such will have two effects: (1) loss of data due to the hidden particles in the coincidences, and (2) spurious data due to the larger pulses formed by partial or complete addition of signals from coincident particles. The lost data are no problem, since they have the same distribution as the acquired data. The spurious data, however, form a pseudosystem of generally larger size, added to the valid data.[11]

This pseudosystem can seriously affect the accuracy of the data when coincidence levels are but a few percent, as shown in Table 6. This is contrary to the claimed permissibility of 10% coincidence, said in COULTER® instruction manuals. The effects are greater on volume distribution data because the $D^3$ weighting, from frequency to volume data, enhances the relative significance of the large end of the distribution. Such clear definition would be impossible for distribution data having only one point for each 2:1 interval of particle volume. Accordingly, care is taken at Particle Data Laboratories to keep coincidence levels near 1% or lower. Data acquisition takes 10 times longer than it would at 10% coincidence, but the avoidance of compromised data makes it well worthwhile.

## VIII. DATA ACQUISITION AND PROCESSING

Three essential priorities for data acquisition are the selection of the orifice for correct size, its high fidelity, and its operational monitoring and handling. Orifice diameter should be 150 to 200% that of the largest particles in the sample, as determined via microscope or electrozone observations. For irregular particle shapes that can more easily block an orifice during measurement, its size should be on the high side.

High fidelity of response is the subject of careful testing in orifice tube manufacture at Particle Data, Inc., using standard samples of particulate materials. The manufacturing procedures are such that the orifices naturally achieve hydrodynamic focusing. Orifices in use at Particle Data Laboratories have the highest quality of response, producing distortion-free measurements without hydrofocusing. If an orifice were faulty or dirty, the ELZONE® system would not produce the typical, sharp resolution of a sub-population as is shown in Figure 4. Instead, data would be skewed toward the higher particle sizes, where the $D^3$ weighting adds most to the distortion of the volume distribution.

For acceptable data, no degree of electrozone orifice blockage can be tolerated. A 1-μm-thick film around the walls of a 48-μm orifice reduces its diameter by 4% and its response, being in inverse proportion to the fourth power of orifice diameter, increases by 17%. Therefore, not only must the analyst be watchful for blockages during data acquisition, but care must be taken during periods of nonuse to keep the orifices clean. It is easiest to keep them wet, preferably in phosphate or detergent solution, whether on or off the ELZONE® stand. Such care will keep the orifice and nearby surfaces, as viewed via the orifice obser-

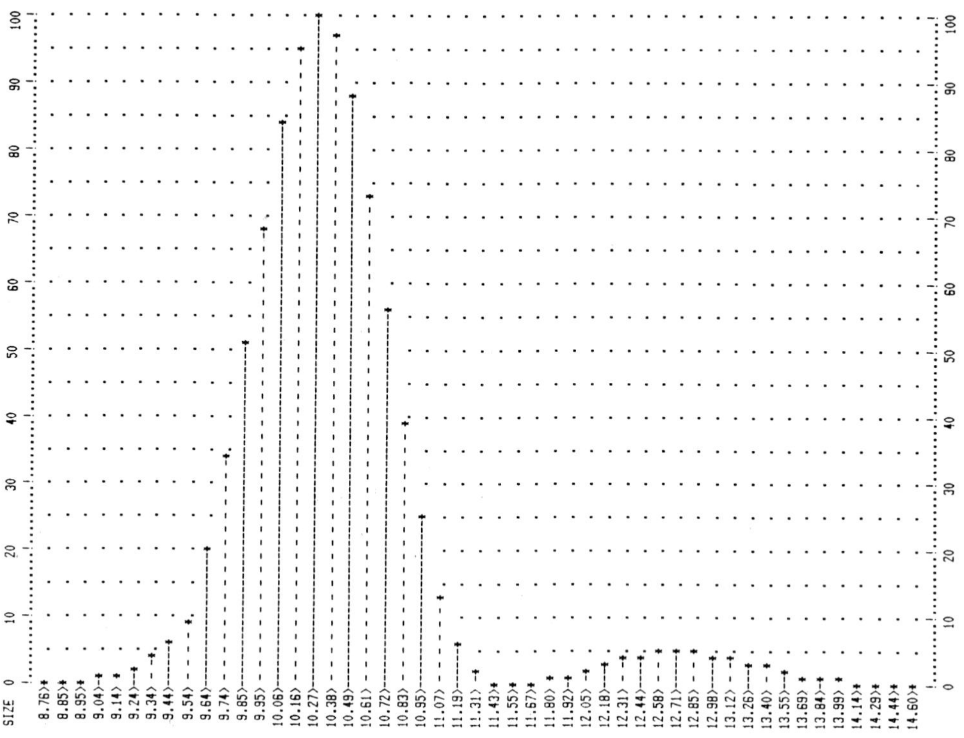

FIGURE 4. Data from properly responding orifice, showing a clear, undistorted symmetry for a latex, with sharp definition of a 5% presence of doublets, at 22 channels per 100% volume increase (26% diameter increase).

vation microscope, free of clinging debris or growth of small bubbles which can interfere with the generation of proper signals. Dry storage of orifice tubes is permissible if care is taken to clean them thoroughly, rinse with distilled water, and then with dry alcohol or acetone. Leaving an orifice tube unwetted on the stand or stored with electrolyte and/or stopcock grease on it must never be done.

Most blockages during operation can be cleared by applying momentary back pressure or high current to the orifice or using a small, clean brush as per the instrument instruction manual. Cases of erroneous storing, difficult blockages, or sticky deposits (usually from samples of biological origin), may require treatment with detergents, strong oxidants, or even (briefly) hydrofluoric acid. Ultrasonic treatment may be required also, however, this must be mild or the orifice may be damaged.

## A. Size Span Settings

Given an orifice diameter at least twice that of the largest particles in the sample, the logarithmic range setting on the ELZONE® unit then determines the width of size span that will be measured. Six diameter ranges are available, from $2^1/_2$:1 to 25:1 full scale (spans of $2^4$, $2^6$, $2^8$, $2^{10}$, $2^{12}$, and $2^{14}$ in volume). The lower three settings are used for narrow distributions, which are generally those having either natural or artificial regulation in their formation, e.g., a finely sized industrial abrasive, biological cells of a single kind, certain river sediments, and particulate products of controlled comminution, classification, or synthesis. Particulate systems of unregulated formation have wider size spans, ranging generally from 10:1 to 100:1 in diameter. For these, a log setting of 12 volume doublings, 16:1 in diameter full scale, is most often used. This avoids the smallest size levels (log 14) for a given orifice, where the signal-to-noise ratio may have low values that can cause problems in the blending of data sets for wide size spans.

The top-of-scale size for a given orifice and log setting depends on the amount of signal amplification, as determined by the current and gain settings; the greater the amplification, the smaller the top-of-scale size. In the setting of signal amplification, two major rules must be observed:

1. To assure capture of all of the largest particles, no signals may exceed top-of-scale. The large particles are far too important in the volume (mass) distribution to allow any to be missed. A single 50-μm particle equals a thousand at 5-μm. For safety, the topmost several channels should have nothing in them when data acquisition has been completed. Of course, this does not apply to the smaller-sized data portion(s) in multirange blending.
2. To assure minimum noise for a given top-of-scale size, the current level must be advanced first, while the gain level remains at minimum. The current level must not exceed that which is accompanied either by partial conductance in the particles (e.g., metals[12] or red blood cells, having an apparent shrinking in size) or by thermal noise from heating of the orifice contents. These phenomena may be observed via trial size analysis; both will skew the distribution curve toward the smaller sizes. The gain setting may then be advanced as needed.

### B. Resolution

The number of channels full-scale may be 64, 128, or 256, as selected via the computer program. Thus, the interchannel diameter ratio (resolution) ranges from 5.2 to 0.36% for channels/doublings settings from 64/14 to 256/4 accordingly. For the frequently used 128/12 settings, the interchannel resolution is 2.2% on a diameter basis.

Whether the size distribution is narrow or wide, 128 channels usually provides an adequate resolution of the distributive details. In some cases, 256-channel full-scale resolution is needed to delineate subsystems. In other cases, 64 channels is enough and saves time in data acquisition. (Each channel accumulates twice as quickly.) This is useful when data sets must be taken rapidly, to minimize or to study distribution changes with time.

Quantitative resolution for a given channel is proportional to the square root of the number of counts accumulated in it, e.g., a channel count of 1600 has an S.D. of 40 counts or 2.5%, yielding a $2\sigma$ "spread" of $\pm 5\%$. In practice, satisfactory data result when the peak-channel count is 4000 or more. Its resolution is then $\pm 3\%$ or better; the total count is then in the region of 50,000 to 150,000. Such data are acceptable for multiorifice range blending, as stated previously.

Quantitative resolution is visually enhanced by the least-squares, polynomial-fit, smoothing command in the computer program. Although this is only a cosmetic effect, it greatly aids human visualization of the distributions. However, presmoothing reduces the effectiveness of any subsequent data conversion and blending operations. Smoothing is, therefore, used only as a "finishing touch" after data processing has otherwise been completed.

### C. Background

Background counts arise from residual particles that have escaped the filtering and rinsing procedures and from stray bits of electrical or sonic noise of various origins. For the usual size distribution data, the steps and techniques that have been emphasized earlier here will easily hold the background to insignificant levels. However, when the particle concentration (contamination) or the number of sample particles available is very low, a background count is taken and stored in computer memory. Later, this is matched with the sample-plus-background data via the overlay display of the computer program and the background then subtracted.

It is possible to perform a similar operation on the "white noise" at the low end of the sensed data. However, white noise is not strictly additive to the real data, thus any "correction" of it is quite questionable. It is far better to use further valid data from a smaller orifice or even a simple Gaussian extrapolation of the data at hand, for which the program is equipped.

### D. Calibration

Each combination of orifice size and logarithmic span requires calibration. Since the electrozone signals are linear with particle volume, calibration data are simply the two numbers required to describe a straight line. The "internal standard particle" reference signal[13] of the ELZONE® unit allows a given calibration to be used for all electrolytes.

Compared to a rod or flake in lengthwise passage, a sphere appears to have an excessive response by about 6% (diameter basis), due to the cusp-like zones at each end where the electrical field "opens" to surround the sphere. This effect lessens as shape becomes more aspherical, virtually disappearing for long rods and thin flakes. Due to the prevailing laminar flow in the sensing orifice these have, in general, lengthwise orientation. Otherwise, some appreciable effects would occur and flakes would have responses more excessive than spheres.

So, to have perfect response to shape, carefully prepared fractions of the sample to be measured would have to be made for use as calibration standards, but that is a lot of work for such a moderate "sphericity effect". As a practical measure, for calibration accuracy in keeping with the electrozone method, estimated adjustments to the calibrations may be made based on the average particle shape for a given sample as judged from microscopic examination.

Thus, very narrow distributions of polymer (latex) spheres are most commonly used for calibration standards. Pollens and spores can also be used, but they can absorb water, increasing in size somewhat. The modal size of such distributions is used in electrozone calibration, since it is readily identified as the peak of the size distribution histogram of a given standard. Likewise, the modal size is readily established via prior, rigorous measurement via microscopy of the highest caliber. For a given orifice and logarithmic scale, three methods for establishing calibration may be used:

1. Given a known interchannel factor (equivalent to the slope of the calibration line), only one size point on the scale must be established, using one known standard
2. Given known, reliable, signal amplification (current and gain controls), two size point on the scale may be established, using only one known standard, by using the amplification controls to place a "known" size point near each end of the scale (within 20% of the end)
3. Two size points on the scale may be established, one near each end (within 20% of scale), using two known standards

In view of the importance of accurate calibration, method 3, above, is relied upon at Particle Data Laboratories. Also, a third standard and/or the other two methods will often be used as checks. Typically, at log 12, a third standard will check within one channel or 2.2% of the modal diameter of the standard.

Calibrations are very carefully done at Particle Data Laboratories. Since orifices must be absolutely clean for highest precision, thorough cleaning of all orifices precedes calibration, i.e., detergent scrubbing and overnight soaking in chromic acid. Well filtered aqueous electrolytes are used, and sample containers are likewise very clean. Suspensions of standards are diluted to effect coincidence levels below 1% and they are given ultrasonic bath treatment prior to usage to assure good dispersion. Each orifice is pretested for cleanliness (no noise or clinging air bubbles, etc.) and then closely monitored during the acquisition of calibration data via the orifice observation microscope.

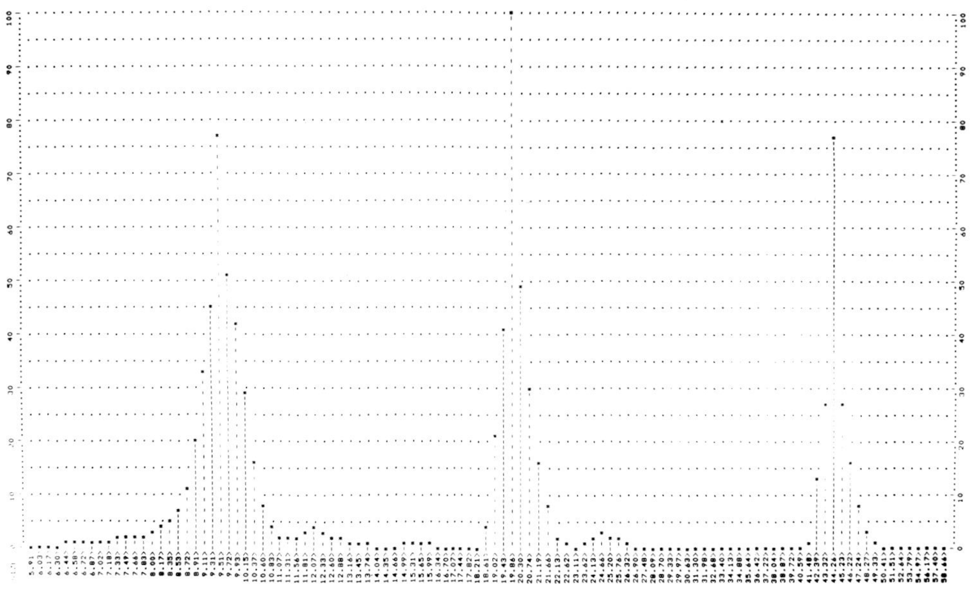

FIGURE 5. Data for calibration of 120 μm orifice, using two latex standards for near-the-end points and showing third standard checking (sizes as labeled).

Calibration data are entered per the computer program promptings. A calibration that deviates too greatly from the nominal log-span setting being used is rejected by the computer program, indicating erroneous entry. For most precise calibration, interpolation is used to determine, to a fraction of a channel, the true peak location on the histogram. Figure 5 is a typical calibration run at log 12, using two standards and showing the fit of a third (middle) standard. Calibration standards should be carefully examined upon purchase (from a reliable source) and then given the best of care. This includes:

1. Storing in a cool, dark place, always avoiding heat
2. Covering tightly always, and avoiding any contamination (bacteria can grow in the standards)
3. Keeping them deagglomerated by occasional shaking, plus using an ultrasonic bath for assurance of dispersion just prior to use (avoid high-intensity ultrasonic probes which can cause irreversible coagulation)
4. Prediluting all latex standards smaller than 2 μm with distilled water, at least tenfold (in concentrated form, they may coagulate upon contact with an electrolyte)

### E. Multiorifice Data Acquisition

For covering diameter spans in excess of 16:1 or 25:1 (12 to 14 volume doublings), multiorifice data sets are required. By far, the majority of samples coming to Particle Data Laboratories have this requirement, and the computer programming is well adapted to it. Adjoining data sets must have sufficient range overlapping to allow their proper blending via the computer program, so that two or more data sets may be combined into a single set of distribution data.

There are two cautions inherent in multiorifice data acquisition. Their circumstances are described below and illustrated in Figure 6.

1. In the overlap regions there must be enough data gathered to provide a sound statistical basis for the subsequent blending of data sets. Each data set will typically have relatively

more particles in the channels at the finer end than at the coarser end. The analyst must be sure, therefore, that sufficient data are in the channels at the large end of the smaller orifice data set. In practice, this means that: (a) the diameter ratio of larger to smaller orifice should not exceed 4:1 (six doublings of distribution span gained from the added data set), and (b) quantitative resolution for each data set must be more than 50,000 particles total.

2. Prior to acquiring the smaller data set, particles larger than 50 to 60% of the diameter of the smaller orifice should be removed from the sample by wet sieving or partial sedimentation. The size distribution will be quite altered in the region of such cut-off, so usage of the smaller data set should not exceed 25 to 30% of the diameter of the smaller orifice. Often, however, the relative frequency of particles that can block the smaller orifice is low enough that the analyst can tolerate them via two convenient features of the ELZONE® system: its data protection programming and its back-pressure push button for orifice clearance. These, in combination with partial settling (turning off stirrer and waiting some seconds), allow the avoidance of sample ''cropping'' by wet sieving.

## F. Data Blending

Multiple sets of electrozone data should be blended in the converted volume (mass) distribution form, rather than in the acquired frequency distribution form. Further, the data smoothing function should not precede the blending. These rules are based on experience as well as statistical factors. Simply put, in the frequency distributions, the data in the overlap region have relatively small numerical values compared to the corresponding values in the mass distributions. The $D^3$ factors in the data conversions are responsible for these large numbers and the resultant improvement in the computer program blending calculations (least-squares averaging). These calculations are more effective if the data have not been previously subjected to similar least-squares treatment via smoothing.

The rules for effective blending of electrozone data sets may be stated as follows:

1. At least 50,000 particles per data set
2. Ratio of orifice diameters 4:1 or lower
3. Wet sieve at 50% of smaller orifice diameter as needed
4. Conversion to mass basis before blending
5. No smoothing of data sets before blending
6. No blending near mode or other high-curvature region
7. No blending if any zero-content channels in overlap
8. At least 20% of each data scale included in overlap

Sieve data are blended with electrozone data by similar overlapping. At best, sieves provide only four data points per diameter doubling. Also, sieves have a characteristic behavior toward aspherical shapes differing from that of electrozone sensing. Such troublesome data overlapping is now carefully fitted via manual graphic technique, but there will soon be added computer programming to perform this task.

## G. Extrapolation

Always, the analyst's goal is to report the complete distribution. However, either single data-set or multiple data-set distributions may be more or less incomplete at either or both ends, with circumstances that make further data acquisition impractical or impossible. For incomplete distribution data, the only statistical index that can be truly expressed is the mode. Unless some arbitrary cut-off size is assumed, a complete distribution is required for expression of the mean and other indexes, to say nothing of the median and other percentiles.

The Gaussian extrapolation function in the ELZONE® programming allows the forming

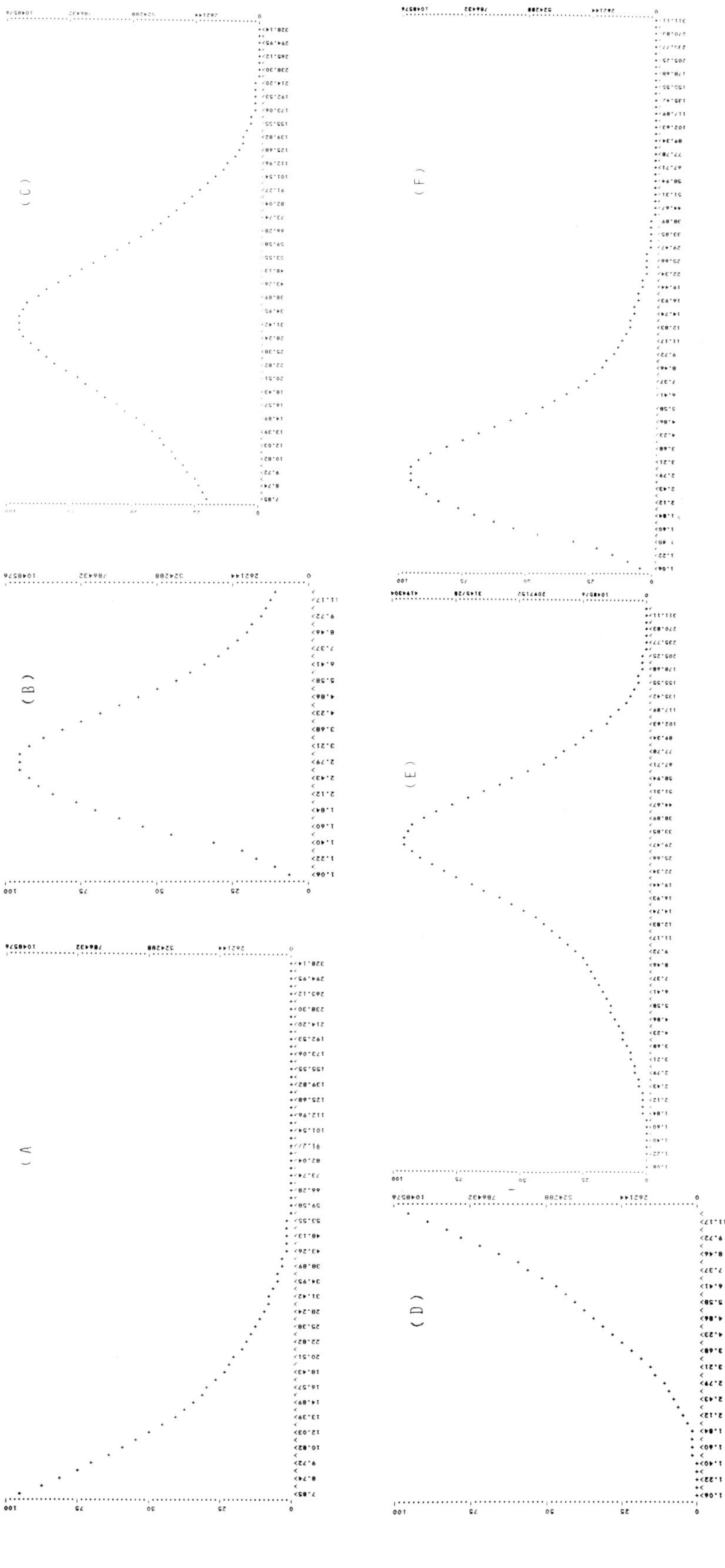

FIGURE 6. Multiorifice data acquisition and range blending. System is a fly ash spanning over 300:1 in diameter, requiring 480-, 120-, and 30-$\mu$m orifices. Graphs show the second stage of blending, i.e., data from the 30-$\mu$m orifice (B) are added to the already formed blend (A and C) of data from the 480- and 120-$\mu$m orifices. C and D are the volume conversions of frequency histogram E. E is then reverse transformed (volume to frequency) to obtain F, which serves as a check on validity of blending (absence of curve discontinuities).

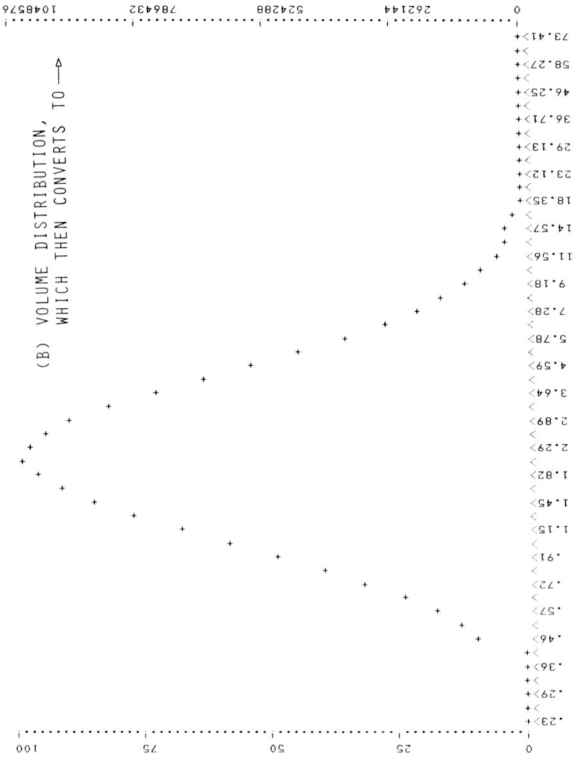
(B) VOLUME DISTRIBUTION,
WHICH THEN CONVERTS TO
0
262144
524288
786432
1048576
0
25
50
75
100
.23
.29
.36
.46
.57
.72
.91
1.15
1.45
1.82
2.29
2.89
3.64
4.59
5.78
7.28
9.18
11.56
14.57
18.35
23.12
29.13
36.71
46.25
58.27
73.41

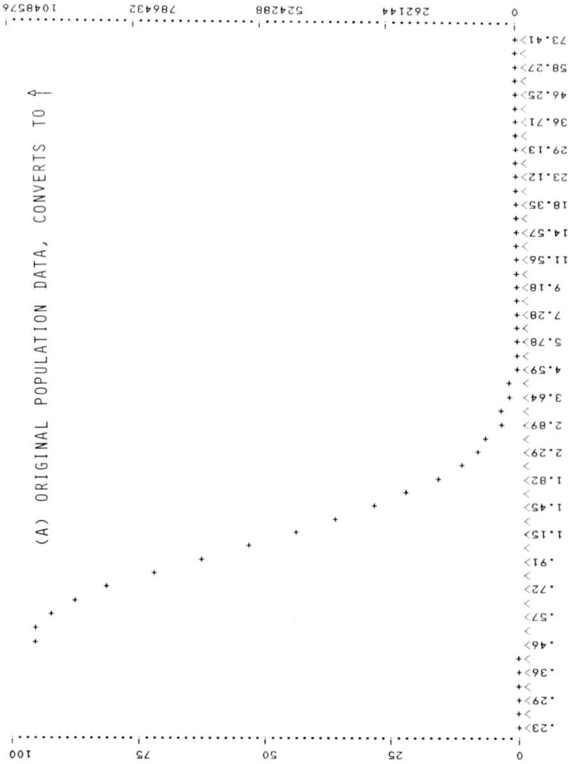
(A) ORIGINAL POPULATION DATA, CONVERTS TO
0
262144
524288
786432
1048576
0
25
50
75
100
.23
.29
.36
.46
.57
.72
.91
1.15
1.45
1.82
2.29
2.89
3.64
4.59
5.78
7.28
9.18
11.56
14.57
18.35
23.12
29.13
36.71
46.25
58.27
73.41

FIGURE 7 A—B.

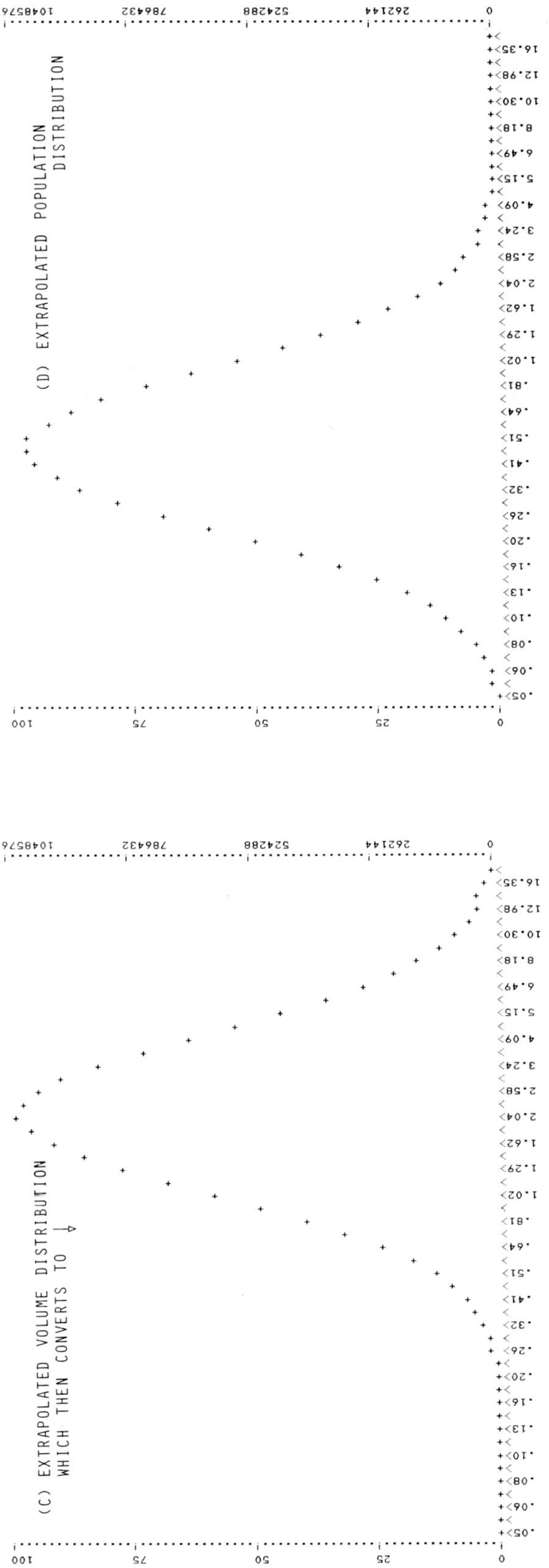


FIGURE 7 C—D.

FIGURE 7. Data conversions and extrapolations: A → B → C → D.

of a well-fitted overlap of actual data and a synthesized Gaussian (log-normal) curve. Accordingly, frequency and/or mass distributions are often made complete via Gaussian extrapolation of either or both ends. Trials have shown negligible differences between data completion by well-considered extrapolation vs. completion by further actual data acquisition.

Often, Gaussian fitting is most feasible on volume distribution data and this is then converted back into an extrapolated frequency distribution. Figure 7 illustrates such a case. Expert judgment is exercised and notations are inserted in the analysis report when such are deemed advisable.

## IX. ANALYSIS REPORTING

Rapid preparation of the analytical report is also computerized. The ELZONE® programming allows report formats to be set up as desired, so it is easy to produce any degree of automatic printout. Because the needs of Particle Data Laboratories' clients vary unpredictably, a comprehensive, "standard" report format has been adopted for general use. Of course, other report formats can be programmed as the need arises.

The elements of PDL's analytical report format are as follows:

1. Source (PDL), Date and ID Number of Report; Operator ID
2. Source of Sample and its ORDER and ID Numbers
3. Orifice, Electrolyte, ELZONE® system settings
4. Modes of Sample Preparation, Dispersion and Data Acquisition
5. For Frequency and Mass Data: Mode, Mean, Median, S.D., Coefficient of Variation and Skewness, on both Logarithmic and Arithmetic bases
6. For Frequency and Mass Data: Listing of quantities at nine convenient percentiles having uniform intervals of 0.77 $\sigma$: 0.1, 1, 6, 22, 50, 78, 94, 99, 99.9, for a useful overview of the distribution that is readily plotted on log probability paper
7. For Frequency and Mass Data: Plotting of distribution in both histogram and cumulative forms, including printing of diameters and readings along the logarithmic size scale
8. For Frequency and Mass Data: Tabulating of each channel size and content and the total of all channels

The above frequency and mass information satisfy most requirements, as they parallel the data obtained from the long-established procedures of microscopy and sieving, respectively. However, the program allows ready inclusion of additional or alternate printouts, as needed. For example, the size scale is sometimes wanted as volume (cubic microns) and the quantity scale sometimes as relative surface area.

Special data transforms and calculations are also available. For the computation of performance parameters for filters and for continuous crystallizers, it is first necessary to convert the quantity axis to be logarithmic on a frequency basis and the size axis to be linear on a diameter basis. Multimodal distributions may require separate quantification of the modes and the comparative figures between modes. Some systems require "before and after" comparisons such as ratios of differences, e.g., a liquid before and after filtering, or a coke before and after oxidation.

The variety of answers requested of the fine particle measurement laboratory is virtually unlimited. The electrozone method provides the majority of them.

## ACKNOWLEDGMENT

This chapter is reprinted by permission of Particle Data Laboratories, Ltd., Elmhurst, Ill.

## REFERENCES


1. **Berg, R. H.,** *Electronic Size Analysis of Subsieve Particles by Flowing Through a Small Liquid Resistor,* Spec. Tech. Publ. No. 234, American Society for Testing and Materials, Philadelphia, 1958.
2. **Kubitschek, H. E.,** Electronic measurement of particle size, *Research Applied in Industry,* Vol. 13, Butterworths, London, 1960.
3. **Spielman, L. and Goren, S. L.,** Improving resolution in Coulter counting by hydrodynamic focusing, *J. Colloid Interface Sci.,* 26, 175, 1968.
4. **Thom, R., Hampe, A., and Sauerbrey, G.,** Die elektronische Volumenbestimmung von Blutkorperchen und ihre Fehlerquellen, *Z. Gesamte Exp. Med.,* 151, 331, 1969.
5. **Adams, R. B. and Gregg, E. C.,** Pulse shapes from particles traversing Coulter orifice fields, *Phys. Med. Biol.,* 17(6), 830, 1972.
6. **Ellison, L. E. and Berg, R. H.,** Particle Pulse Analyzing Apparatus Employing Linear Amplification and Logarithmic Conversion, U.S. Patent 3,626,166, 1971.
7. **Karuhn, R. F. and Berg, R. H.,** Particle shape analysis via electrolytic sensing zone, Proc. Tech. Program, Int. Powder Bulk Solids Handling Processing Conf., Chicago, May, 1978.
8. **Berg, R. H. and Karuhn, R. F.,** Particle Length and Volume Comeasurement with Controlled Orientation, U.S. Patent 4,290,011, 1981.
9. B.S.I., Electrical sensing zone method, *Methods for the Determination of Particle Size of Powders,* BS-3406, Part 5, British Standards Institute, London, 1982.
10. **Karuhn, R. F. and Maret, A. R.,** Size characterization of liquid toners, TAPPI Print. Reprogr. Test. Conf. Proc., Rochester, N.Y., November, 1979, 169.
11. **Berg, R. H.,** Sensing zone methods in fine particle size analysis, *Materials Res. Stand.,* 5(3), 119, 1965.
12. **van der Plaats, G., Herps, H., and Willeams, L.,** Size determination of conductive particles with a Coulter counter, Proc. Particle Size Analysis Conf., Loughborough Tech. Univ., Loughborough, England, September, 1981.
13. **Haigh, G. T.,** Current Normalizer for Particle Size Analysis Apparatus, U.S. Patent 3,745,455, 1973.

Chapter 9

# PARTICLE SIZE INSTRUMENTATION — COULTER® COUNTER

**Shepard Kinsman**

## TABLE OF CONTENTS

## I. INTRODUCTION

Most industries today have a need for quality instruments for measuring concentrations and particle size distributions. The COULTER® Counter is such an instrument and can examine particles one by one at rates up to 3000/sec. The information thus obtained is read or printed out as concentration and size distribution information. Particles down to 0.5 μm and below, and up to 1200 μm and above, can be examined.

## II. PARTICLE DETECTION PRINCIPLE

Sample particles are suspended in a conductive liquid such as saline. This suspension is rather dilute and is confined in a round-bottom beaker. A small aperture with a vacuum on one side is submerged in the suspension and the suspension is sucked through. A platinum electrode on each side of the aperture causes an electric current to pass through the suspending liquid in the aperture. Each time a particle traverses the aperture an electronic pulse is generated, the pulse amplitude being a measure of the particle volumetric size. All the pulses are classified electronically for their amplitude into 16 size interval channels.

## III. INSTRUMENT CALIBRATION

Near-monosize reference microspheres of known diameter are used for calibration of the adjustable size threshold dial. Any given calibration procedure takes about 5 min. Calibration at more than one size level is easily done, but the high degree of linearity makes this unnecessary. The operator also can use the microspheres to monitor the instrument stability once a week.

## IV. DATA PRESENTATION

It is common to present data in both tabular and graphic form. The size parameter is in particle diameter (equivalent spherical diameter) or a volumetric value (cubic micrometers). The quantity parameter can be either relative or quantitative and can be expressed as volume or weight percent or count percent. Concentration is frequently shown as the number of particles per milliliter of sample suspension. The 16 size channels are in geometric intervals of double the volume size, or 1.26 times the equivalent spherical diameter size.

The chosen size parameter intervals can be manipulated for shape factor correction if the particle system being examined has some degree of shape uniformity throughout.

## V. MEASURING DYNAMIC PARTICLE SYSTEMS

Many users have found this unique device very useful for graphing particle systems undergoing changes. It is a simple matter to study dissolving rates, swelling rates, settling rates, floccing and agglomerating rates, precipitation and crystallizing rates, and the stability of emulsions. When the particle system to be studied is not compatible with water, non-aqueous electrolyte systems can be used.

## VI. SAMPLING AND DISPERSING PARTICLE SYSTEMS

The quality of any particle size analysis result is greatly influenced by the accuracy and reproducibility of the sampling technique and by the accuracy and reproducibility of the sample dispersing technique. Because of its simplicity and speed, this device has proven to be a great asset when studies of sampling and dispersing are undertaken. These must be done carefully or the whole effort becomes inaccurate and variable. For instance, our laboratory has done comparative studies of sampling techniques.[1] The speed of the instrument allowed the generation of statistically valid data with only a small amount of lab time required. Various dispersing techniques can quickly be compared in a similar manner.

## VII. INTERFACED CALCULATORS AND COMPUTERS

The model TA II COULTER® Counter is interfaceable to calculators, microprocessors, and computers.[2] These devices are capable of (1) collecting and storing data, (2) computing, converting, and manipulating data, and (3) presenting data in tabular or graphic forms. Listed below are a few of the useful data variations obtainable from these devices, with no extra laboratory time:

Quality control — go or no go — decisions
Curve fitting for extension or extrapolation of data
Interpolation of data points
Combining data from two apertures for wide size ranges
Averaging data
Computing surface area
Computing mean, mode, quartile, and S. D.
Computing ppm
Combining of data from different methods such as COULTER® Counter and sieves
Adjustments for shape of particles

## VIII. OPERATIONAL VERSATILITY

Even without interfaced computers, the Model TA II has several useful capabilities. The quantitative data can be the particle count or the totalized particle volume in each channel. A selector switch determines whether data will be cumulative or differential, or both. Operator-selected curve editing for background or extrapolation can be stored. For many particle systems, extrapolation into submicron sizes can be validated by a mass balance procedure between the sample presented to the instrument and the sample measured by the instrument. In this case particle density must be known.

## IX. SPECIAL TECHNIQUES

Some particle systems require special techniques. Particles as small as 0.35 μm may be examined by using small, 20- or 30-μm diameter apertures. Cleanup of everything contacting the sample is required. Suspending and flushing liquids should be carefully filtered through 0.22–μm membranes and all glassware should be amply flushed. With a little practice, high degrees of cleanliness can be maintained.

A modified electrolyte system may be used to keep large dense particles properly suspended. Up to 30% by volume of glycerin may be added to the electrolyte. This retards the gravity effect on the particles. The stirring blades must be very near the bottom of the beaker and a round-bottom beaker must be used.

Some organic cells, when suspended in an electrolyte, apparently absorb the electrolyte and become partially conductive. Such materials can be examined with the COULTER® Counter but the particle size data may be small, such as equal to the nucleus. In some cases a factor can be applied to the size read by the instrument to give a final size more nearly equal to the cell outside diameter.

Most metal conducting particles tend to resist conducting current during analysis, probably due to an interface layer such as the Helmholtz double layer or an oxide film. Exceptions to this have been found. By increasing the electrolyte conductivity, or by reducing the voltage applied, interface layer breakdown can be avoided.

Most water-soluble particles can be measured by changing to an organic electrolyte or by overnight saturation of the electrolyte followed by filtering.

Airborne particles can be captured on membranes and released into an electrolyte for analysis on a COULTER® Counter, according to Rosinski and Nagamoto.[3] Contamination particles in hydrocarbons also can be captured on membranes for COULTER® analysis.[4] In this case, residual oil is removed by flushing the membrane with petroleum ether before backwashing into the electrolyte. In both of these cases, the releasing of particles from the membranes is aided by ultrasonics.

Narrow-size particle systems which may not be well defined by the standard 16-channel geometric progression interval system can be recorded on a multichannel C-1000 device. This has narrow channels with linear interval progression.

Wide-size particle systems, in some cases, require two differently sized apertures to cover the full-size range. It is easy for the operator to analyze the sample on one aperture and then change to the other to complete the analysis. However, when analysis time is important, two sample stands may be connected into the COULTER® Counter through an accessory switchbox so that the two required apertures are permanently in place, this avoiding aperture changing.

## X. MEASURING PARTICLE SYSTEMS DOWN TO 0.01 MICRONS

A complimentary instrument is available for examining particle systems down to 0.01 μm. This device reads out the average particle size and a polydispersity factor. It does not give a complete size distribution. A suspension of particles is placed in a laser beam and a photodetector reads a 90° scatter signal. This signal is dependent upon the Brownian motion speed of the various particles colliding and moving. The averages of these speeds is a good indication of the particle size involved.

## REFERENCES


1. **Lines, R. W.,** Some observations on sampling for particle size analysis with the Coulter Counter, *Powder Technol.*, 7, 129, 1973.
2. **Kinsman, S., Hoff, E., and Wood, M.,** A new data reduction unit for the Coulter Counter, Proc. Tech. Program Powder Bulk Solids Handling Process., Rosemont, Ill., May 16 to 18, 1978.
3. **Rosinski, J. and Nagamoto, C. T.,** Water-affected fraction of natural 1.5-9 μ diameter aerosol particles, *J. Colloid Interface Sci.*, 40(1), 116, 1972.
4. **Schmitz, J. E.,** Particle size analysis of oil samples with a Coulter Counter, *Filtration Sep.*, July/August, 426, 1973.

Chapter 10

# FINE PARTICLE CHARACTERIZATION METHODS IN LIQUID SUSPENSIONS

**Alvin Lieberman**

## TABLE OF CONTENTS

# I. ABSTRACT

Fine particles are characterized in terms of concentration, particle size, and size distribution or composition for many reasons. Particle size information can be obtained for particulate materials either in a dry state or when suspended in liquid. Concentration data are of concern when particles are present as an interference or contaminant in liquid, or have been collected in a liquid suspension from an aerosol or powder stream. Particle composition is of concern where identification of a contaminant source, or in situations where size-related particle composition information is important. For composition analysis, the particulate material is usually removed from liquid suspension. To obtain the greatest detail on particle size distribution the particulate material is frequently examined after thorough dispersion into a suitable liquid.

The following discussion covers some of the reasons why fine particles are characterized in terms of size, composition, or concentration. Problems in handling particles before and during the characterization processes are discussed, including some of the procedures in common use for sampling and the physical phenomena which affect materials in the sample. The methods and instruments for characterizing fine particles are described in terms of their operating principles and procedures, particularly for commercially available devices. Both the capabilities and limitations of these instruments are mentioned. Some descriptions of necessary ancillary elements for practical instrument operation are given, ranging from sample feeding devices to computer interface devices. Some typical applications are described along with problems and their solutions as have been encountered in the particle size measurement area.

# II. INTRODUCTION

## A. Objectives

Particles must be characterized for a variety of reasons. Some of the reasons for characterization are primarily associated with manufacturing problems based upon product quality, product manufacturing convenience and cost, and for product handling convenience. For example, pharmaceutical products are often produced in tablet form. The tablets are prepared by agglomerating powders with a suitable binder. The tablet strength and tablet formation requirements vary markedly as particle size and particle size distribution changes. In addition, pharmaceutical product activity, once the tablet has been ingested, is a function of active-agent surface area related to particle size.

Mineral manufacturing processes often involve comminution. The energy required for grinding ores or other mineral products is related to the desired particle size as well as the size of the material before grinding. Transport and chemical reactivity problems are also functions of particle size. Food product quality, as for spray-dried materials or ground food products such as sugars, flour, and spices will vary markedly with size of the material.

The importance of particle size in the several examples above can be understood better when one considers that the color, gloss, and appearance of many materials depend upon the basic particle size; light scattering, absorption, and reflection effects can be definitely associated with size of particles making up the surface of the item being observed. In a similar manner, surface area controlled reactivity will increase markedly as particle size decreases, since surface area varies inversely with the square of the particle diameter. Transport of products, combustion rates, and reaction rates depend upon surface area to a great extent. Thus, if manufacturers can produce powders and fine particle product precursors in the desired size range at minimum cost, a savings in manufacturing cost can be obtained.

Where the liquid substrate in which particles may be suspended is the product, then the particles may be considered as contaminants. In this situation, particle content must be limited insofar as is economically and practically feasible. Particle size and concentrations are of concern for the liquids while particle composition is a problem area for specific materials. Particle composition is of concern where chemical reactions occur, or pharmaceutical products or edible liquids are being manufactured or processed.

For many liquid products a maximum particle concentration is specified. The maximum concentration may be defined in terms of a maximum allowable weight or volume fraction in the liquid, or in terms of a maximum number of particles present in the liquid. In the latter case, particle size distribution is also specified. For most liquid products the appearance of the liquid is a matter of concern. If a suspension of small particles is present in excessive concentration then the liquid which should appear clear and transparent will be opalescent or opaque, thus detracting from the attractiveness of a product. This is an area of particular concern for beverages and other edible liquids.

In many mechanical systems liquids are used for cooling, lubricating, cleaning, or power transmission. Most of these mechanical devices contain pumps, valves, hydraulic actuators, or other moving components with small clearances. If particulate material is present, either in excessive quantity or with excessive particle size, then interferences with mechanical operation can occur and mechanical failures can result.

Many pharmaceutical products must be prepared in such a way so that even nonviable inert particles are not present in quantities greater than allowed by standard requirements. This is of particular concern with parenteral and injectable liquids. Even so, one may note that the U.S. specifications for allowable particle content in parenteral liquids permit particle concentrations appreciably higher than those allowed in many hydraulic control units.

Electronic semiconductor manufacture requires reaction with corrosive chemicals for photolithographic purposes, along with flushing and cleaning of semiconductor component surfaces during and following fabrication. The reactant liquids and the flushing and cleaning fluids must be scrupulously clean. In many cases, feature-to-feature spacing on microelectronic devices is less than 1 μm. If conductive particulate contaminants are present in the reactant or flushing liquids, then failure of a specific component at the point of the particle bridge can occur. For this reason, electronic fluids must be cleaned scrupulously and should be characterized to ensure that particulate contaminants have been eliminated from the liquid. In several mechanical systems such as valves, pumps, hydraulic components, mixers, crystallizers, etc. which operate with liquids, particle content generation and particle type in the liquid can be used to characterize the operation of the mechanical devices. Indication of the particle size distribution and concentration in hydraulic oil from a high-pressure pump will serve to indicate the wear rates and efficiency factors for the pump; valve wear and hydraulic component wear can be indicated by particle generation rates in fluid over a period of time. Particle concentration vs. location data can be used to define mixer efficiency.

Liquid-borne particle control devices are typically particle-size sensitive, that is, the control efficiency is a function of particle size. Larger particles are removed with greater efficiency than smaller particles. For this reason, performance of these devices is defined by challenging a device with either polydisperse or monodisperse suspensions of particles and recording removal efficiency for each of the particle sizes in the challenge material. For monodisperse particle challenge, it is necessary only to determine particle concentration by count, volume, or mass in the challenge and effluent suspensions. For polydisperse challenge materials, it is necessary to determine particle size distribution as well as concentration in the challenge and effluent suspensions.

Another reason for characterizing particulate material is to fulfill requirements of in-house and imposed specifications and standards. Definitions of materials often contain requirements for maximum or minimum concentration of particles in various fluids or may contain requirements for particle size and/or particle size distribution. These requirements may be for suspensions of material or for the particulate material itself. Specifications may exist in the form of in-house requirements, where a manufacturer has become aware that product quality depends strongly on the nature of the particulate material either making up or contaminating his product. The in-house specifications may be based upon experience from a particular organization or may be derived from perusal and consideration of standards produced by voluntary standards groups such as American Society for Testing and Materials (ASTM), Society of Automotive Engineers (SAE), or other foreign groups engaged in standards preparation. At the same time, many organizations must follow government agency requirements such as the *Good Manufacturing Practices for Large-Volume Parenterals* promulgated by the U.S. Food and Drug Administration in which particle contents are limited by the *U.S. Pharmacopeia* for large-volume parenteral fluids. Similarly, many military specifications for lubricating or hydraulic oils or fuels indicate concentration and particle size distribution specifications.

### B. Measurement Protocol

The first step in characterizing fine particles in liquids is to specify the particular parameter of size, concentration, or composition which is of concern for the specific measurement. Next, one should attempt to indicate the overall range of variation that can be expected or is within the range that will be measured. Once these two broad levels are defined, then it is possible to select optimum devices and the optimum protocols from the wide range of techniques that are available. Next, one should select the measurement method that will provide maximum information on a particular parameter being measured. The method may produce data on mass, number, or volume concentration of particles; it may produce data on the size of individual particles using any of the applicable particle size parameters, and it may produce data on particle composition or specific chemical element content. Next, considering the procedures that will be carried out during the measurement process, one then determines the method of particle collection or observation that will produce a sample that can be most easily handled by the measurement method. As one obvious example of the basis for such choice, consider the requirement for examination by the scanning electron microscope for particle morphology vs. the requirements for a sample to define mass loading or turbidity. In the first situation, one would deposit particles on an electron microscope screen from liquid, and for the second, one would deposit particles on the surface of a preweighed substrate or filter medium. In the third case, one may not require any collection techniques but would simply measure visibility through the fluid.

Following measurement of the particle parameters of concern, the best advantage should be taken of the data obtained. In many cases a simple mathematical description of particle size distribution can be used; in some cases mass concentrations are required, in other cases fractional compositions for various materials must be known. At the same time, it is necessary to keep in mind those situations where measurements are carried out to ensure that the maximum level of a particular component is never exceeded. In this case, it can be adequate to record no data until the level is exceeded. Data can be handled by direct observation, hard copy preparation, or computer memory storage.

## III. PARTICLE SYSTEM DEFINITIONS

In order to aid in defining selection of an optimum measurement system, it is helpful to consider some of the particle systems that must be observed.

### A. Particle Products, Sources, and Particle Types

One large class of particulate materials that is characterized in liquid suspension includes product materials that are formed by chemical or physical reactions. A second class of product particle includes those which are formed by comminution or size reduction. The first class of reaction product materials includes such common items as polymer latices, inorganic salts, food products, sugars, crystals, flavor compounds, refined metal powders, encapsulated medicinal or cosmetic products, etc. For the most part this type of material will have a blocky structure approaching a cubical or spherical format. However, some reaction product materials can have a rather high aspect ratio, depending upon crystal structure of the original material. Particles formed as a result of vapor phase reaction can often have extremely convoluted structures as individual substructures coalesce to form chains of particles. It is necessary to characterize such reaction particle materials in terms of size and composition so that manufacturing efficiency can be optimized, and application and utilization of the particles can be carried out in the best possible manner.

The second class of product particles includes the particulate material formed by comminution or subdivision. Raw materials, frequently encountered in massive structure, must be subdivided so as to permit ease of handling, processing, or subsequent reaction. This type of product includes ores that must be milled to provide maximum surface area availability for chemical reaction and/or heat transfer. It includes cement which is milled to produce a powder with optimum handling and reaction characteristics. It includes grains for food product materials that are ground to size for optimum reaction and sensory characteristics. It includes solid fuels designed for combustion in special burners whose efficiency is a function of fuel particle size and size distribution. For the most part comminuted particles are characterized in terms of particle size and size distribution along with particle concentration. Details of the comminution process and of typical size distribution produced by comminution will not be discussed here. They can be found elsewhere in detail.[1-4]

In addition to characterizing particle product material, quite frequently it is necessary to define contaminant particles suspended in liquids or that have been collected from other environments into a liquid carrier for characterization. Contaminant material may be collected from assembly areas, where products are being manufactured or devices being operated, that should be free of contaminants. The contaminant particles may be collected from mechanical, electrical, or optical devices that should be clean or from liquids that should be free of particles. The particulate contaminants may result from carryover of manufacturing debris such as machining fragments, fragments of plastic from a molding operation, paint chips, wear particles from moving parts, etc. These particulate materials may be in almost any physical form or configuration depending upon their source. They can be composed of metal, metal oxide, and inorganic or organic compounds including hydrated salts, polymer, plastic fragments, etc. The contaminant particles may consist of chemical reaction products produced by oxidation of lubricants or hydraulic fluids and their additives. The contaminants may consist of ingested debris drawn into or collected by devices during normal operation. The devices can include moving mechanical parts or liquid system reservoirs that can breathe during temperature changes. The ingested debris can include dust, fibers, and soil particles reentrained from the ground or floors of assembly and operating environments. The ingested debris can include airborne dust and animal and vegetable particles that are floating in the atmosphere or diverted into systems by manmade movement and activities. The ingested debris can include bacterial material growing within lubricant, hydraulic, or cooling liquids. Shape, structure, and particle size distribution of contaminant particles can take any form imaginable since the sources, introduction processes, and physical phenomena occurring after introduction are so varied. It is, therefore, of concern to determine particle size and particle size distribution, concentration, and/or composition of the contaminant particles present in and on mechanical devices or product fluids.

### B. Particle Descriptions and Definitions

In some of the previous comments, bases for selecting parameters for characterization have been referred to in a very broad sense. It is now time to define some of the parameters that are of concern in particle characterization and indicate measurements and ranges that are in common use.

The first parameter to be discussed is the particle shape factor. Shape is a fundamental characteristic of fine particles and powders. One may characterize individual particle images, or the particles themselves, by assuming a fairly simple particle structure such as a sphere, ellipsoid, cube, or parallelopiped. In this situation an aspect ratio, a ratio of particle perimeter or circumscribing circle perimeter to particle area, a ratio of a maximum to a minimum dimension, etc., can be used as a shape factor for the particles, considering a two-dimensional image.

In recent years, many studies have been carried out on shape factor description systems which can be computer stored and processed. For example, vector magnitudes from the profile centroid to its perimeter as a function of angle can be used to generate a wave-form which can be subjected to Fourier analysis or other computations.[5]

If one follows the earlier work of Heywood,[6] then shape factors can be related to ratios between selected orthogonal parallel planes tangential to opposite sides of the particle. Similarly, some simple comparisons to ideal solids such as sphere, cube, or ellipsoid can be accepted for shape factor definition.

Shape factor is of importance in defining packing fraction, bulk density, or void fraction of a mass of particles. Pressure drop for fluid flow through a packed bed, heat transfer, and particle transport phenomena are also affected by particle shape factors. In a similar manner, surface structure is of importance in defining particle-fluid systems, particularly if one is concerned with suspension rheology, particle reactivity, or other surface-related functions. Practically speaking, particulate material is never homogeneous. There is always a surface layer differing from the interior of the particles. The surface layer can contain adsorbed gases, particularly water, and may consist of an oxide coating on a base metal or a hydrated layer on soluble salt particles. In the case of manmade artifacts, the surface can be similar in nature to the gelatin capsule in which many liquid medicines are provided. For the most part surface structure is called out as "rough, smooth, having reentry characteristics, convoluted", etc. Perhaps the closest approach to a quantitative definition of surface structure is the fractal dimension definition for characterizing particle boundaries.[7] The fractal dimension is a measure of the ability of a curve to fill space so that an extremely convoluted line that fills a two-dimensional space completely has a dimension of two, whereas a straight line has a dimension of one. Thus the more convoluted a particle surface may be, the larger its fractal dimension will become. As with the shape factor, surface structure affects rheology and friction, thus relating to particle and particle suspension transport along with fluid flow, pressure drop, and heat transfer and reactivity as affected by the nature of the particles.

Particle size can be defined in terms of a linear, area, or volume dimension for a particular material. The dimension is usually stated in terms of the value for an equivalent sphere. Equivalency can be based upon a physical function if such is being measured. Thus, diameters which represent length, surface area, volume, visibility, or settling rate of an equivalent sphere can be defined for individual particles. The choice for a specific diameter selection frequently depends upon the process variable of concern. For example, a diameter based upon equivalent area would be of more use in describing paint or pigment particles. Deposition in the respiratory tract or in curved ducts is related to the mass and aerodynamic diameter of a particle; reaction rates, droplet growth, heat transfer, etc., are related to particle surface area. Many size definitions are based upon a simple geometric factor such as the length of a chord traversing the particle in a specific direction, the diameter of a sphere of equivalent projected area, or other such geometric factors. At the same time, since physical properties are of great importance, many diameter specifications are based upon just those physical properties. For example, an equivalent optical diameter is defined as the diameter of a sphere with equivalent reflectivity, scattering, or absorption characteristics; equivalent aerodynamic diameter is defined as a diameter of a sphere of unit density having the same terminal sedimentation velocity as the particle of concern (note that a Stokes diameter is the

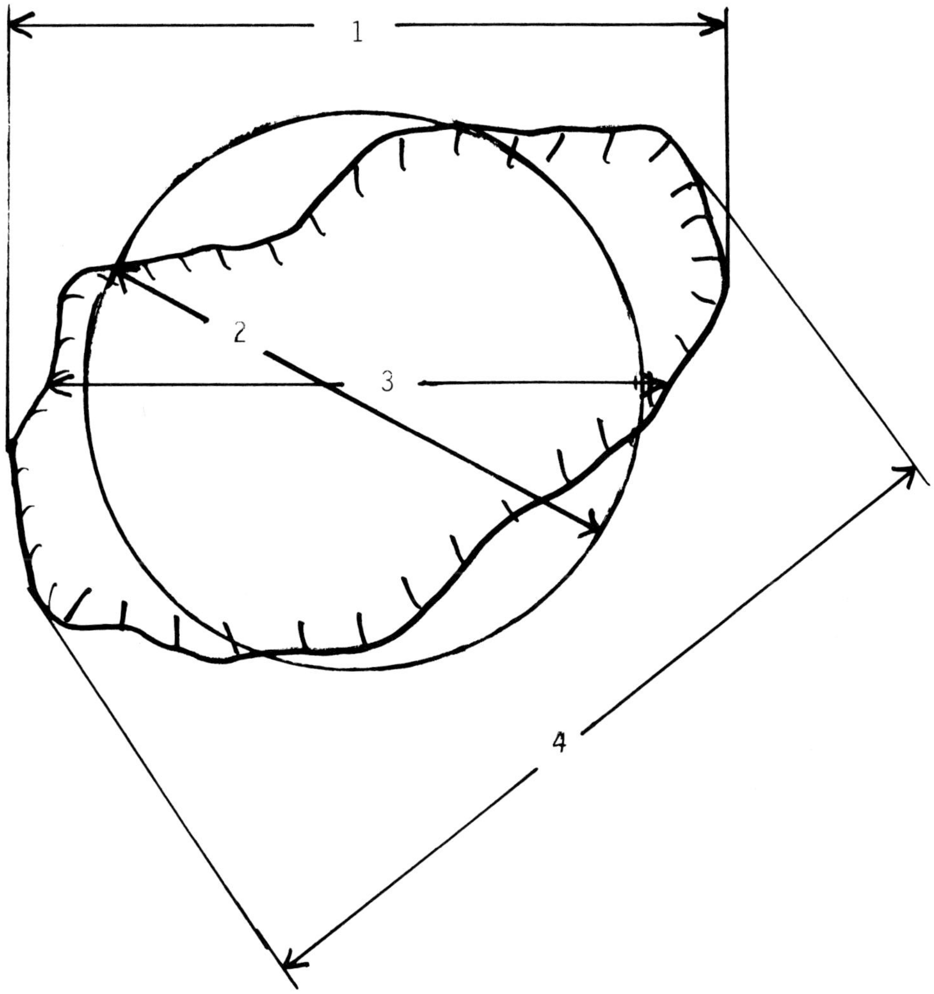


FIGURE 1. Some geometric diameter definitions: (1) Feret's diameter; (2) equivalent area diameter; (3) Martin's diameter; (4) longest dimension.

diameter of a sphere having the *same* density and sedimentation velocity as the particle of concern). Some typical geometric diameter definitions are shown in Figure 1.

In the "real world", individual fine particles are seldom if ever encountered. In the mass of particles that are seen, the size of each single particle will differ to some extent. Therefore, it is necessary to describe a particle size distribution. In many cases the distribution is monomodal and can be well described by a logarithmic probability function, particularly if the dimension of concern is an arithmetic mean rather than a surface or volume mean. The arithmetic mean is defined as the sum of all particle diameters divided by the total number of particles measured. The surface mean is the diameter of a hypothetical particle with the same surface area as the average surface area of all particles. The volume mean is the diameter of a hypothetical particle having the same volume as the average volume for all particles. Since the arithmetic or geometric mean diameter has no real physical significance, it is common to define terms relating to definite physical properties such as surface, volume, or mass. If surface is of concern then more consideration should be given to the smaller particles, while if volume or mass is of concern then the larger particles will have more

**Table 1**
**STATISTICAL AVERAGE DIAMETERS**

| Diameter | Definition | Description |
|---|---|---|
| Arithmetic | $\Sigma\,(nd)/\Sigma n$ | Average sphere diameter |
| Length | $\Sigma\,(nd^2)/\Sigma\,(nd)$ | Diameter of sphere with average projected surface area |
| Surface | $\sqrt{\Sigma\,(nd^2)/\Sigma n}$ | Diameter of sphere with average total surface area |
| Volume | $\sqrt[3]{\Sigma\,(nd^3)/\Sigma n}$ | Diameter of sphere of average volume |
| Volume-surface | $\Sigma\,(nd^3)/\Sigma\,(nd^2)$ | Diameter of sphere of average volume per unit surface |
| Weight | $\Sigma(nd^4)/\Sigma\,(nd^3)$ | Diameter of sphere of average weight |

effect on the volume or mass of the total particle population. Table 1 shows some average diameters and defines the size parameter of concern.

Probability distributions, both linear and logarithmic, have been found to be very helpful in defining particle size distributions. It is possible to describe a particle size distribution using only a size and standard deviation (S.D.). This technique is extremely useful for defining materials; however, in "real life", data deviate from the calculated values at the two tails of the distribution. Even though the error is small, since the number of particles at the tails is also small, some problems exist and other distribution functions have been described to aid in typifying particular situations. For example, the Rosin-Rammler equation,[9] shown as Equation 1, was developed to plot size distribution of ground friable materials:

$$n = 100 \exp\left(-\frac{d}{c}\right)^{b} \tag{1}$$

where n is the population percent of particles smaller than diameter d, and b and c are constants.

The Nukiyama-Tanasawa distribution[10] was developed for defining spray droplet size distributions and is shown as Equation 2:

$$d = \frac{k_1}{v}\left(\frac{\sigma}{\rho}\right)^{1/2} + \frac{k_2\,\mu^{0.45}}{(\sigma\rho)^{1/2}}\left(k_3\,\frac{L}{G}\right)^{1.5} \tag{2}$$

where $\sigma$ is liquid surface tension, v is the relative velocity of gas and liquid, $\rho$ is liquid density, $\mu$ is the liquid viscosity, L is the liquid volumetric flow rate, and G the gas flow rate.

Roller's equation[11] is a size distribution equation, relating weight percent and size for ground powders. It is shown as Equation 3:

$$w = (a\sqrt{d})\,\exp\,(-b/d) \tag{3}$$

where w is the weight percent of particles smaller than d, and a and b are constants.

Again, the objective is to produce a simple equation that can be used with minimum effort to adequately describe a particular class of material produced by a specific generation mechanism. Many systems are present where particles are generated or grown and removed at the same time. Bacterial suspensions, crystallizer populations, and aged aerosols are a good example of this type. For this reason, the concept of a self-preserving size distribution was developed by Friedlander.[12] The self-preserving size distribution function is one where the concentration varies inversely with a power (usually 3 to 4) of the particle radius.

Further characterization of particulate material may require definition of physical and/or chemical composition. When defining the chemical or physical composition of particles, the procedures often require collection of the particulate material on a suitable substrate in order to concentrate the particles to a level where conventional wet chemistry or absorption or emission spectroscopy can be used to define chemical composition of the particulate material. If physical measurements are to be taken on the particulate material, then the particles should be present in sufficient bulk as to permit the required physical measurements to be carried out. Physical measurements can often be carried out on particles almost individually. Such measurements include optical effect measurements, moving particle velocity determinations, and other procedures wherein individual particles can be observed or measured. For most chemical composition determinations the mass of an individual particle is normally so small as to require concentration to the point where analytical system sensitivity is adequate. Consider that a unit density micrometer diameter particle weighs less than 1 pg.

In addition to the simple definitions of surface roughness, the integrity of the particle structure near the surface is also of importance. Rigidity or deformability of the surface of particulate material can be of great importance in determining packing or transport phenomena. A good example of the importance of this feature is the capability of blood cells to deform so as to pass through capillaries much smaller than the normal dimension of an at-rest blood cell.

### C. Particle Concentration Definitions

Particle concentration measurement is primarily of concern in defining the content of materials in liquid suspensions. The number or mass of particles in a given quantity of liquid is of importance in characterizing the cleanliness or acceptability of the liquid. Similarly, if a mixture of materials is present within a dry powder then the concentration of one fraction within the total is also of concern. In this respect, one may define fractions as composition controlled or size controlled. In the latter case, then, the particle size distribution function can be used to define the concentrations of any size fraction within the total particle sample. The concentration can be stated as a cumulative or differential fraction. The cumulative fraction can be either the fraction equal to or greater than or the fraction equal to or less than any stated size. Figure 2 shows the size distribution data for a near-monodisperse material plotted differentially on the basis of particle population and particular volume. Note that the median diameter is somewhat larger in the latter case; note further how the population data differ from the volume data, particularly at the extremes of particle size ranges. The sample illustrated was carefully prepared by wet sieving so that essentially all particles should be larger than 6 μm in diameter and smaller than 12 μm in diameter. Indeed, the volume population shows more than 90% of the particles larger than 6 μm and 95% smaller than 12 μm. The population data, however, show nearly 35% of the particles smaller than 6 μm and more than 99% of the particles by number smaller than 12 μm. This shows the importance of matching the analytical procedure to the preparation process for particulate material.

In determining concentration within the substrate liquid sample, then the measurement procedure is selected on the basis of the desired definition, e.g., population or volume/mass. If population concentration in terms of number of particles of specific size per unit volume of liquid is concerned, then one determines the number concentration in a suitable liquid sample. The method of determining the number concentration may involve observation of the particles within the liquid by optical or electrical resistance techniques. Alternately, one may remove the particles from a known quantity of liquid to a suitable examination substrate and determine the number of particles that were present in a unit volume of liquid. In most cases, particles are removed by filtration through a suitable medium whereupon they can be observed at a convenient time. Comments on the technique and problems in both measurement methods will be provided later in this discussion. In determining the concentration of particles suspended within liquids, the same requirements exist in terms of concentration and collection

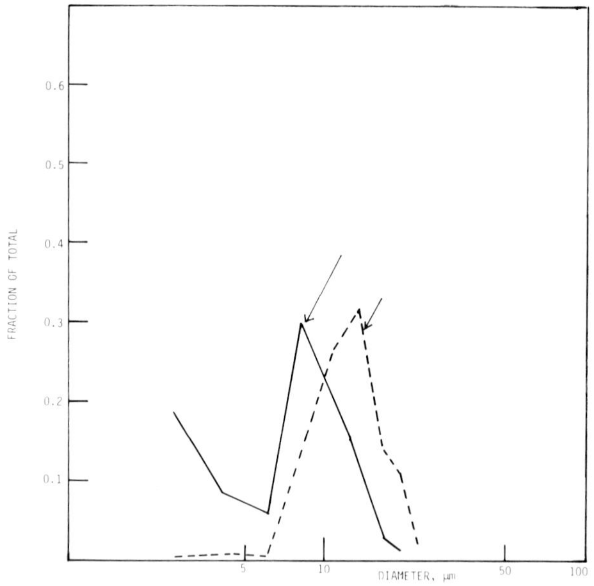


FIGURE 2. Monodisperse particle size distribution.

of an adequate quantity for measurement as were mentioned in the previous section on requirements for determining physical and chemical composition.

## IV. SAMPLE HANDLING

### A. Dry Powder Handling

Practical application of powders in use may require that the powders be received in the application area, stored for various periods of time, and transported to a system, reactor, or container where the powder will be used. For optimum use of materials that may be costly, it is advisable to characterize the materials so as to avoid overuse, underuse, or incorrect application of the particulate material. Sample acquisition for characterization is best done with convenient size samples that may range from microgram to kilogram levels. Because of the quasistable state of any powder system, it is necessary to insure that samples obtained are truly representative of the powder batch or powder system. Nonrepresentative samples can be the result of aggregation during storage or fine percolation during transport, as well as the more frequently blamed inadequate procedures.

For powders flowing in a line or duct, mechanized systems have been developed which intermittently remove a sample from the line to permit accumulation of adequate sample for examination.[13] Such devices can be used for acquisition from screw conveyors, pneumatic conveyors, or fluidized material transport systems. Typically, the intermittent sample collection devices can remove kilogram quantities of material in a relatively short period of time. When a bulk sample has been obtained from the outlet of a transport line sample then further quantity reduction is required before any analysis is carried out. In this way temporal variations in the nature of the material flowing down the line are minimized.

Sampling of particles stored in a container is often carried out by using a sample thief. With this device a hollow rod is inserted into the container to a suitable location and an aperture at the end of the thief is opened for a time long enough to allow particles to flow into the aperture. The aperture is then closed and the thief removed from the container. Use of this type of device is based upon subjective decisions as to the best location for sample removal. For this reason it is wise to procure several samples from locations throughout the interior of the container. When containers are filled, particle size segregation occurs due to percolation of fine particles downward through the spaces between the larger particles. The degree of segregation depends upon the size of the container, the filling rate, and the particle size distribution within the powder before filling.

After the samples have been removed from the line or container, then further sample size reduction is frequently required. If good characterization is to be accomplished, then the analytical sample handling from the bulk of material should be carefully carried out. Several methods are available for such sample size reduction or subdivision. The least effective is probably the most widely used; that is, straightforward scoop sampling from a conical pile of powder. Again, subjective decision is typical for selecting the scoop sample size and the location for removal. A more representatve sample can be obtained by using the cone and quartering technique, wherein a conical pile of powder is physically subdivided into four sections by vertical quartering. The retained quarter can be further subdivided in the same manner until a sample size suitable for analysis has been obtained.

Samples can be further subdivided by using a chute riffler. A stream of powder is dropped onto a sharp edge and the particle sample will be subdivided reasonably well into two fractions; each fraction should be nearly identical. The fractions can be further subdivided until a usable sample size is obtained. A spinning riffler, if properly used, is more convenient and provides more uniform subdivision.[14] In this device, the powder sample is dropped over a period of time into compartments spun slowly in the stream of falling powder. By ensuring that each compartment is exposed to the stream many times reasonable subdivision of the entire sample is obtained. In any case, sampling techniques require that the technician carrying out the work does it with extreme care. The state of the art for design and operation of these mechanical devices is not yet to the point where a skilled operator can be eliminated.

In applying any sample acquisition technique, the first requirement is a decision as to the magnitude of the sample to be obtained for characterizing the sample. The objective is to produce data which can be extended to the population from which the sample was taken. The procedure implies that the characteristics of the sample are, at best, identical with the characteristics of the population from which the sample was obtained or at least can be related somehow to the characteristics of the entire population. In order to specify the relationship between the sample and the base population, it is necessary to apply statistical techniques. For a complete discussion of small particle statistics the reader is referred to the text by Herdan.[15] For the purposes of the present discussion, no derivations will be made or justified. Only conclusions, as stated by Herdan and other authors in the field, will be given.

General procedures for statistical experimental planning involve: (1) stating the nature of the value to be characterized and an acceptable error probability, (2) collecting the data that define the value, and (3) computing a test statistic such as chi square and comparing that statistic to a tabulated critical value. The confidence with which we may accept the value of the stated characteristic depends upon how the computed test statistic compares to the tabulated critical value. Thus, one may note that the number of samples that must be obtained for any lot size is proportional to the intersample variability. In other words, if one wishes to characterize the particle size distribution of a shipment of powder and each sample provides wildly varying data, then the number of samples should be increased greatly over a situation where the samples appear almost identical. For random sampling, the square of the S.D. ascribed to a characteristic of the population can be expected to be equal to or greater than the square of the S.D. between samples divided by the number of samples. For completely random sampling conditions the inequality becomes an equality.[15]

The above illustrates one simple example of statistical experimental design. Where characteristic distributions in a population are Gaussian in nature, the S.D. is equal to the square root of the average value of that characteristic. Thus, in defining the true value for the characteristic, one may determine the average value and an S.D. for that average value; one may further state confidence limits by using tabulation of areas under the normal curve and selecting a specified percentage confidence level. We may then define the number of S.D. to be added to the average value so as to define the fraction of the sample that will have characteristic values that fall within acceptable bounds.

Statistical experimental design must be used wisely before maximum use can be made from such design. It is necessary that the technician ensures that the samples are truly random or representative of the population from which the samples are obtained. At the same time some "common sense" application has to be made to the sampling technique. For example, a 5-$cm^3$ thief sample cannot be used to characterize a 1000-$ft^3$ bin of powder. This is particularly the case if one suspects that segregation or percolation has occurred in a bin. In the same way, a 1- to 10-mℓ/min sample abstracted from the center of a tube carrying supposedly pure flush water at the rate of 100 gal/min will not provide samples whose data can be accepted easily. When protocols are planned for sample number and frequency, one should consider normal and abnormal variations in a production schedule. If production is uniform and if processing is constant, sample intervals can be scheduled easily. If a process is cyclic in nature or perturbations occur frequently then the sample scheduling must take into consideration the cyclic nature of the process.

For optimum characterization, this step should be carried out immediately after samples have been procured. Storage or transport can affect particles. Unfortunately, immediate analysis is frequently inconvenient. Samples may be procured from field installations or in remote plant areas and require transport to a laboratory in another area for analysis. Work load in the analytical laboratory may be such so that samples must be queued for processing. For this reason some care must be given to storage conditions for dry particulate material. Problems in sample storage for analysis can be considered small-scale examples of those with bin storage of process materials. The same problems of percolation and fractionation can occur with sample storage as can occur with bin storage. Differences of scale exist, of course. These problems are primarily those of agglomeration due to forces of gravity, fractionation due to percolation as affected by particle size distribution, and one additional problem. The last problem is that of physical and/or chemical reaction with atmospheric gases and vapors occurring during long storage periods. The problem of percolation and size fractionation can be handled by careful remixing or by ensuring that the entire sample is used for characterization. If agglomeration occurs, particularly at the bottom of the sample container due to the weight of larger particles, or if agglomeration occurs because of inter-particle reactions, then particle redispersion can seldom be carried out effectively. The question of differentiating between deagglomeration and particle fracture cannot be easily answered. If particles are friable or if reaction is possible, it is best that the sample be analyzed as soon as possible or that the sample be stabilized by dispersing in a suitable fluid where each particle will be separated by an inert liquid film. If particles must be stored in a humid environment or one in which reactive gases may be present, then it is necessary that the sample container permit hermetic sealing; under extreme conditions the sample can be covered with an inert gas such as dry nitrogen or carbon dioxide.

Samples as received should be dispersed as soon as convenient. Even when dispersed in liquid, problems can occur. The forces which cause particles to agglomerate when dry can still be effective even after the particles have been dispersed in a liquid. Since particles are usually more dense than the dispersion liquid, eventually they will settle to the bottom of the container during storage. If the concentration is sufficiently high, one particle will deposit on top of another and eventually the intervening liquid film will be removed. The two particles will then achieve intimate contact. Combination of forces due to mechanical linkage

Table 2
DISPERSION COMBINATIONS

| Powder | Liquid | Dispersant |
|---|---|---|
| Alumina | Water | Triton® X-100 |
| Barium cromate | Mineral spirits | — |
| Calcium carbonate | Water-methanol | Triton® X-100 |
| Calcium oxide | Mineral spirits | — |
| Carbon dust | Water | Triton® X-100 |
| Coal dust | Mineral spirits | — |
| Clay | Isopropanol | — |
| Collagen solids | Isopropanol | — |
| Copper | Water | Photoflow 200 |
| Coffee (dried) | Mineral oil | — |
| Diamond powder | Water | Aerosol® OT |
| Diamond powder | Mineral oil | — |
| Glass | Water | Triton® X-100 |
| Granite dust | Water | Triton® X-100 |
| Guar gum | Mineral spirits | — |
| Iron powder | Water | Triton® X-100 |
| Kaolin | Water | Ethoneen C-15 |
| Paint pigment | Water | Aerosol® OT |
| PVC powder | Isopropanol | — |
| Pumice | Mineral spirits | — |
| Quartz | Water | Triton® X-100 |
| Silica | Mineral spirits | — |
| Silicon carbide | Water | Tween® 20 |
| Sodium nitrite | Mineral spirits | — |
| Taconite | Water | — |
| Teflon® | Isopropanol | — |
| Tungsten powder | Water | Triton® X-100 |
| Vanadium powder | Isopropanol | — |
| Zirconium silicate | Water | Triton® X-100 |

for irregular particles, van der Waals forces, and electrostatic forces will retain the particles in intimate contact with one another once the intervening liquid film has been removed. In general, deagglomeration forces will vary inversely with particle size; that is, smaller particles will be more difficult to redisperse than larger particles. In dispersing dry powders, it is necessary that the powders be thoroughly wetted by the dispersing liquid — the contact angle between the powder and the liquid should be as close to zero as possible. For this reason powders of specific chemical content should be dispersed in specific liquids; wetting agents or surface active materials may be very helpful. Table 2 shows some examples of material-liquid combinations that have been found useful in preparing good dispersions.

## B. Liquid Suspension Sample Handling

### *1. Powder-Liquid Interactions*

Dry powder samples are frequently dispersed in liquids for analysis as a liquid suspension. Suspensions are convenient to analyze since the particles can be well dispersed and transported through and into a variety of analytical systems. Suspensions of particles in liquid are also analyzed, since the liquid itself may be the material of concern and the particles are present as a contaminant whose quantity and nature is to be defined. In either case, some possibility exists for particle-liquid interactions that may affect the integrity of the data produced. The first and most obvious area of concern is to ensure that the particles to be analyzed are not soluble in the liquid in which they are to be dispersed. In situations such as definition of crystallizer nuclei size distribution the analyst has no control over choice of liquid or particulate material; the particles are present within a saturated solution. This situation requires that the analyst control sample temperature during the entire handling process so that the particles neither dissolve and become smaller nor act as nuclei and grow in size as the solubility shifts with temperature. Another obvious chemical reaction problem to avoid is the possibility of the particulate material reacting with the liquid so as to cause any particle size change.

If particle suspensions are to be retained, or have been retained for any period of time prior to analysis, the analyst must be aware of the possibility of bacterial growth. Bacterial growth will result in the generation of erroneous particle size information due to bacteria causing individual particles to appear larger than they actually are and to the generation of clumps of bacteria within liquids that are normally not considered nutrients. Such liquids include deionized water and mineral oils. Another problem that can occur is changes in the physical nature of the liquid due to the particle being present. Viscosity of the substrate liquid may change because of the presence of large quantities of subanalyzable particles. Gel formation can occur, resulting in a very large viscosity increase.

In this case, the operation of stream analyzers can be affected due to the change in liquid velocity passing through the analyzer and the resultant change in output pulse structure. If an optical device is used for particle sizing and for concentration measurements, then it is necessary that the analyst choose a liquid in which the particles will be visible, i.e., one with a refractive index different from that of the particles. Otherwise, a conventional non-imaging particle counting and sizing device will not detect the particles at all. Similarly, the conductivity of the particle should be different from that of the substrate liquid if an electrical analyzing device is to be used. Fortunately, response can be obtained even with refractive index differences as little as 0.02 for the optical instruments and essentially all particles rapidly acquire an insulating oxide layer.

If feasible, the specific gravity of the particulate material and that of the dispersant liquid should be as close as possible. If not, difficult handling problems will occur as particles settle to the bottom of the liquid reservoir and agglomerate, or as particles float on the top of the liquid surface and flocculate during storage, or coat the surfaces of storage vessels preferentially and do not pass through the particle characterization instruments. Another, perhaps obvious, requirement is that the interfacial tension between the particle surface and the liquid be adequate to permit good wetting, otherwise particles will not be dispersed to any appreciable extent.

### 2. Particle Condition Considerations

When characterizing fine particles in liquid suspension, the particle condition requirements can be reiterated; the particles should be sufficiently separated so that no interparticle or intraparticle reactions occur and the particles should be present in sufficient quantity for measurement, but not so many as to result in interference with the normal operation of the characterization system.

First, consider the spatial distribution of particles in a liquid suspension under normal circumstances. A good dispersion of particles in a suspension should be uniformly distributed with sufficient space between each particle so that the characterization device can examine a particle, report its condition, and return to quiescent state and readiness for the next particle to be examined. In actuality, particles are not uniformly distributed but are present in random spatial distribution. Since many particle characterization instruments observe individual particles as they pass through a sensing volume, the random distribution results in a finite probability that more than one particle will be present within that volume at any specific instant. As a result, information that is reported for a single particle may be due to more than one particle present at any time. Coincidence errors can be calculated, based on accepted statistical theory.[16] If one assumes the statistical probability where no particles, one particle,

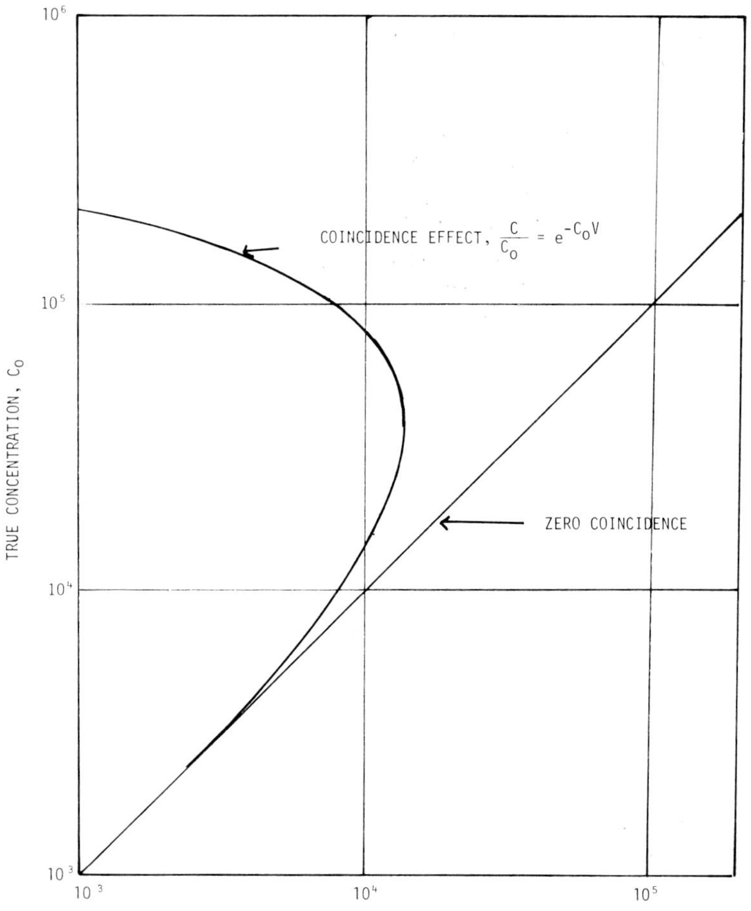


FIGURE 3. Coincidence effect on concentration measurements with single-particle counting instruments.

or more than one particle is present in the sensing volume at any time is given by a Poisson distribution, the coincidence error can be expressed by a simple exponential relationship: $N_m = N_{te} - N_tV$ where $N_m$ is the reported number of particles in the sensing volume V, and $N_t$ is the actual number. As the concentration increases, the reported number of particles continues to decrease, as shown in Figure 3. Eventually, at very high concentration, the population of individual particles approaches zero and the sensing volume is always populated, thus allowing no dead time for the electronic system to operate. In addition to the statistical error produced by more than one particle being present in the sensing volume at any time, the saturation limitations of the sensing system electronic reporting and/or counting circuits must be considered; that is, sufficient dead time must be allowed for the electronics to recover from information from a previous particle.

Particle suspensions which are stored in sample containers or in quiescent transport lines will be characterized by loss of particles which have settled on the lower surfaces of these containers due to gravity. If the container has been stored for a long period of time, or if the transport line has been in a zero flow condition, in all likelihood the particles which have settled to the walls of the container will be very difficult to redisperse since the liquid film between the particle and the wall will have been squeezed out by forces due to gravity. Further, if particles are in contact with each other then agglomeration will result. If particles are to be characterized for composition, then examination of agglomerated particles is satisfactory. If particle size distribution or particle concentration by number is required, then agglomeration of particles in the sample will result in measurement errors. Settling effects can be controlled by stirring or adequate mixing by other means during storage.

Wall deposition can occur in transport lines where particles will settle on the walls, even though liquid is flowing, due to the gravitational forces exerted on the particles, particularly if their specific gravity is significantly greater than that of the liquid. Once the particles have settled upon the walls of a tube then they will tend to remain, since the liquid velocity profile even in turbulent or plug flow is such as to produce very low velocity directly at the tube wall. Over a period of time, however, additional particles will settle in the low velocity zone and eventually the particles will be removed from the wall by the following processes:[17-18] (1) a small particle will settle upon the wall surface; (2) another particle will settle directly behind or on top of the first particle; (3) additional particles will settle to form a cluster or a filament that will extend into the higher velocity regions of the flow stream lines; (4) the clump will be lifted from the tube wall and swept down the length of the tube; and (5) the clump, because of its large size, will descend rapidly to the tube wall where it will either settle or strike another particle with sufficient force to entrain the other particle back into the flow — the entrained particle may be an individual or a secondary agglomerate. This reentrainment process, known as saltation, can result in particles much larger than the original being brought into the measurement or characterization device.

In most cases, liquid samples which are stored will be exposed to cyclic temperature situations. Settling will be decreased by the presence of thermal eddy currents within the liquid and resultant particle transport and mixing. The effect of such thermal eddies will be significant for small particles. For particles much larger than a few micrometers in diameter, thermal eddies will have little effect.

In process and sample lines where samples are to be taken, the wall loss problem can be reduced by maintaining turbulent flow at a high enough rate so that the quiescent layer of liquid near the wall is of minimum dimension. In other words, the Reynolds number should be maximized for good particle suspension. As long as the Reynolds number exceeds approximately 3000, then plug flow will eventually be established in the liquid line. As a safety factor, however, it is suggested that sample liquid flow lines should be operated at a Reynolds number of at least 5000 in order to ensure good distribution of particulate materials for sampling.

*3. Liquid Sampling*

Sample size and sample frequency considerations are related to the nature of the sample that is being measured, to operation of the characterization instrument, and to capabilities of personnel carrying out the work. As an example, consider some basic requirements for particle counting. As concentration of particles in a sample decreases, the counting accuracy will also decrease. In general, if particles are distributed in a population with a Gaussian distribution, the variance of data for a series of samples will be equal to the square root of the mean values for those samples. Thus, if a sample taken from a large population has a count of 100 particles, the variance is $\pm 10\%$. If the sample had a count of 1000 particles then the variance is $\pm 3\%$. Figure 4 shows how uncertainty of data varies inversely with particle sample size.

In normal sampling of material to be characterized one must also take into consideration the process of concern; that is, is the process continuous and uniform or is the process one that has intermittent factors? If the process is uniform and variations are cyclic, a constant sample interval frequency can be established. If the process is not uniform and variations in the nature of the process can be tied to specific operations then the sampling frequency should be selected on the basis of operational variations. When acquiring samples from flowing lines, valves are often used to control sample flow to the particle characterization

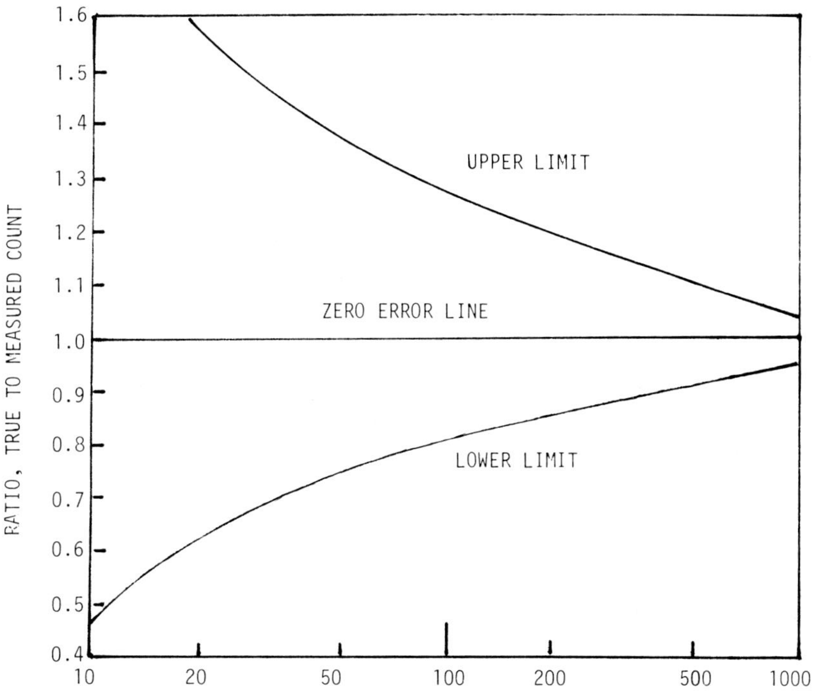


FIGURE 4. Statistical error limits at low count levels.[19]

system. Acquisition of data from the characterization system should take into consideration the fact that valve operation frequently generates artifact particles by frictional wear of the valve components, therefore, sufficient time for valve flushing should be allowed before the sample is collected.[20] Further, if a line to be sampled is under any significant pressure then it may be necessary to reduce pressure between the sample line and the characterization or analytical device. Invariably, some gas will be dissolved in the liquid being analyzed and the pressure reduction may result in generation of gas or vapor bubbles. If particle counting or sizing data is desired, then it may be necessary to vent the line to the atmosphere so that a place for bubbles to escape is provided. In a similar manner, sudden variations in the sample line dimensions at constant flow under pressure may result in cavitation that generates very stable bubbles. For this reason no sudden changes in sample or flow line dimensions should occur. Design of valves, gasketing, O-rings, retainers, etc., should be examined carefully. Cavities where eddies may deposit particulate material that can be reentrained at a later time to contaminate subsequent samples should be avoided.

Sample acquisition from batch containers is another problem area requiring some planning. For example, clean liquid is often stored in a reservoir from which material is drawn as needed. From time to time, sample withdrawal may be advisable to define the nature of the material in the container or reservoir. In the same way as line sampling, any artifacts introduced by valve operation at the beginning of the withdrawal process can be eliminated by discarding that part of the sample obtained during initial valve operation. In sampling from reservoirs or containers, it is usually not necessary to assure uniformity of particle distribution within the container since the sample which is being drawn from that container will be equivalent to the material that is being used from that container. On the other hand, if a group of smaller containers is to be used and samples are selected from the group, then it is indeed necessary to ensure thorough mixing of the material in the sample containers. Before a sample container is opened the exterior surfaces should be cleaned carefully, particularly around the closure. Whenever possible, screw-type closures should be used that have been gasketed with materials impervious to the liquid in the sample and made of material that will not abrade easily. Inert polymer liners are preferable to the lined paper gaskets often found in commercial bottle closures.

Withdrawal of a sample from the sample container is usually a matter of pouring or aspirating from the container. If aspiration is used, the material in the liquid should be stirred prior to the aspiration process. When pouring, the technician should take care to minimize bubble generation and splashing caused by pouring too rapidly. On the other hand, pouring should not be at so slow a rate so that a film of liquid adheres to the exterior upper portion of the container where it may inadvertently wash debris from the container exterior. If batch samples are to be transported in secondary sample containers, some recommendation for the sample or transport containers can be made.[21] First, it is necessary to be certain that the sample containers are adequately clean. A suitable container size for measurement of so called "clean" liquids is usually in the range of 100 to 500 mℓ with a popular intermediate level of approximately 250 mℓ. Containers can be procured that are made of glass, polypropylene, polyethylene, or metal. Glass is the material of choice since it can be easily cleaned and its cleanliness verified. With suitable packaging, it can be shipped even under hostile conditions of temperature and vibration. The glass can resist mildly corrosive conditions and will not normally add artifacts that may be confused with original particulate material. Polypropylene is also useful, but can be easily confused with polyethylene, which shreds very easily under mildly abrasive conditions. Polyethylene is less resistant to acid corrosion than glass or polypropylene. Metal containers usually should be avoided; stainless steel containers are suitable but costly. The very convenient drawn aluminum or lined mild steel containers corrode very rapidly and can add significant debris to any sample.

A convenient closure and seal system for sample containers consists first of a 1-mil nylon or Mylar® film that is wrapped around the container top and held loosely in place with an elastic band. The cover for the container is typically a screw cap. By using the film, it is possible to provide a gasket and seal that will be nonreactive with most liquids and will not add artifacts to the sample. For cleaning, new containers should be cleansed either by scrubbing with detergent solution or in a commercial dishwasher. Following adequate rinsing with tap water, a further rinse with deionized water is recommended. The deionized water can then be removed from the interior of the container and interstices of the closure by rinsing with filtered isopropanol or methanol. The alcohol can then be removed by a stream of filtered dry air. If large quantities of sample containers are to be used, a mass cleaning procedure should be used that includes scrubbing, rinsing, fluid exchange, and drying followed by cleanliness verification.[22]

Sample size and frequency considerations are seldom defined adequately. For the most part, analysts have depended upon subjective and intuitive procedures for defining sampling protocols. However, without a good sampling plan, the relationship of analytical results from the sample to properties of the system from which the sample was procured becomes uncertain or even impossible to interpret meaningfully. Statistical techniques are available to improve validity of sampling.[23] Sampling programs can be designed so that a meaningful identification of the population being sampled can be produced and legitimate and efficient selection of number and frequency of the samples can be made. A random sampling process should be established, rather than any sampling program which can be subject to bias. Random sampling is best carried out by first subdividing the population to be sampled into a number of segments. For example, a container can be conceptually subdivided into zones either horizontally and vertically or radially and vertically. The zones to be sampled are selected by assigning a number to each zone in a definite sequence. A selection of zones to be sampled is made by starting in an arbitrary place within a random number table and choosing numbers according to a predecided pattern that may consist of a consecutive sequence, an alternate number selection, or an interval selection, and sampling those zones whose numbers are chosen until a satisfactory quantity of samples have been obtained. Results can then be analyzed to determine whether systematic variations occur within the population being sampled that may be a result of systematic trends or cyclic processes. Measurement plans can be designed to identify and optimize these effects. Note that random sampling is not equivalent to representative sampling. Representative sampling is possible when the population has been homogenized before sampling. Relatively high-concentration particle samples within a liquid can be so homogenized as to permit justifying abstraction of samples that can be identified as representative.

In the case of powder characterization, samples received by an analyst will require subdivision where replicate measurements or observation by more than one technique for different characteristics should be carried out. If the sample received has already been homogenized, caution is necessary to avoid demixing or segregation by percolation or other processes during the individual sample extraction.

With these considerations in mind one may define an acceptable sampling protocol that includes details of sampling periods, sampling location, and sampling procedure. In taking samples, one should decide the number of samples to be taken in each size and location. One may further consider whether individual samples should be analyzed separately or a composite prepared from individual samples for further subdivision or direct analysis. In the cone and quartering process, sample composite preparation is frequently used. When the properties of a material to be characterized are totally unknown it is frequently good practice to collect a small number of samples for preliminary analysis.

With regard to sampling, in general one may conclude that sampling requires decisions that are frequently difficult to formulate. Development of optimum sampling protocols for long-term operation frequently require investigations with presampling protocols. These require acquisition of ''preliminary'' samples and measurements taken to determine some ranges of averages and relative S.D. Once these quantities are defined, definition of statistical sampling errors, uncertainties, and protocol can be carried out with some confidence. Procedures can be established that will provide an optimum sampling plan. However, the analyst should never replace experience and common sense with sampling theory. Sampling theory can only be used to complement these factors for best results, but sampling theory should be considered as part of the experiential background with which the analyst lives.

Sample dilution problems occur in determining particle size distribution and concentration of particles in liquid suspensions, particularly when single-particle counting and sizing instruments are used. In these devices, the liquid sample is passed through a sensing zone or volume of finite dimensions. By definition, the signal from any individual particle requires that only one particle be present in the sensing volume; further, electronic system operation also requires that between each signal pulse there exists a finite ''dead'' time for zero basing purposes. As particle concentration or liquid flow rate increases, two phenomena occur which can cause difficulty: coincidence and saturation. Coincidence is defined as the presence of more than one particle simultaneously in the sensing volume, following the treatment of Jaenicke.[16] It is interesting to note that this effect produces a curve as shown in Figure 3.[24] Note that this curve shows the possibility that extremely high concentrations may be indicated as zero measured concentration. It should be emphasized that the coincidence error is not affected by flow rate through the sensing zone. The number of particles present within the sensing zone at any time is a function only of particle concentration and sensing zone dimensions and does not increase or decrease as flow rate changes. Only the residence time for a single particle and the residence time intervals change with flow rate. The effect of changes in the flow rate at any concentration are twofold: the pulse counting rate will increase as flow rate increases at any fixed concentration and the pulse duration and pulse rise time will decrease as flow rate increases since the particles pass through the fixed sensing zone at a faster rate. Saturation of the electronic counting circuits in the particle counter occurs since the electronic system is observing discrete pulses. Each pulse must rise above the zero level base line and pulses must be separated by a finite period of time in order for electronic circuits to recover from a previous pulse. If two pulses occur too close together, the electronic system will record the two pulses as a single pulse. Note that spatial distribution of particles in the liquid is random, therefore the pulses will also occur at random time intervals. As a general rule of thumb, it has been observed that for any electronic pulse counter to record pulses that are randomly spaced, it is necessary that the counter be capable of counting at a uniform rate approximately ten times the long-term average rate for the randomly spaced events.

Thus, coincidence will result in several particles producing a single pulse corresponding to a single larger particle which reduces the particle count and distorts the particle size distribution. Saturation will result in several pulses being counted as a single pulse with amplitude that depends upon the counting circuit logic. If peak reading logic is used, then the multiple pulse signal will be recorded as equivalent to the largest pulse. Again, indicated concentration will be reduced and particle size distribution will be distorted.

In such a situation, it will be necessary to dilute the sample to a level where the particle counting device will be able to measure at the concentration in the diluted sample. Where fixed ratio dilution can be established, then a relatively simple procedure can be used. For dilution ratios larger than approximately 25:1, a sequential procedure is recommended since volumetric measurement accuracy for very small quantities of liquid is low: (1) establish the desired dilution ratio; (2) fill a clean container to a preset marked level, e.g., 100 m$\ell$, with clean diluent liquid; (3) remove a quantity of clean diluent equal to the volume of sample to be diluted (for example, if a dilution ratio of 25:1 is to be used, remove 4 m$\ell$ of clean liquid from the 100-m$\ell$ volume); and (4) the original sample container should be well mixed and a sufficient sample poured into the marked diluent container to bring the total quantity back to the original 100-m$\ell$ level. In this way 4 m$\ell$ of sample will have been added to 96 m$\ell$ of clean diluent so as to produce the desired 25:1 dilution ratio. This procedure does not involve direct volumetric measurement of the concentrated sample material with possible retention of particles on the volumetric container and it minimizes possible contamination of the clean container with concentrated sample.

If, however, a required dilution ratio is not known but the original sample is suspect because of an observed atypical particle size distribution, the following procedure can be used: first, determine the particle concentration in a sufficient number of size ranges so as to produce a cumulative size distribution and assume that a logarithmic or power function distribution should have been produced. Examine the data and observe particularly the first two or three particle size ranges, counting up from the smallest range. If these do not follow the anticipated size distribution function format, dilution is indicated. Note the particle concentrations recorded in all ranges for future reference. Second, make either a 2:1 or 10:1 dilution, using the procedures described in the preceding paragraph. Reexamine the indicated particle size distribution that is produced following the first dilution. If the first dilution is adequate, the ratio of particles measured before and after dilution will be essentially equal to that of the dilution ratio. If the dilution is not adequate then the particle concentration, in the smallest size ranges where the highest particle count data are expected, will not decrease to any noticeable extent; however, the particle concentration in the larger size ranges may decrease. Repeat the dilution process as many times as necessary to insure that particle concentration in the smallest recorded size range decreases in accordance with the dilution ratio last used. Record particle concentrations in each of the size ranges in accordance with dilution ratio effective for that size range. Note that some care is required at this point. Sequential dilutions of the type described here will result in particle concentration decreasing in the larger size ranges to the point where data are statistically insignificant. The analyst should remember that, even in the largest size range where data are recorded, sufficient data should be available so that S.D. anticipated for a series of samples are acceptably low. That is, if the number of particles in the largest size range is less than 10 the relative S.D. will be approximately 33%. Precautions are necessary anytime liquid is handled, as in making dilutions. It is possible that artifacts can be introduced during the dilution process from handling, from inadequately clean diluent, and from containers and measuring devices that are used in the dilution procedure. It must be remembered, further, that in handling any suspension good mixing is required throughout the procedure since large dense particles can be lost on walls easily during the handling process. Finally, the requirements for extreme care in handling clean liquids and measuring small quantities of liquid cannot be overemphasized.

Installation and operation of characterization devices in-line occur quite frequently. Such installations are used for routine monitoring of deionized water lines in electronic manufacturing operation, of cleaning water lines in pharmaceutical and medical device manufacturing, and for composition characterization primarily in many chemical process component transfer lines. As in any particle characterization system, certain considerations are necessary to optimize sample acquisition and handling as well as overall system design and operation. Problems exist for in-line particle sampling that require unique solutions and techniques. One of the first problems that must be considered is that of liquid system interfacing. In other words, how does one best connect the system from which the sample is obtained to the device or system where the sample is characterized? No single all-inclusive answer can be provided for every system; some variations may always be required from one system to another.

Problems of sample acquisition from liquid systems have been addressed for many years and many sampling system problems are well known. The desirability of sampling from a turbulent rather than laminar flow line in order to obtain the most representative samples has been known since approximately 1960. Isokinetic sampling is considered most desirable, however, if the Reynolds number in the liquid system is less than approximately 75,000, isokinetic sampling is not an absolute necessity. For most process lines, superkinetic sampling is considered superior to subisokinetic sampling, if isokinetic sampling cannot be obtained.[25]

Sample point locations depend upon the nature of the system being sampled, obviously. However, one may look at some possible liquid system operations and recommend some rule-of-thumb locations. For example, if a liquid system involves a filter for cleanup, one should sample shortly after the filter to ensure integrity of filter system operation. If the liquid transport system involves mixture, addition, or reaction due to chemical or thermal effects, a sample point could be installed shortly after that process has been completed. One may wish to install a sample point at a reservoir inlet or outlet line. In addition, consideration must always be given to the possibility of sample acquistion at "the only available location due to operating and system layout". For this reason, data may be viewed with consideration of abnormal or nonrepresentative conditions existing at or directly upstream of the sample point. Figure 5 shows an example of sample point choices for a simple hydraulic process loop.

In designing sample probes or acquisition systems, a fundamental decision must be made on the selection of either permanently installed probes or removable types which can be inserted into the liquid systems through suitable valves and packing glands. The removable probe can be easily inspected and cleaned between samples, but well-designed fittings and seals must remain at each sample point. Note that permanently installed probes, although subject to particle deposition from the line, provide less opportunity for ingression of atmospheric contaminants and debris into the liquid system being sampled and are somewhat less complicated to manipulate.

In either case, the probe inlet should face into the liquid flow unless a very low flow rate

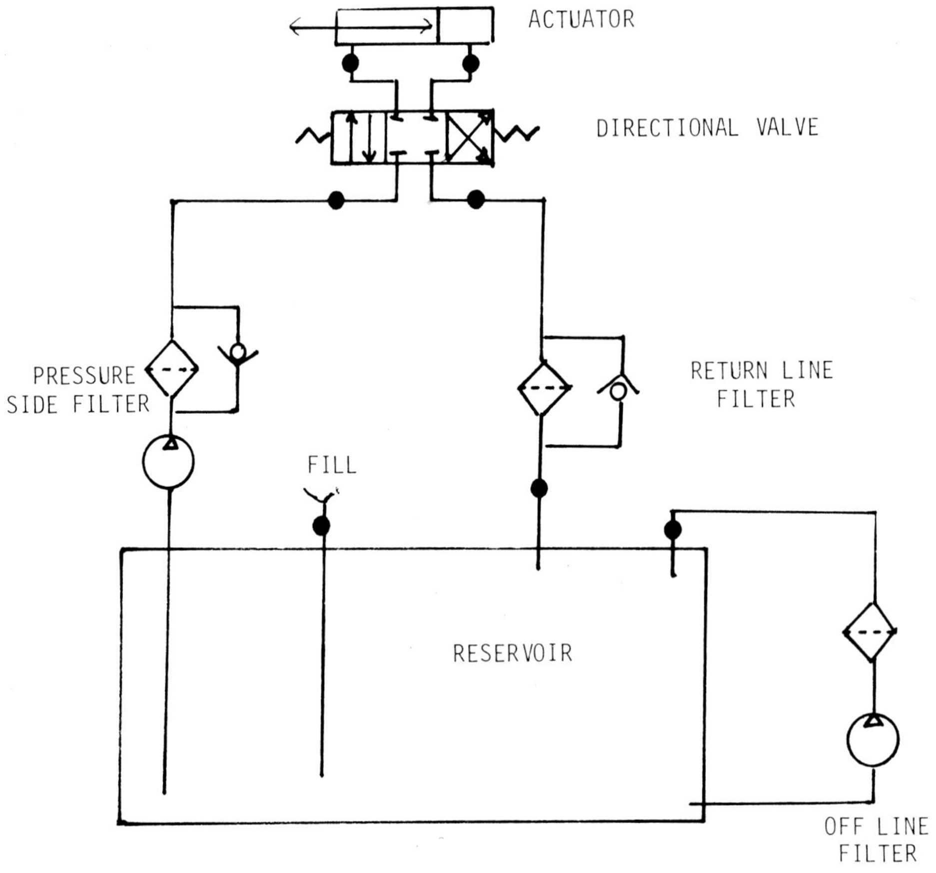


FIGURE 5. Hydraulic loop sample point locations. (●) Candidate sample point.

is in use. In this situation, a simple Tee sampler may be used. For most in-line particle counting and sizing systems, a flow rate of 10 to 200 mℓ/min is used. Sample probe inlet diameters from 2 to 6 mℓ will result in adequate liquid velocity to minimize deposition and settling in the probe. During sample transport from the probe to the instrument, the sample line should be as short and direct as possible, whether the line sample point is used for permanent installation or for intermittent connections. Note further that connections to the liquid system, to the sample line, and to the characterization instrument should be as smooth as possible. Use compression fittings rather than flare fittings; use ball valves and tubing with smooth inner surfaces. If sample lines longer than a few meters are needed, no data should be recorded until sample flow in a newly connected line has been stabilized and the line from sample acquisition point to the measurement or collection point have been thoroughly flushed. A point of interest in sample line design for particle counting and sizing operation is the necessity for maintenance of a constant flow rate and pressure to ensure operating stability of the particle counting and sizing instrument. If liquids containing gas are to be sampled, provision for degassing or venting to the atmosphere may be required as part of the sample line design. The problem of flow stabilization can be handled by using a protected pressure-compensated flow control valve. The problem of possible gas release requires special treatment for each installation.

If dilution is required for an in-line particle characterizing device, some precautions are necessary. These arise from the fact that in changing particle concentration the particle size distribution that is reported can also be changed. Many particle size distributions can be described with a power law function; that is, the concentration is an inverse power function of size with the exponent having a value of $-3$ to $-4$. Thus, if one wishes to reduce concentration of smaller particles to the point where the particle counting device will not be overwhelmed by excessive particle concentrations, then the large particle concentration may be reduced to the point where data are no longer statistically valid.

In-line dilution systems require treatment such that a recirculating fluid is passed through a filter that will remove essentially all particles in the size range of concern. Clean liquid is mixed with sample, usually with a simple Tee mixing chamber or, if necessary, motionless mixers can be used. The diluted material is then passed through the particle characterization system. Effluent can then be pumped back through the filter for continued use as diluent. The hardware and controls for such a dilution system require accurate flow measuring and control systems so as to permit accurate dilution ratio definitions. If dilution ratios greater than 100 to 1 are required it will be necessary to use a two-stage dilutor since flow measurement accuracy limits the practical dilution ratio levels.

If particle counters are used, then sample flow rate control is necessary. This requirement arises because of the nature of electronic system limitations for such devices. Further, many counting and sizing instruments measure sample time and assume continuous known and constant flow rate as the basis for defining sample volume. Flow rate control can be accomplished through application of a pressure-compensated flow control valve immediately downstream of the particle counter, and immediately after its own filter that is required to protect the flow control valve element from plugging when particles pass through. A second means of controlling flow rate would be to use a metering pump with motor speed control. The particle size analyzers in common use require fixed flow rates at a specific level for optimum operation; thus the flow rate and sample size selection must be matched to the operation of the particle analyzer.

## V. PARTICLE CHARACTERIZATION INSTRUMENTS

When particles in liquid suspensions are characterized, then the characterization process either examines the particles while they are suspended within the liquid or removes the particles from the liquid and characterizes them after collection and removal. A variety of instruments can be used for characterizing dry powders or discrete samples that have been removed from a liquid suspension. In addition, many instruments are useful for characterizing the particles within the liquid.

### A. Dispersion Problems

When characterizing powders, one must realize that a sample of dry powder may be in a condition where individual particles are strongly agglomerated due to a variety of adhesion mechanisms. The mechanisms may include electrical forces, van der Waals forces, interfacial tension (when a second phase is present), mechanical interlocking, etc. The nature of the force that may cause particles to become agglomerated is of no particular concern to this discussion. However, the requirement for dispersing the agglomerates to permit examination of discrete particles is important. If particles are agglomerated, then the particle size distribution reported, whether it is based upon Stokes' diameter or any other physical characterization base, can be incorrect. For this reason, it is necessary to achieve good dispersion of the particles within the fluid in which they are being examined.

As a general rule of thumb, particles smaller than 5 μm in diameter are frequently found difficult to disperse in gas. Considering the surface-to-mass ratio of such small particles, one often finds that electrostatic and van der Waals forces frequently exceed available shearing force that would be required to disperse the particles adequately. Even when alpha sources are used to achieve charge equilibrium in the gas stream, the remaining surface charges on the particles prevent good dispersion. For adequate dispersion, particles are frequently handled in liquid. The first requirement is that good contact be achieved between the particle and the liquid in which it is to be dispersed. The first step in dispersion is consideration of the history of the powder to see if surface contamination has occurred that may make dispersion difficult in a particular liquid. For example, the presence of an oil film or stearate grinding aid may result in an organic coating upon a powder that will make it very difficult to be wetted by water. Many metal powders may have an oxide surface film of varying composition depending upon the nature of the powder, atmospheric moisture, and storage conditions. Thus, consideration of the nature of the powder and its history should be the first step in choosing a suitable dispersing liquid. All of the powder characterization instrument manufacturers have accumulated listings of compatible liquid/powder systems for best dispersion. Where it is necessary to use an electrolyte liquid, nonionic dispersion aids are recommended. Note that surface active agents are frequently used as dispersing agents. However, a dispersing agent is any general additive that is suitable for maintaining maximum interparticle distance within a suspension while a surface active agent is designed to achieve intimate contact between the surface of the particle and the liquid in which the particle is dispersed. When using surface active agents, it is necessary to ensure that agents do not cause a charge to remain upon the particle that will result in potential gradients within the suspension. Electro-viscous effects can arise that will distort results.[26]

In nonpolar fluids, electrostatic effects can interfere with transport and dispersion. Thus, when oils or aliphatic hydrocarbons are used for dispersing liquids it may be necessary to add an antistatic additive such as ASA-3 (Shell Oil) to reduce electrical charge density within the liquid. Quantities of some 20 ppm have been found effective in controlling charge density in tests carried out with hydraulic oil.

The technique for dispersing powders in liquids is relatively simple. It is common practice to start by preparing a rather concentrated slurry of the powder in a small quantity of liquid. High shear forces are applied to the slurry by placing it in a container where a high shear mixer can be used or by exposing the slurry to ultrasonic energy. After powder has been thoroughly wetted by this procedure, with or without dispersant addition, then additional dilution with careful mixing can be carried out. Care is required in applying this technique. The investigator should try to use shearing forces no greater than those to which the system will be exposed in use. Thus, if a powder to be characterized is a carbon black or colorant that will be dispersed with high shear forces into a viscous medium, the investigator can use extremely high shear forces for characterization analysis. However, if the powder consists of fragile capsules that will be gently dispersed in a liquid substrate, mild dispersing forces should be used when preparing to characterize that type of powder.

### B. Dry Powder Particle Size Analyzers

Screens and sieves have long been used as a simple, reproducible, and inexpensive method for particle size analysis as well as for process material separation. Standard woven wire sieves can be used for defining particles with dimensions ranging from approximately 20 μm up to 125 mm. Electro-formed or micromesh sieves can be used to define particles as small as 5 μm. Woven wire sieves can be fabricated with openings that are essentially square. Electro-formed or perforated plate test sieves can have openings that take any desired shape.

Problems exist in sieving, as with any test method. These problems include uniformity of sieve opening dimensions and shape and sieving technique and measurement. Sieve opening variability is primarily associated with woven wire sieves. Aperture width distributions generally follow a Gaussian distribution about the mean width dimension. Relative S.D. up to 15% of median dimensions can occur.[27] Electro-formed or machine-perforated plate sieves are usually more uniform in their opening dimensions.

Operating problems are primarily those due to operator technique and data manipulation.

The first problem that usually occurs is overloading, with resultant blinding (blocked sieve openings) and agglomeration of soft particles, comminution of friable particles, inadequate dispersion of flocculated particles, etc. Inadequate shaking action can result in unsatisfactory exposure of all particles to the screen surface. Finally, damage to sieve surfaces, particularly for sieves with openings less than 200 μm in diameter, can occur if mechanical cleaning methods are used. It is recommended that ultrasonic baths and/or high-velocity low-pressure air streams may be best for cleaning.

Note that passage of particles through the openings of a sieve is possible if, and only if, the sieve surface and particles move relative to one another. The relative motion may be produced by mechanical agitation of the sieve or by hydraulic/pneumatic transport of particles relative to the sieve opening.

If one observes the powder mass passing through a given sieve as a function of time, then three regimes of material passage can be noted.[28] These are associated with the mix of particle sizes in the residue on the sieve. The first regime represents a preponderance of subsieve particles on the sieve; the second regime exists when the residue contains a mixture of sizes; the third represents the situation where the residue contains mainly particles that are near-mesh or larger in size.[28] Considering the nature of this curve, the question arises of determining the exact time for stopping the sieving operation to define the mass of particulate material as a function of sieve size. It is generally agreed that the time for sieving to stop is when the operation is in the third regime. For the most part, an arbitrarily chosen mass rate change as a function of time is accepted as defining conclusion of the sieving process.

A series of standard methods have been developed for sieve analysis of powdered materials. One class of procedure involves passing the dry powders through a "nest" of sieves; a second class requires that the powders be suspended in a fluid that is flowing through the sieves. Alternately, one may use a single sieve and determine the fraction of the total material presented to that sieve that passes through the sieve. One may use a nest of sieves as prepared by selecting sieve mesh opening sizes from the largest to the smallest and placing the sieves in sequence so that the largest opening is seen first by the mass of powder. Powder is induced to pass through the sieve openings by causing relative horizontal and/or vertical motion between the powder particles and the sieve. Eventually particles larger than the sieve openings will be retained on the surface of that sieve while the smaller particles will fall through to the next sieve or to the collecting pan. Relative motion between the powder and the screen can be obtained by mechanical agitation of the sieve, by local air jets passing up through the sieve to disperse powder and allow it to fall on a new location, by producing vertical oscillation in an air column within the nest of sieves, or by controlling momentum in a liquid stream flowing through the sieves.

Relative motion between particles and the sieve can be imparted by an apparatus which either vibrates or shakes the material. Mechanical devices include the "Rotap" sieve shaker which imparts a horizontal rotary motion combined with vertical jarring produced by a hammer. High-frequency vibration can also be used to induce particle passage through the screens. For sieving down to approximately 35 μm, an air jet sieve is quite useful. With this device, the powder is placed upon the screen surface and fitted with an air-tight cover. Using a vacuum pump, air is drawn upward through the sieve along one radius by a set of rotating nozzles. Air leaving the area above the sieve is drawn downward through another segment of the sieve and exhausted through a filter. When powder is placed on the screen the upward-flowing air jets tend to break up the flocs on the sieve and to minimize blinding of the openings. Note that the upward air flow cleans the sieve apertures immediately above the jets and the return air flow aids passage of fines through the sieve.[29]

Another automatic sieving device is the auto sieve.[30] With this device, the powder to be sieved is sequentially passed through four preweighed screens and the surface of the screen in use is continuously washed while the screen is oscillated. An automatic weighing is made after each oscillation, and when the change in weight has stabilized then the weight is entered into the device memory. The next sieve is then fed with the residual powder to define the next step in the size distribution. In this system, the sieves are contained within a rotor assembly where three sieves are placed along with one section free for oversize passage. Mesh sizes can be changed or shifted on rotors as desired. This device is designed for analysis of powders with particles in the size range from 38 to 500 μm.

When using screens or sieves to characterize the size distribution of particulate material the data produced are reported in terms of a size distribution by weight. The size distribution may be cumulative or differential and is reported in terms of percent of the sample in the several size ranges represented by the sieves in use. For the most part, the particle size range that is determined by sieving runs from approximately 10 μm up to several millimeters.

When using analytical screens it is necessary to be sure that the screens are in good condition with no rips, tears, or distorted apertures in woven wire screens. If micromesh or photo-etched screens are used, the surfaces should be examined to ensure that damage has not occurred to these relatively fragile devices. When making a sieve analysis it is wise to ensure that the powder is not flocculated or agglomerated because of excessive moisture or electrostatic charge. If this is the case, heating may be required before analysis. When carrying out wet screening, the operator should make sure that the powders and liquid are compatible and that the suspension liquid is clean in the size ranges of concern. Following an analytical screening operation, it may be wise to examine the surface of the screens to ensure that excessive blinding has not occurred during the screening operation. Sieving as a characterization process finds greatest use in preliminary evaluation of powders for applicability to specific purposes, e.g., fluidization, pelletization, pipeline transport, etc. Further, sieving as an analytical method has a long history of data which can be used as benchmark bases.

Gas elutriators have been used to determine powder size distribution by fractionating the powder sample into several classes based upon aerodynamic size. Either gravitational or centrifugal elutriators can be used. Both operate on the basis that a directed laminar flow stream of air carries particles in one direction, while centrifugal or gravitational forces carry particles in the opposite direction. By balancing the air velocity against the centrifugal or gravitational force, one may selectively remove particles below any present size by adjusting the air velocity in the elutriator to obtain equilibrium between the opposing forces. The small particles are removed with the transporting air stream, while the larger particles are retained in the elutriator. For a vertical elutriation chamber where gravity is the opposing force, the procedure is to disperse a sample of the powder so that an aerosol is formed containing suspended particles in all the size ranges. At the lowest air velocity the smaller particles will be carried upward in a laminar flow air stream and can be collected on a filter surface. After the smallest particles have been removed from the aerosol the action can be stopped and the filter removed and weighed. Next, the air velocity is increased until the next larger size range is removed from the dispersed aerosol. The procedure is continued until the powder has been satisfactorily categorized in terms of the percent by weight in each of the aerodynamic size ranges corresponding to the air velocity. Size determination can be made by carrying out each step sequentially or by using a cascade procedure where up to eight elutriation chambers are arranged in series. The data produced are in terms of mass distribution percentages defined on the basis of aerodynamic diameters. For gravitational separators, particles in the size ranges from approximately 5 μm up to 100 μm can be handled. Centrifugal devices can carry out fractionation for particles down to at least 1 μm in diameter. Samples of 5 to 20 g are required for good operation. (It is worth noting that these devices can also be used effectively to produce reasonable quantities of fractionated powders for experimental purposes.)

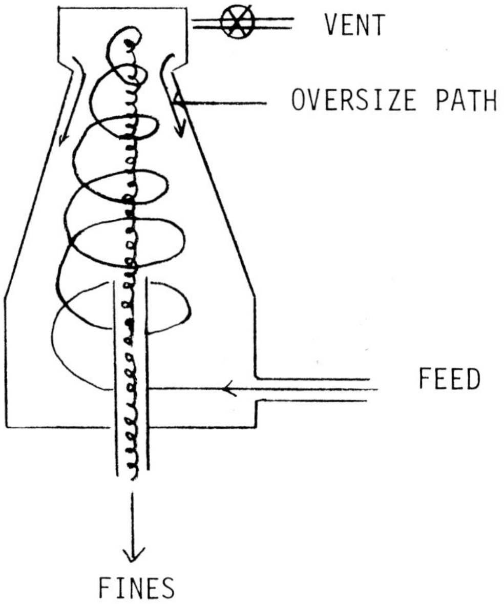


FIGURE 6. Cyclo Sizer operation.

When using an elutriator for powder characterization it is necessary that the powder be relatively free flowing and that the powder not generate large static charges, otherwise losses on the walls of the elutriator may become excessive. The sharpness of cut that is obtained with the gravitational elutriator is typically less than that for a centrifugal elutriator since larger flow passages are required with the smaller gravitational forces. In this situation the gas velocity profile is typically laminar and may cover a large numerical range, particularly for the smaller size classes.

It is worth noting that the basic operating principles of a gas elutriator can be applied to define performance of systems where the gas is replaced by liquid. To reiterate the operating principle, these devices operate by noting that the point where the forces causing a particle to settle are balanced by opposing forces. The forces resulting in sedimentation or settling of the particles are due to either gravity or centrifugal force. The opposing force can be due to gas flow or to buoyancy in a liquid.

Some of the devices which have been used for gas elutration will be described briefly. A modification of the traditional cyclone collector was reported by Kelsall and McAdam.[31] The basic system, termed a "Cyclo Sizer" consists of an inverted conical cyclone collector operating as shown in Figure 6. Since the stream of fluid carrying particles is directed into the bottom portion of the cyclone with tangential motion, then the helical fluid flow travels to the top of the Cyclo Sizer with centrifugal force directing the larger particles toward the periphery of the Cyclo Sizer. Fine particles remain with the main liquid stream and are passed out through the central vortex tube.

The Haultain infrasizer uses an air stream to lift particles through cones of increasing diameter, as shown in Figure 7.[32] Since each succeeding cone is somewhat larger than the preceding one, then it will retain material of somewhat smaller size than the preceding cone. A special feature of this device is the use of a spinning ball check valve that acts to disperse powder at the beginning of each elutriation chamber. As the air stream enters at the bottom of the chamber, the ball forms an annular passageway where relatively high-velocity gas causes turbulence at the bottom of the cone to aid in dispersing particles.

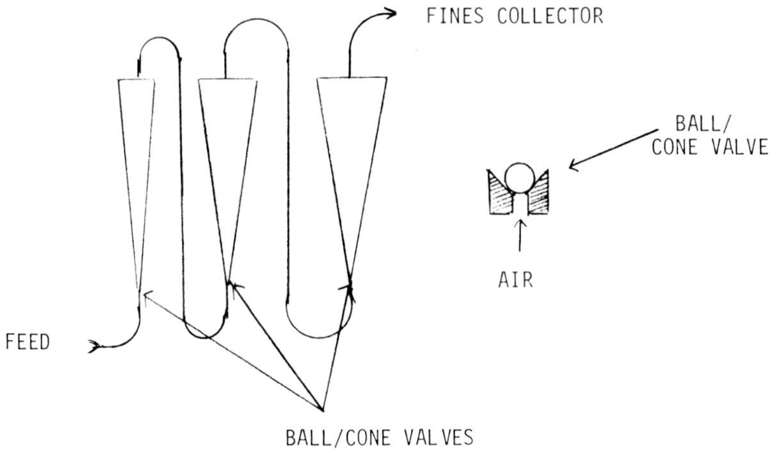


FIGURE 7. The Haultain infrasizer.

The MSA-Whitby centrifugal sedimentation equipment is used for size distribution analysis.[33] A glass centrifuge tube of approximately 1.5 cm inside diameter is used with a graduated capillary of 1 mm diameter at the bottom of the tube. In operation, the tube is filled with clear sedimentation liquid and a suspension of material to be analyzed is placed in liquid of slightly lower density than the sedimentation liquid. Upon centrifugation for a specific time, particles will sediment through the sedimentation liquid and form a packed column in the capillary at the base of the tube. By measuring the height of sediment column in the capillary one can determine the rate of sediment accumulation and, therefore, the quantity of material which will sediment a given distance at the centrifuge speed used for the measurement.

A number of gravitational and centrifugal sedimentation devices have been developed for determining size distribution of powders dispersed in liquid. At this time, some popular applications use optical or X-ray scanning of the sedimentation fluid. The scanning level is adjusted and timed so as to derive information that will permit definition of the Stokes diameter distribution of particles in the fluid. Normally the particles are in a well-mixed, homogeneous suspension at the beginning of the measurement. As the particles are sedimented, changes in the energy of the radiation beam passing through the monitoring zone of the suspension are used to define concentration at the point of monitoring. Either light or X-ray radiation is used. The Lambert-Beer law will define the relationship of beam intensity change and suspension concentration for a fixed beam path length system. By determining passage and system constants for a given instrument, it is possible to directly relate the intensity loss to the Stokes diameter.[34]

In X-ray sedimentometers, the transmittance of the X-ray beam is directly related to the mass of particles within the beam path by a simple exponential relationship similar to the Lambert-Beer law.[35] In this device, the X-ray beam path is moved down the sedimentation column to achieve rapid information production while material is settling under the influence of gravity.

Several gas elutriators have been used for fractionation as well as size definition. For example, the Roller elutriator was described in 1931.[36] A sample of powder is placed in a U-tube and a high-velocity jet used to disperse and aerosolize the powder. Once the powder is dispersed the resulting aerosol is transported upward through a conical section into the separator tube by the gas originally used to disperse that powder. The air flow rate is selected to provide a given velocity in the cylindrical section of the separator tube. The given velocity is defined by Stokes' equation: $V = 10^{-8} g\rho d^2/18 \nu$ where V is the terminal velocity of a

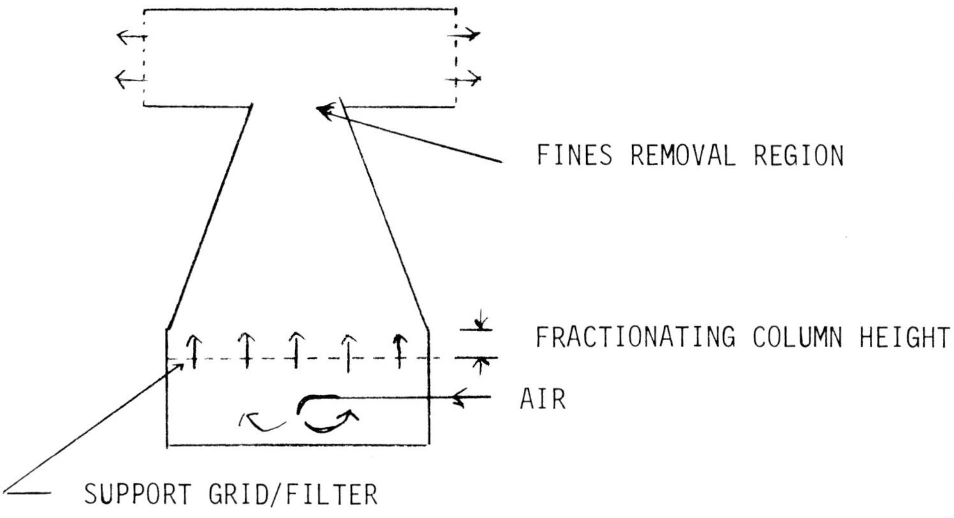


FIGURE 8. Short column elutriator.

given particle (equal to the vertical upward velocity of the air flow), g is the gravitational constant, ρ is the absolute density of the particle, d is the Stokes' diameter in micrometers, and ν is the viscosity of air. This device is limited in terms of making narrow distribution since the diameter of the cylindrical section of the separator tube can be several tens of centimeters and the laminar flow air velocity profile is parabolic, therefore sharp fractionation is not possible. The lower size limit for this instrument is about 5 μm and fractionation time can be up to hours. In addition, electrostatic charge can cause problems in transporting particles even when conductive metal separator tubes are used.

A sample of powder can be fractionated or characterized with this system, given sufficient time. Characterization is carried out by placing a powder batch in the U-tube, defining terminal velocity for the smallest particle size of concern, and selecting a separator tube diameter and air flow rate to permit generating that velocity. At the end of a fixed time the material in the filter is removed. Next, terminal velocity for the next largest size is chosen and the air velocity increased with a second filter in place. Following sufficient time for equilibrium, the particles in the size range defined by the difference in the two terminal velocities will be in the filter. This procedure is repeated until sufficient batches have been removed to adequately characterize the particle size distribution by Stokes' diameter. Definition is made by weighing the particles collected in each of the filters.

An air elutriator with a short fractionating column is shown in Figure 8. In describing the system, the developers state that this device is practical only for powders larger than 10 μm Stokes' diameter.[37] For particles in the size range from 2 to 5 μm, a centrifugal classifier was later developed.[38] This device is shown in Figure 9. It consists of an axially symmetrical chamber to which a tangential stream of classifying air is supplied. Material is dispersed within the air stream and accelerated to the speed of the transporting air before it enters the classifying chamber. After entering the chamber the air flow spirals toward the center of the classifying chamber and transports the fine material. Coarse particles are forced against the periphery of the chamber as a result of centrifugal force and rotate in the form of a stable ring of material. Fine material is transported out the through the center of the chamber and collected by a filter.

An instrument popular for characterizing powders and ores is the Bahco classifier.[39] Its operation is shown in Figure 10. Except for the powder feeder, the entire system rotates. Entering air flow is smoothed through a stack of radial disks which are configured so as to

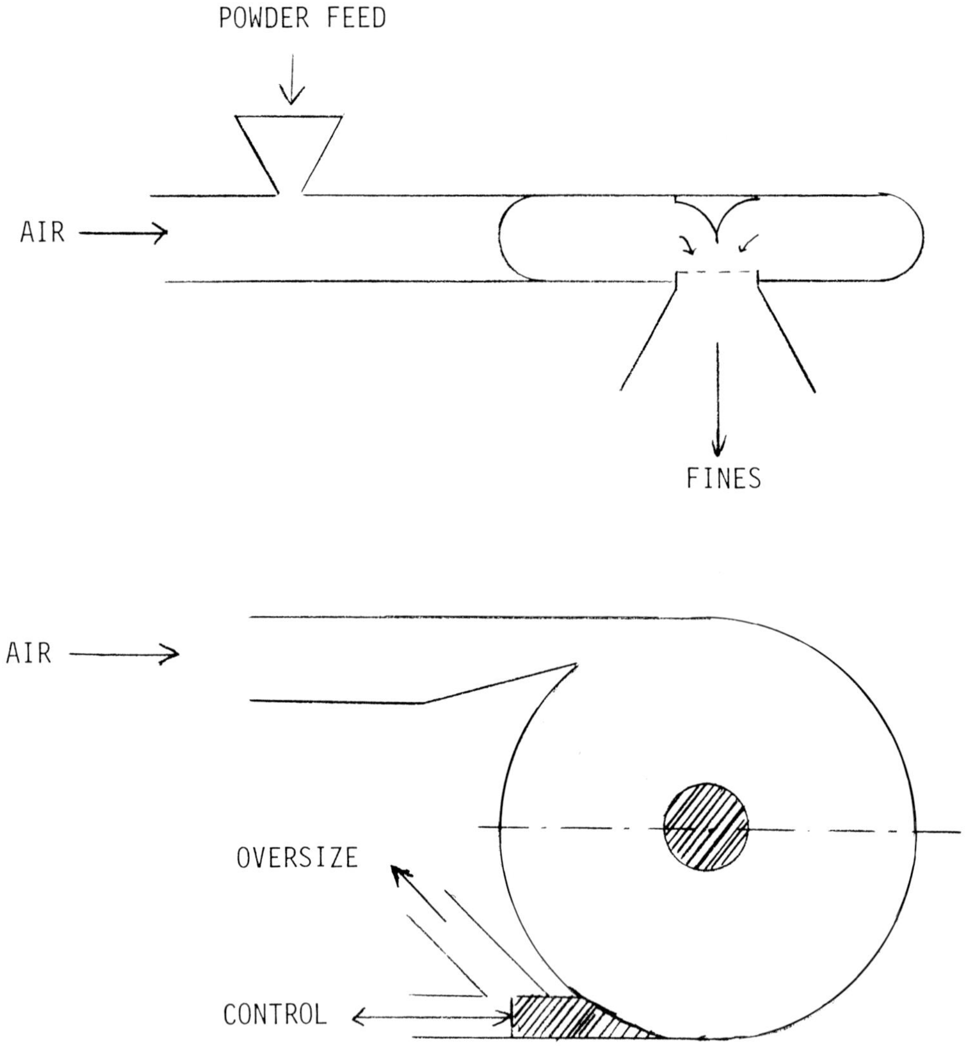


FIGURE 9. Tangential elutriator.

ensure that the air spirals as it moves into the fractionation chamber immediately above the disks. Powder is fed into this chamber and particles are exposed to the opposing centrifugal and air drag forces. The flow conditions at the point where the particles enter the fractionation chamber determine the cut size. Air flow rate decreases outward from the orifice and centrifugal force increases, ensuring that particles already moving outward will continue to do so. Particles which start to move inward are in an accelerating air flow and experience smaller centrifugal force. When the finer particles reach the large angular space at the edge of the top rotor plate the air flow rate drops sufficiently where the centrifugal force is maximum. Thus, finer particles are deposited at the upper periphery as shown in the figure. As can be seen, the powder is split into two fractions by this instrument, with the cut size being determined by air flow conditions. A series of fractions, either for characterization or classification, can be obtained by operating the instrument at a range of flow conditions.[6]

A variety of elutriating systems can be used for characterizing or fractionating powders. They are capable of defining particles as small as 0.1 μm for the X-ray sedigraph system or up to several hundred micrometers for some of the gas elutriating devices. The instruments

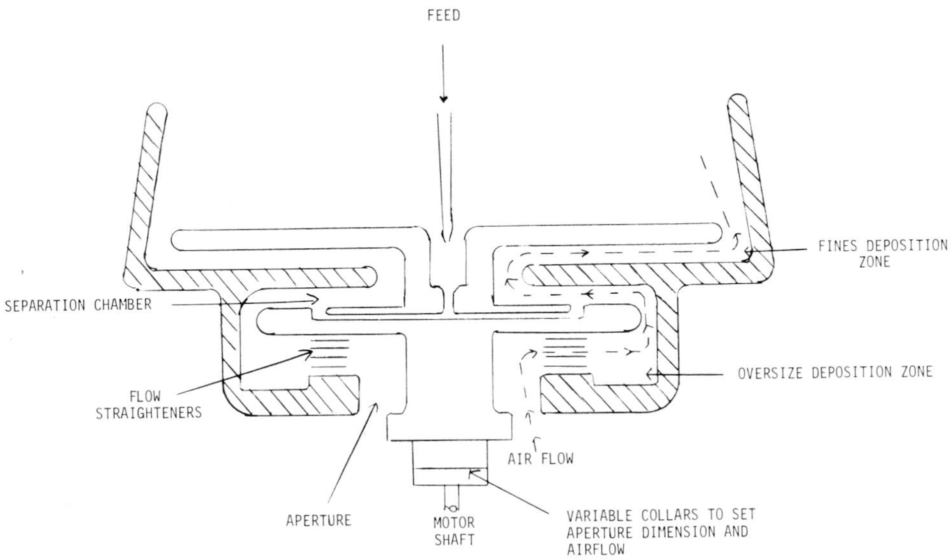


FIGURE 10. BAHCO classifier.

are characterized by dependence on fluid flow uniformity. They operate and produce data in terms of a Stokes' diameter. This discussion has barely touched the description of some of the many elutriating and sedimentation devices commercially available and described in the literature. However, the range of available operating systems has been discussed.

### C. Collecting Devices for Suspension Analysis

Once the powder sample has been satisfactorily removed from the overall mass of powder and has been transported to a point where it can be retained or deposited for analysis or observation, a choice of collection systems is available. As indicated in previous portions of this discussion, the powder may be present in a storage container such as a hopper, or it may be stored in a conical pile on a convenient surface. Alternately, the powder may be flowing in high concentration in a fluid suspension. Means of sample acquisition from any of these systems have been discussed previously. Once the sample has been obtained, however, a number of choices exists for arranging retention of the sample for subsequent characterization. Some examples of these choices will be given following.

If one considers a free-flowing powder moving horizontally, a simple sample collection system is an enclosure where the only force affecting the particles is gravity. In this situation the powder particles settle on a surface that can be removed and handled for subsequent examination. If desired, adhesive coatings can be used to ensure retention of the powder at the point of first contact. Practically speaking these sedimentation collectors are seldom used for powder particles smaller than approximately 10 μm in diameter. This type of sample collector should not be confused with the horizontal elutriator used for sampling respirable particles in industrial atmosphere,[40] since those samplers are used solely for aerosol sampling and are operated at flow conditions that permit separation of particles as small as 1 μm in diameter.

The general subject of sample subdivision is not part of this discussion. However, as stated previously, the analyst concerned with powder characterization must be reasonably assured that his samples are valid. When one wishes to analyze samples from a relatively large powder batch, cone and quartering methods, sample splitters, or spinning riffler devices are used. For further discussion of the effectivity for such devices the reader is referred to work by Kaye.[41]

When powders are suspended in fluid, the particles can be easily removed from the fluid and deposited upon any one of a variety of substrates by using filtration. Filtration devices involve a means of sealing a section of filter medium to a suitable support and drawing fluid at a known flow rate or of a fixed volume through a measured cross-sectional area of that filter medium. The filter medium can be chosen from a wide range of materials with a wide range of flow rate, retention, and surface characteristics. The holder for the filter medium is primarily required to provide a positive seal at the edge of the medium so that the area and flow rate of fluid are definitely known and uniform.

A variety of filter media are available. For example, cellulose fiber papers are used for sampling where composition of the particulate material is to be determined. This medium has the advantage of low ash content and good strength. Glass fiber filters are used where high collection efficiency is desired and ease of extraction for subsequent chemical analysis is of importance. Filter media consisting of porous membranes can be procured with a wide range of pore size distributions. The pore shapes are very tortuous in cross-section through the membrane, thus resulting in high retention efficiency. Membrane materials can be chosen from a wide range of soluble and low-ash content organic materials or can be selected from relatively insoluble materials such as silver or Teflon®. Thus, analysis is made quite easy. In addition, particle collection takes place at or near the surface of the membrane and optical examination is easier. These comments also apply to the Nuclepore® filters. While most membranes are produced by formation of a gel containing pores, the Nuclepore® filters are made by placing thin polycarbonate sheets in a neutron flux which penetrates the polycarbonate plastic. Chemical etching subsequently penetrates and dissolves the neutron-damaged plastic in such a way that uniform pores with diameters from 0.2 μm to approximately 10 μm can be produced. Because of the extremely smooth surface the Nuclepore® filters are often used for collection of particles to be analyzed by visual or electron microscopy. For collection of larger quantities of materials, sintered glass fiber or Alundum® thimbles are often used, particularly in the elutriator fractionator-analyzer devices. Application of electrostatic fields to relatively low-efficiency filters will improve their collection efficiency without significantly affecting pressure drop or flow resistance through the filter.

Most filter media used for collection of materials for microscopic observation or chemical analysis are retained in a holder from 25 to 100 mm in diameter. Fluid flow rates from 0.1 to 50 ℓ/min can be used. Following collection of the powder on the filter surface, the filter can then be placed upon a microscope slide or on the examination stage of an electron microscope for determination of particle size distribution, total particle number, or chemical composition of the powder sample that has been collected. Powder sample composition can be determined by a variety of means that will be described briefly at a later point.

It is also possible to collect powders from a suspension by imposing energy gradients on the powder-fluid suspension. This can be done in a centrifuge or cyclone where centrifugal forces result in deposition of the powder particles on the interior surfaces of the cyclone while the fluid continues to rotate. Cyclone collectors are used for powder particles larger than a few micrometers in diameter suspended or gas or liquid. Centrifuges are capable of collecting submicron particles. Sampling cyclones with tangential entry are normally from 1 cm in diameter to 10 cm in diameter with lengths 5 to 8 times the diameter. Fluid flows range from 1 or 2 ℓ/min for the 1-cm cyclone to several hundred liters per minute for larger cyclones.

The cyclone collector is only one form of the energy gradient devices used for particle separation in both gas and liquid systems. For removal of particulate materials from air suspensions, then, any of several energy gradient devices can be used. The simplest type is the inertial collector. It can be used to collect particles significantly smaller than those affected primarily by gravitational forces. The inertial collector uses the phenomenon that particles moving in a gas streamline tend to continue in the original direction when the streamline is deflected by an obstacle. The particles impact upon the obstacle and can be retained by a suitable adhesive coating or by electrostatic or van der Waals forces. Most inertial collectors in use today accelerate the particle stream by passing it through a nozzle which faces an impaction plate. By careful selection of the nozzle-to-plate distance and nozzle design, it is possible to collect particles equal to or larger than a preselected size with constant efficiency. If one passes the remaining particles through a following nozzle at a higher gas velocity, then smaller particles are collected. In fact, when one defines airborne aerosol particles, a series of nozzle-plate combinations are arranged sequentially so as to fractionate the particles in the series of size ranges. The nozzle-plate combination can be arranged as single nozzles with circular or rectangular cross-sections, or can be arranged as a number of parallel nozzles for larger total sample passage.

An alternate procedure is to use a single nozzle collection plate system which is submerged below the surface of a liquid. Particles which impact upon the collection plate will be washed into the liquid and retained for analysis. Particles smaller than 1 μm in diameter can be collected at near-atmospheric pressure.[42]

Energy gradient devices are used primarily for particle collection from gases. For example, analytical electrostatic precipitators are used for several reasons. The electrostatic precipitator is capable of collecting airborne particles with constant and relatively high efficiency over a very wide particle size range. The sampling and efficiency are not strongly affected by the concentration of particles nor by the amount of material that has already been collected. Material can be deposited in a form convenient for recovery and/or examination by a number of means.

Most analytical electrostatic precipitators are based upon a design developed in 1938.[43] This device is normally operated with a negative corona wherein negative high voltage is applied to a central electrode located within a grounded tube through which the sample is passed. Typically the grounded (conductive) tube is 2.5 to 3.5 cm in diameter and 15 to 25 cm long, while the central corona wire is approximately 0.5 mm in diameter with a sharp point and 2.5 to 5 cm long. Good collection efficiency is obtained for particles ranging from submicrometer up to approximately 100 μm in diameter where concentrations vary from a few micrograms per cubic meter to several hundred milligrams per cubic meter. Sampling flow rates up to 2 $m^3$/min can be used.

When one wishes to examine extremely small particles in dilute aerosol streams, a thermal gradient device can be used. A thermal precipitator removes particles from an aerosol stream by passing it through a channel where a high temperature gradient exists. Temperature gradients ranging upwards from 750°/cm are frequently used. These are obtained by using a relatively small aerosol passage. Thus aerosol sample flow rates are typically less than 10 mℓ/min. A thermal precipitator will quantitatively deposit particles from approximately 0.01 μm in diameter upward.[44] Because of the low precipitation velocity in this device, shattering or breakup of agglomerated particles does not occur during sampling. On the other hand, poor adherence of collected material is often noted, and thermally unstable materials can be adversely affected. For examination of collected materials, an electron microscope grid or a thin glass cover slip can be placed in the thermal gradient on the cold surface for subsequent analysis by electron microscopy or visual microscopy.

Magnetic forces can also be used to separate particles from gas or liquid suspension. If a container is packed with strands of a ferromagnetic material and placed within a strong external magnetic field, high-gradient magnetic forces are produced near the edges of the strands. As the suspension flows through the packing, magnetic and paramagnetic particles are captured with high efficiency. Although the process has been used mainly for fluid clarification, the particles that have been collected on the strands within the container are available for examination and characterization. Mass capture efficiencies of over 99% in gas streams have been reported for particles 3 μm and up.[45]

### D. Measurement Devices for Collected Particles

#### *1. Light Microscopes*

A microscope can be used to provide magnified images of particle deposits on a suitable substrate. A great deal of information on particle size, morphology, shape, and composition can be obtained by such examination. The techniques, procedures, and equipment to carry out such examinations are dealt with in great detail in the journal *Microscope* and in standard optical texts. Some general comments will be given here.

The optical microscope is a convenient and useful instrument for characterizing collected materials. It provides visual data and both subjective and objective observations can be made quickly. Typically, a light microscope can be used up to magnifications of approximately 1000 times the numerical aperture of the objective lens. Thus, the smallest particles that are typically defined are approximately 0.5 μm in diameter. With the light microscope, one can define the refractive index of particulate materials by simple comparison techniques. For checking identity of unknown materials, dispersion staining can be used.[46] Both light and electron microscopes are used for determination of particle size and particle size distribution. Use of an eyepiece reticle that has been calibrated against a stage micrometer is common. Eyepiece reticles can be calibrated in terms of linear dimensions or areas for whichever particle size parameter is of concern. Further details on techniques of measuring particle size in a light microscope are given in the British Standard Method.[47] For particles below 1 μm in diameter, generally a transmission electron microscope or scanning electron microscope is used. Calibration of the electron microscope is typically based upon comparison with monodisperse latex particles in sizes ranging from approximately 0.09 μm up to 5 μm in diameter. For particles smaller than 0.1 μm in diameter, calibration is based on crystallographic determination of lattice parameters.

In general, measurements of microscopic images by manual observation can be carried out with reasonable accuracy for short periods of time. After an hour or so, operator error begins to enter into the measurement. Thus, automated systems involving sophisticated image analysis have been developed. The microscope is still used, but the image itself is analyzed by electronic means. Particles can be counted and measured at very high rates of speed with programing for any diameter, area, or specific shape factor parameter. Thousands of particles can be counted and sized in seconds. A video camera is used to acquire the image through the microscope and computer processing permits rapid measurement of a number of factors. Typically, up to eight basic dimensional measurements can be obtained directly, and a wide variety of algebraically derived additional measurements can be prepared. Results can be graphic, in terms of alphanumeric output, with statistical information. Histograms can be presented, hard copy outputs can be provided, and digital or analog information can be produced.[48]

If a skilled microscopist is carrying out the work, good sizing and counting accuracy can be expected. For routine measurements with manual observation, the accuracy of sizing and counting is frequently degraded. For example, ASTM method F312-69, *Microscopic Sizing and Counting Particles from Aerospace Fluids on Membrane Filters*, indicates procedures for size and quantity measurement determination of particles larger than 5 μm on membrane filters. It is of interest to point out that acceptable repeatability of ±25% is specified in this standard procedure. For this reason, automated image analyzers are frequently used in defining the size and number of particles measured microscopically.

#### *2. Electron Microscopes*

Transmission and scanning electron microscopes are frequently used for observation of particles, and a transmission electron microscope is used where particles are quite small and detail as to their size or projected shape desired. Resolution of particles as small as 10 Å can be obtained in a transmission electron microscope. This can be compared with the maximum resolution capability of a light microscope (approximately 0.5 μm). Note, however, that an electron beam as used in an electron microscope can cause degradation of fragile particles. For this reason an evaporated metal coating is typically used to protect the material, particularly if delicate organic structures are to be observed.

A scanning electron microscope can be used to provide more detailed structural information in particulate material. A scanning electron microscope uses a different approach than that used in transmission microscopy. In the latter, an image is formed when the object interferes with the electron beam so that the electrons which remain unaffected by the object reach a fluorescent screen, forming a shadow of the object. In the scanning electron microscope, the object is scanned by a fine electron beam which sweeps over the surface of the object. Because of the beam interaction with the object surface, a sequential emission of electrons is generated depending upon physical characteristics of the object. The emitted electron beam is collected and, after suitable amplification, produces a voltage that modulates the brightness of a display tube whose screen is scanned synchronously with the object scan. Thus, a sequential image is obtained. For sizing, the scanning electron microscope uses a beam of some 50 Å, permitting image resolutions of approximate 100 Å. Further, the scanning electron microscope offers possibilities of analysis since electron emission is induced by the probe beam. Four collection modes may be used for information gathering. The first is secondary electron imaging, used to determine size, shape, and surface characteristics; the second is back-scattered electron imaging, used to permit composition differentiation between organic and inorganic materials; the third is X-ray fluorescence, used to determine elemental composition; and the fourth is electron diffraction, used to define crystalline structures.[49]

### E. Composition Analysis for Collected Samples

When information on chemical composition of particles is required, it is normally necessary that the particles be immobilized before analytical techniques are carried out. There are essentially no instruments that can provide useful information on chemical composition of single fine particles *in situ*. One should consider that a typical particle of 1 to 10 μm in diameter has a mass of approximately $10^{-10}$ to $10^{-12}$ g and is very difficult to retain in place for a period of time long enough for most analytical beam interpretation devices to operate satisfactorily. For this reason, particles are deposited on suitable substrate for concentration, manipulation, handling, or insertion into a suitable analytical system.

A number of analytical procedures can be used for definition of particle composition. These range from classical wet chemical analytical procedures through spectroscopic and electrochemical methods. It is beyond the scope of this discussion to describe all of the methods in detail. Some of the more useful and appropriate techniques will be discussed briefly.

Sampling and concentration for composition analysis is carried out primarily with filters. Once the material has been collected on the filter surface, the filter can be transported or stored and the material remains stable. In addition, the filter surface can be subdivided to permit using a number of measurement techniques — some of which may be destructive — on a single sample. Some care, therefore, is required in selecting the filter medium for the particular analytical procedure and material to be analyzed. When a particular compound or element is to be analyzed, the filter medium should be selected with a low background level of that material; if the analytical procedure is one that observes primarily surface structure, the filter medium should be chosen so that particulate material is collected on the surface of the filter rather than within the interstices of that filter medium. For analysis of reactive compounds, it is necessary to ensure that the filter medium does not sorb or retain reactive gases.

In general, water-soluble anions are analyzed by conventional wet chemical, spectroscopic, and electrochemical methods while water-soluble cations are analyzed by those methods along with X-ray fluorescence and activation techniques. Mineral components are typically analyzed by emission spectroscopy and X-ray fluorescence, while metals are analyzed by a number of spectroscopic and fluorescence methods. Organic compounds are typically analyzed by gas chromatography, mass spectrometry, and infrared or ultraviolet analysis. Note that these general technique comments are not hard and fast. A number of variations and preferable exceptions always exist.

Frequently, in carrying out chemical analysis, it is desirable that the material collected on the filter surface be concentrated even further. In this case, a fibrous medium filter is used, and the material is concentrated along with some elemental separation carried out by using the ring oven technique.[50] In this device, a solvent is added to the sample on a filter paper and the paper is placed in a ring oven. The solution migrates under the action of capillarity, and heat causes the solvent to evaporate. The material is then analyzed by standard spot test techniques. Appel et al.[51] used a technique of solvent extraction in specific materials, followed by carbon determination to identify elemental carbon primary and secondary organic materials.

A series of tests that can be applied to aerosol particles of masses down to $10^{-16}$ g are described by Bigg et al.[52] Precoated electron microscope screens are used. The screens are coated with a thin film of reagent and reaction is caused to proceed by exposure to the vapor of a suitable solvent. Reaction products are then examined in the electron microscope and compared with similar particles captured on a nonreactive surface. For detection of individual nitrate particles as small as 0.5 μm, an electron microscope measurement is made of the reaction spot produced by reaction of nitrate ion with nitron.[53]

A method using gas chromatography is suggested by De Maio and Corn.[54] In this device, a single benzene extraction of particles collected on filter paper is followed by use of a conventional packed column. Dulka and Risby discuss general sampling techniques particularly applicable to collection of material for micro-trace analysis.[55] They describe current methods of microanalysis, including detection limits and areas of applicability. Similar procedures specific to particles that may be present in normally clean liquids such as found in parenteral solution is described by De Luca and Boddaputi.[56]

Application of an automatic technique for determining carbon, hydrogen, and nitrogen simultaneously in size-fractionated ambient aerosol is described by Lee and Hein.[57] In this technique, a sample is burned in pure oxygen at high temperature, converting the elements to carbon dioxide, water, and nitrogen.

In the hands of a skilled operator, morphological analysis, preferably with the aid of an optical or electron microscope, is a valuable and useful tool. McCrone[58] discussed the procedures for morphological analysis of particulate samples. A number of procedures and laboratory systems are outlined. McCrone discusses techniques for morphological analysis based on refractive index. This identification technique is based on recognizing the difference in dispersion of the refractive index of particulate solids and that of the liquid medium in which the solid is imbedded. Particle boundary color changes are used to identify the specific materials.

X-ray fluorescence analysis of aerosols is described by Adams and Van Grieken.[59] The use of the method is described along with a procedure for correcting absorption by the filter medium. Birks[60] describes the application of X-ray spectroscopy to analysis of airborne pollutants and indicates cost effectivity of the technique. A novel multielectrode flame ionization detector is described by Altpeter.[61] In this device, the individual particle appears to provide its own event marker as it is consumed in the flame. The marker pulse is followed by another current pulse which may be due to the actual charge carriers delayed by their time of flight from the flame to the electrode. A discussion on flameless atomic absorption spectrometry for metal analysis is given in Begnoche and Risby.[62] A complete description of the method for sampling particles from ambient air in volumes ranging between 20 and 300 ℓ, and the analytical procedures, are given with reference to a number of metals. An example of the use of neutron activation for analysis of specific radionuclides is given by Benson and Gleit.[63] The use of infrared spectroscopic analysis is described by Cunningham et al.[64] Cunningham shows clearly the variation in the chemistry of airborne particles as a function of particle size and time by use of the infrared technique.

A number of mass spectrometric devices are available. One is described by Davis.[65] In this device the particles impinge on a heated ribbon, and the resulting burst of ions produced by each particle is analyzed in a 3-in.-radius magnetic sector mass spectrometer. Particles containing 1000 atoms of a particular element in particle concentrations of $<1/cm^3$ can be detected. The method, in combination with pyrolysis, can be used to characterize organic particles, as well.[66]

Proton-induced X-ray emission analysis (PIXE) is useful in measuring changes with particle size distribution of a number of compounds.[67] This technique has found very wide use in measurement of general composition of collected aerosols. Convenience and ease of measurement for the technique are outstanding.[68] Gamma-ray emission induced by inelastic scattering of protons has been described by Macias et al.[69] for simultaneous determination of carbon, nitrogen, and sulfur in collected samples. Improved mass spectrometric techniques are described by Cronn et al.[70] By use of computer-controlled, high-resolution mass spectrometric thermal analysis, sample handling was minimized since portions of the fiber filters could be inserted directly into the spectrometer and molecular information was obtained simultaneously for organic and some inorganic pollutants in submicrogram quantities.

Ion and electron microprobes are useful for analyzing individual particles or areas on specific particles as small as 1 μm in cross section. Application and capabilities of ion and electron microprobe instruments for characterizing particulate matter are described by Gavrilovic and Majewski.[71] Resolution limits, sample handling requirements, and some typical data are given. Real-time composition and size data have been obtained for aerosolized particles.[72] A beam of particles is injected into a vacuum chamber and flash-vaporized on a hot filament. The vapor pulse is ionized by electron impact and ions analyzed by a quadrupole mass spectrometer.

Consideration has been given to the use of Raman spectroscopy for particulate analysis, as described by Rosen and Novakov.[72] Although this technique has not been widely exploited, laser-induced mass spectroscopy has become widely used.[74] High sensitivity, particularly for elements with low mass numbers, is easily achieved along with spatial resolution to 1 μm. In some cases, surface analysis of collected particulate material is of particular interest, as indicated by Keyser et al.[75] They point out a number of spectrometric techniques that are particularly applicable for surface analysis. These include electron microprobe analysis, X-ray emission spectrometry, electron spectroscopy for chemical analysis, Auger electron spectrometry, and secondary ion mass spectrometry (SIMS). These techniques are often useful for analysis of individual particles ranging in size down to 1 μm in diameter. In some of the techniques, ion bombardment to remove surface material permits analysis of the composition of several layers of the particulate material.

In general, the summary of analytical methods given is not a complete discussion of the methodology, nor does it discuss in detail exactly which elements and compounds can be analyzed by which methods. This discussion is designed primarily to give a rapid overview of techniques that have been used in analysis of particles and to indicate sources where additional information can be procured. At the same time, the listing of methods is not complete. Even as this discussion was written, new methods were being published.

## F. Flow and In-Place Methods

### *1. Acoustic Response*

If liquid is contained within a cylindrical vessel, a pulsing ultrasonic transducer can be mounted in the wall of the vessel so that the inner wall of the tube serves as a plano-concave lens to focus the ultrasonic beam within the liquid.[76] Each burst of ultrasonic signals is transmitted into the fluid in a conical beam and reflected back to the transducer by the walls of the vessel and by particles in the fluid. Reflected signals are received by the ultrasonic transducer and can be converted into information that defines the number and size of particles within the liquid. By analyzing only those pulses which return within a given time period, then an electronic window can be defined to specify the bounds of the sensing zone. Reflected pulses which originate within this window are counted and sized as particles only if the resulting response is above an adjustable threshold voltage. The threshold voltage is then used to define the size of the particles present.

The smallest particle that can be seen by this device is reported to be approximately 5 μm.[77] Concentrations up to several hundred particles per milliliter can be defined with this device since the transducer can be mounted in the wall of a tube or pipe with relatively high liquid flow rates. The technique is quite convenient. Sample sizes and volumes are defined by the dimensions of the window used in the electronic system. The procedure has the advantage that it is nonintrusive except for the necessity of mounting a component that will contain the ultrasonic transducer in the flow system. In addition, it is necessary to ensure that artifacts due to air bubbles or other debris are not present in the sample being analyzed.

*2. Electrical Resistance Methods*

Particles must be suspended in a conductive liquid for measurement by these methods. For the most part, a 1% salt solution in clean water is used. However, additives can be provided for most organic liquids to permit conductivity to be achieved. Where this is necessary, the conductive liquid is passed through a small aperture located in an insulating barrier. Typical aperture diameters range from 30 μm to several hundred micrometers in diameter. The resistivity of the liquid is measured continuously while it is flowing through the aperture. If a particle is present within the aperture zone, then the measured resistivity will change dependent upon particle volume. By determining the number and amplitude of the electrical pulses produced when particles pass through the aperture then the particle number and size distribution can be determined.[78] With this device, sophisticated electronics can be used to investigate the shape and duration of a series of pulses. In this way, the technologist can obtain information on the shape and size characteristics of each particle passing through the aperture in use. Particles can be sized with reasonable accuracy where the particle diameters range from approximately 3 to 50% of the orifice diameter. Auxiliary techniques to reduce the smallest size that can be measured involve such procedures as hydrodynamic focusing and flow straightening, so as to cause particles to pass only through the central portion of the aperture. In this way, resolution at the smallest particle size is improved to permit better sizing accuracy.

One may question the validity of the concept if metallic or conducting particles are being measured. In practice, it is found that essentially all particles have sufficient oxide coatings or chemisorbed layers of miscellaneous nonconducting materials so that essentially all particles react as do insulating materials. Precautions are required to minimize the effect of obscuration — either total or partial — of the aperture. If partial blockage occurs, changes in the Reynolds number of the flow through the orifice may initiate turbulent flow eddies that result in pulse distortion and erroneous size information produced.

Since this device is a single-particle counter, it is subject to the same coincidence limitations as any other single-particle counting system. If particles are present in excessive concentration, more than one particle will be counted as a single particle with resultant concentration errors. Normally, concentration capabilities range up to several tens of thousands of particles per milliliter of liquid for apertures smaller than 50 μm or so. It is reported that on-line capability is present, however, this requires operation of two or more apertures

$$I_V = \frac{16\pi^2 a^6}{r^2\lambda^4}\left(\frac{m^2 - 1}{m^2 + 2}\right)\sin^2\phi$$

$$I_H = \frac{16\pi^2 a^6}{r^2\lambda^4}\left(\frac{m^2 - 1}{m^2 + 2}\right)\cos^2\theta\cos^2\phi \qquad (4)$$

a = particle radius
m = refractive index
r = observation distance
λ = wavelength

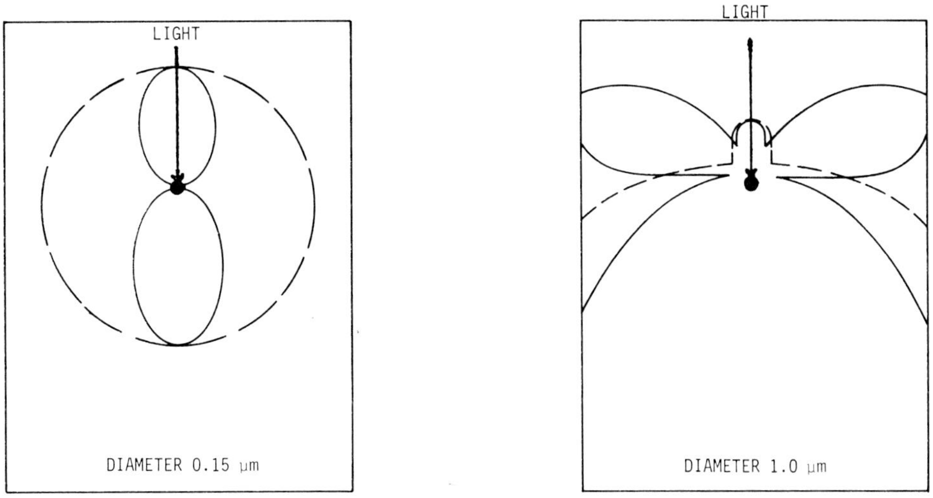


FIGURE 11. Rayleigh scattering (a<≪λ).

with back-flushing plumbing so that if one aperture becomes plugged or blocked by debris, flow can be transferred to the second aperture while the first is back flushed.

The data produced by a resistive pulse analyzer are in terms of number and volumetric size of particles per unit volume of liquid. By adding fairly simple data processing devices, the particle size can be shifted from a volume base to a diameter or area base quite easily. Particle size distribution can then be defined in almost any desired statistical terminology. Computer interface systems are common for these instruments so as to carry out a wide range of data manipulation.

*3. Photometric and Turbidometric Devices*

Operating principles of these devices are quite simple: a light beam is passed through the sample and the total amount of light scattered out of the sample at specific angles, or absorbed by opaque particles within the sample, is measured and related to the quantity of particulate material within the liquid. Empirical calibration to a turbidity unit scale is normally used to define the quantity of suspended solids, although changes in scattered light are dependent on size, shape, and concentration of the suspended particles rather than on any individual factor.

Photometric analyzers are used primarily on-line to indicate the turbidity or concentration levels in process liquids or in water lines. Back-scattering systems are used for measurement of concentrations up to 10 to 15% solids. Angular scattering systems can be used for measurements of solids in concentrations to 50,000 ppm. Absorption analyzers can be used for concentrations to approximately 1000 ppm.

For optical detection, most particles of interest for characterization have dimensions that are within two orders of magnitude of the wavelength of visible light. When a particle or particles is illuminated by a beam of such light, the amount of light that is scattered by the particle varies roughly with the projected area of the particle or particles larger than the light wavelength in dimension. As particle size decreases then the amount of scattered light varies with the sixth power of the scattered light.

$$I_v = \frac{\lambda^2}{4\pi a^2}\left|\sum_{m=1}^{\infty}\frac{2m+1}{m(m+1)}\left\{a_n\frac{P_n(\cos\theta)}{\sin\theta} + b_n\frac{d}{d\theta}P_n(\cos\theta)\right\}\sin\phi\right|^2$$

$$I_H = \frac{\lambda^2}{4\pi a^2}\left|\sum_{m=1}^{\infty}\frac{2m+1}{m(m+1)}\left\{a_n\frac{d}{d\theta}P_n(\cos\theta) + b_n\frac{P_n(\cos\theta)}{\sin\theta}\right\}\cos\phi\right|^2 \quad (5)$$

Where $a_n$ & $b_n$ are the amplitude scattering functions.

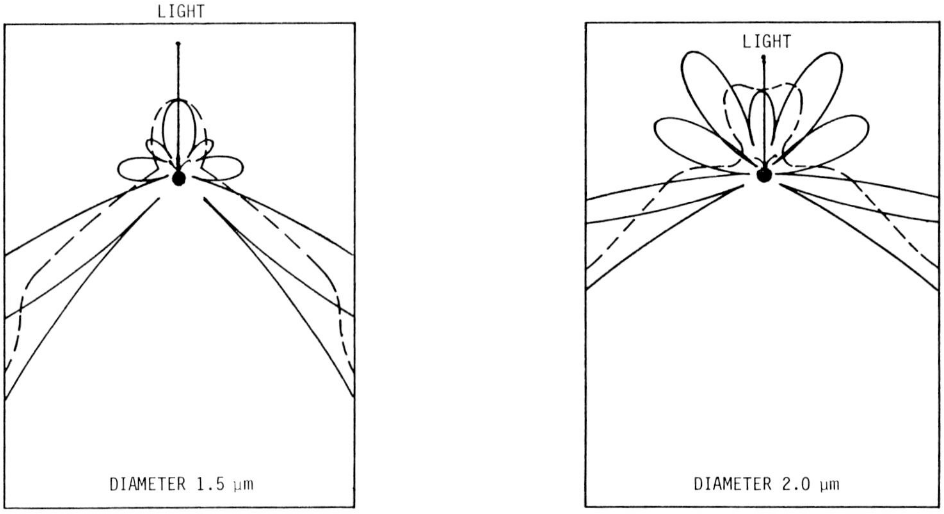


FIGURE 12. MIE scattering (all particle sizes).

Equation 4, the Rayleigh scattering equation for small particles, is shown in Figure 11 for $I_v$, vertical polarization, and $I_h$, horizontal polarization. Figure 11a shows the vertically and horizontally polarized scattering patterns for particles that scatter in the Rayleigh regime ($a \ll \lambda$), while Figure 11b shows the scattering patterns for particles in the transition size range where Mie scattering is becoming important. Equation 5, the Mie scattering equation that describes light scattered from larger particles is shown in Figure 12 for vertical and horizontal polarizations. Figure 12 shows how the complexity of the scattering pattern increases as particles become larger. For much larger particles, geometrical scattering occurs. Equations 6, 7, and 8 shown in Figure 13, describe diffraction, reflection, and refraction effects, respectively. Figure 13 shows the beam paths through and around a particle that can be expected as a result of each of these effects.

Particles may be passed through a brightly illuminated volume sufficiently small so that only single particles are present within that volume at any time. It is possible to collect light scattered by each particle over a specific range of angles. The amount of light and the angular distribution of that light, as seen by the above equations, depends upon particle size, shape, and refractive index of the particle. The total light flux in each pulse can be measured, and particle size distribution and concentration data obtained, once the flow rate of the liquid stream in which the particles are suspended is known. Alternately, one may pass a liquid sample through an illuminated sensing volume and observe the light removed as the absorbing and scattering particles traverse the sensing volume. The light is removed by a combination of the absorption and scattering out of the direct beam. The amount of light remaining can be correlated to the size of a particle in terms of its projected area and optical characteristics.

Commercially available instruments will handle particles over a dynamic range of approximately 50:1, from 1 μm in diameter upwards, in any single measurement when ob-

1. DIFFRACTION

$$I_D(\theta_1\ a) = \frac{\alpha^2}{4\pi}\left(\frac{\lambda J_1\ (\alpha \sin \theta)}{\alpha \sin \theta}\right) \tag{6}$$

2. REFLECTION

$$I(\theta) = \frac{1}{8\pi}\left\{\frac{\sin \theta/2 - (m^2 - 1 + \sin^2 \theta/2)^{1/2}}{\sin \theta/2 + (m^2 - 1 + \sin^2 \theta/2)^{1/2}}\right\} \tag{7}$$

3. REFRACTION

$$I(\theta) = \frac{2}{\pi}\left(\frac{m}{m^2 - 1}\right)\left|\frac{\{m \cos (\theta/2) - 1\}^3\ (m - \cos \theta/2)^3\ (1 + \sec^4 \theta/2)}{\cos \theta/2\ \{m^2 + 1 - 2m \cos \theta/2\}^2}\right| \tag{8}$$

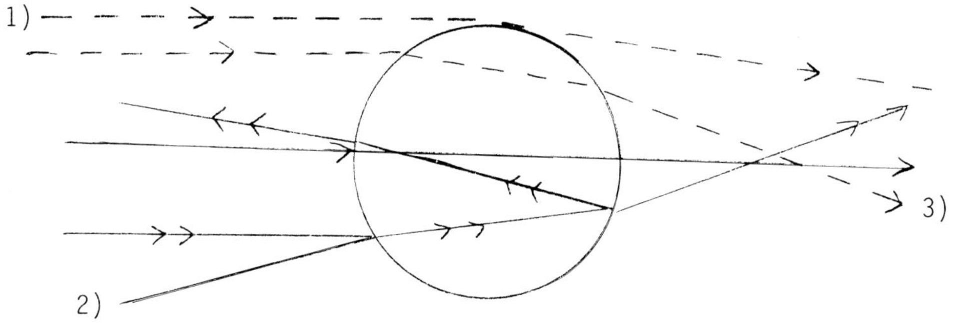


FIGURE 13. Geometrical scattering.

scuration techniques are used. For light-scattering measurements, single-particle counting devices are capable of sizing and counting particles over the same dynamic range, with a lower sensitivity limit of approximately 0.5 μm.

In addition to the theoretical response of optical instruments, one must also consider the effect of instrument construction limitations. These are primarily due to buildup variations in manufacturing tolerances that range from 0.1 to 5% in mechanical dimensions, lens focal lengths, electronic component drift, etc. As one example of the effect of tolerance variations, consider uniformity response when the viewing volume is defined by an incandescent filament. The viewing volume dimensions are often defined by an aperture illuminated by a condensing lens. The aperture is imaged in the sample stream by a relay lens; light scattered or absorbed by particles is collected and measured by a photodetector. If a laser is used, then the Gaussian distribution of light across the laser beam requires limitation of the beam so that variations of no more than 10% exist. In either case, light intensity across the sample stream will vary by up to 10% in one dimension and approximately 5% when a relay lens, even with good depth of field, is used. In addition, consider that the liquid in which the particles are suspended will be in laminar flow at the point of observation. The resulting parabolic velocity results in particle dwell times that may vary by a factor of 2, since the maximum stream velocity at the central line of flow is twice the average velocity. This dwell time variation results in variation in pulse rise time and duration, requiring wider electronic system bandwidths which will then increase the electronic noise level. The result of the several variations in light level, flow, electronic component variation, and so forth can therefore result in a distribution of pulse heights resulting from particles of identical size. For a very good commercial instrument, these variations may range up to 10% of the mean depending upon instrument design parameters. The net result is that pulse height distribution may have an S.D. of 5% of the median pulse height, even from monodisperse particles. The result of such variations is that instrument resolution is degraded. Both sizing and counting accuracy will then be affected.

Any time that single-particle counters are used, the normal concentration of particles within the liquid should not exceed the maximum capabilities of the device. Since the particles

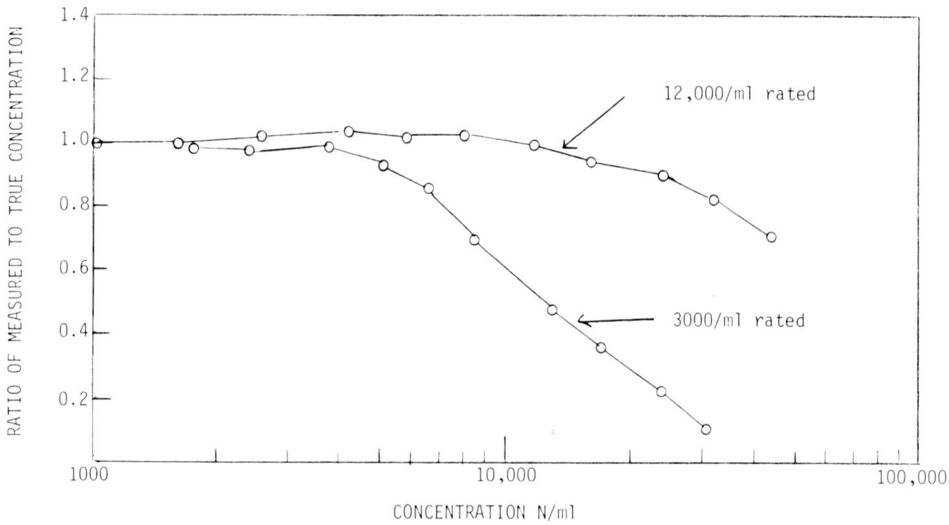


FIGURE 14. Measured coincidence losses.

are randomly distributed within a sample of fluid there is always a finite probability that more than one particle may be present within the sensing volume at any time. Since the particle counters are not imaging devices, but measure light level when a particle is present, then the instrument cannot differentiate between the light flux produced by several particles in the sensing volume or by a single particle with somewhat larger size. Thus, the effect of coincident particles within the sensing volume is an exponentially increasing error in the indicated concentration as the actual concentration increases. For this reason the indicated concentration will always be less than the actual concentration. The errors can be calculated based upon an assumed Poisson distribution. The errors were described by Jaenicke.[16] Figure 14 shows the effect of coincidence error on reported concentration data. Note that eventually the single-particle counter will report zero concentration when the actual concentration is so high as to result in at least one particle larger than threshold being present within the sensing volume continuously.

## VI. SUMMARY COMMENTS

### A. Standard Methods and Materials

In the preceding discussion, methods and devices for sampling and characterizing particles have been discussed. Although not stated explicitly, analysts have long noted that problems exist in correlating measurements made with different types of devices and even interlaboratory measurements made with the same type of device on referee materials can differ. For this reason, work is in progress to develop standard operating methods and standard reference materials for use in calibrating the analytical instruments. Most of the standards are under the jurisdiction of voluntary standards agencies. These include such groups as the ASTM, the British Standards Institution, the Association of German Engineers (VDI), the Japanese Standards Association, and the International Standards Organization. In addition, many government-controlled agencies and trade organizations provide both procedures and standard reference materials.

A variety of standard monodisperse particles are available for calibration purposes, mainly for particle size specification. Care is required in using these materials, since satisfactory dispersion is frequently difficult to accomplish. Controlled, reproducible concentrations are almost impossible to achieve after any significant storage unless very good procedure for redispersion, mixing, and sampling are used.

It is emphasized that in the final analysis, accuracy of sampling, sizing, or measuring any aerosol parameter depends upon the experience and skill of the operator in adjusting the instrument, carrying out the procedure being used, and interpreting the data developed by the system.

### B. Data Processing

In characterizing samples, particularly for concentration and/or size data, it must be remembered that a sample is procured from a much larger population. If the sample is too small, then variability from point to point will result in wide differences between measurements. Thus, if very small samples are obtained, the variance may be unacceptably large. On the other hand, if the sample is too large or too concentrated, instrumental operation limits can introduce errors.

If one wishes to convert data from one instrument type to another it is necessary to convert, for example, particle number data from one instrument to particle volume data (with specific gravity correction) for another instrument, or vice versa. In addition, the effect of the breadth of the particle distribution, as well as the more mundane problem of good operating and sampling protocol, must be included in the conversion process.

Once the particular problems have been defined and the objectives for sampling stated clearly, a protocol must be selected in accordance with desired data presentation. Then, one should consider the objectives of data collection and select optimum procedures. Obviously, for characterization of a specific sample, one procures the largest possible fraction of that sample and carries out the necessary analyses on the material. On the other hand, to characterize the material in a larger parcel, one should select sample sizes and sample locations that are representative of the overall system, and determine a sampling schedule that will represent the typical operation within the area or the environmental factors that may affect the material. On the other hand, sometimes only abnormal conditions may be of interest. In this case, the question of records, their disposition, and/or storage arises. One may consider the possibility of storing data for a fixed period of time and continuously cleaning out and discarding data that were stored for a given time period. Alternately, one may consider the possibility of acquiring and storing data only when an out-of-specification condition occurs and ignoring the acquired particle characterization data the rest of the time. Even with the present computerized, high-density data storage systems, the quantity of data that can be acquired and recorded from particle measuring systems can be enormous. It is possible to encode and store large quantities of data that describe routine conditions in great detail but do not justify more than a simple description of "satisfactory". That type of information should be recorded and stored only if there is a requirement that the details of such data be maintained.

A variety of instruments, devices, and procedures have been described for the sampling and analysis of particulate material. A question always asked when one begins such a program is the selection of the optimum system for the particular problem. Unfortunately, there is no simple answer. One must first define the problem in terms of the objectives for sampling and analysis, the accuracy of data required, the averaging time and sampling frequency (for transient and unstable systems), the type of data desired and the volume of data that can be handled, the allowable response time for data production, the scale of effort that can be applied, the skill level of available operators, funding available for purchase/lease of equipment, funding available for consumables, service and maintenance requirements, choice of laboratory vs. field instrument types, portability of samples and/or of instruments, environmental constraints, and system capabilities. In addition, it may be necessary to include some intangible factors, such as the possible necessity for correlation with historical data for coordination with participating organizations in a group study or because enforced inheritance defines available equipment. In selecting a system for sampling and analysis, one examines the details of the problem and chooses an instrument system, a sampling and analysis protocol, and data processing procedures in accordance with those details. The device performance characteristics discussed in this chapter should be of assistance in making these choices.

## REFERENCES


1. **Snow, R. H. and Luckie, P. T.,** Annual review of size reduction — 1975, *Powder Technol.*, 23, 31, 1979.
2. **Austin, L. G.,** A review; introduction to the mathematical description of grinding as a rate process, *Powder Technol.*, 5, 4, 1971/1972.
3. **Holt, C. B.,** The shape of particles produced by comminution, a review, *Powder Technol.*, 28, 59, 1981.
4. **Meloy, T. P. and Gumtz, G. O.,** The fracture of single, brittle heterogeneous particles — statistical derivation of the mass distribution equation, *Powder Technol.*, 2, 107, 1968/1969.
5. **Meloy, T. P.,** A hypothesis for morphological characterization of particle shape and physiochemical properties, *Powder Technol.*, 16, 233, 1977.
6. **Heywood, H.,** Symposium on particle size analysis, *Trans. Inst. Chem. Eng.*, 22, 214, 1947.
7. **Kaye, B. H.,** Specification of the ruggedness and/or texture of a fine particle by its fractal dimension, *Powder Technol.*, 21, 1, 1978.
8. **Green, H.,** The effect of non-uniformity and particle shape on average particle size, *J. Franklin Inst.*, 204, 713, 1927.
9. **Rosin, P. and Rammler, E.,** The laws governing the fineness of powdered coal, *J. Inst. Fuel*, 7, 29, 1933.
10. **Nukiyama, S. and Tanasawa, Y.,** Experiments on the atomization of liquids in an air stream, *Trans. Soc. Mech. Eng. (Jpn.)*, 61, 1939.
11. **Roller, P. S.,** Law of size distribution and statistical description of particulate materials, *J. Franklin Inst.*, 223, 609, 1937.
12. **Friedlander, S. K.,** On the particle-size spectrum of atmospheric aerosol, *J. Meteorol.*, 17, 373, 1960.
13. **Jordison, F.,** Tips on automatic sampling, *Chem Eng. (Jpn.)*, 5, 62, 1939.
14. **Kaye, B. H.,** Investigation into the relative efficiency of various sampling procedures, *Powder Met.*, 9, 213, 1962.
15. **Herdan, G.,** *Small Particle Statistics*, Butterworths, London, 1960.
16. **Jaenicke, R.,** The optical particle counter: cross sensitivity and coincidence, *J. Aerosol Sci.*, 5, 95, 1972.
17. **Cleaver, J. W. and Yates, B.,** Mechanism of detachment of colloidal particles from a flat substrate in a turbulent flow, *J. Colloid Int. Sci.*, 44, 464, 1973.
18. **Fletcher, B.,** The incipient motion of granular materials, *J. Phys. D:*, 9, 2471, 1976.
19. **Michels, D. E.,** Sample size effect on geometric average concentrations for log normally distributed contaminants, *Environ. Sci. Tech.*, 11, 2, 1977.
20. National Fluid Power Association, Method for extracting fluid samples from the lines of an operating hydraulic fluid power system (for particulate contamination analysis), *Am. Natl. Stand. Inst.* B93.19, 1972.
21. **Lieberman, A.,** Automatic particle measurement system and liquid-handling techniques for clean liquids, *J. Test. Eval.*, 3, 398, 1975.
22. National Fluid Power Association, Procedure for qualifying and controlling cleaning methods for hydraulic fluid power sample containers, *Am. Natl. Stand. Inst.* B93.20, 1972.
23. **Kratochvil, B. and Taylor, J. K.,** Sampling for chemical analysis, *Anal. Chem.*, 53, 924A, 1981.
24. **Lieberman, A.,** Flow rate and concentration effects in automatic particle counters, Proc. Natl. Conf. Fluid Power, Chicago, October 1975.
25. **MacInnes, W. F.,** *Summary Report on Development of Sampling Techniques*, Rep. 1049, AFFTC-TR-60-44, Fluor Corporation, Irvine, Calif., 1960.
26. **Elton, G. A. H.,** Electroviscosity in dilute heterodisperse suspensions, *Br. J. Appl. Phys.*, 3 (Suppl), 60, 1964.
27. **Leschonski, K.,** Sieve analysis, the Cinderella of particle size analysis methods, *Powder Technol.*, 24, 115, 1979.

28. **Whitby, K. T.,** The Mechanics of Fine Sieving, Ph.D. thesis, University of Minnesota, Minneapolis, 1945.
29. **Lauer, O.,** Instruments for particle size determination by screening, Proc. Powder Bulk Solids Conf., Chicago, May 1978.
30. **Orr, C., Davis, D. K., and Camp, R. W.,** Particle size analysis by automatic siever, *Powder Technol.*, 24, 143, 1979.
31. **Kelsall, D. F. and McAdam, J. C. H.,** Design and operating characteristics of a hydraulic cyclone elutriator, *Trans. Inst. Chem. Eng.*, 41, 84, 1963.
32. **Haultain, H. E. T.,** *Trans. Can. Inst. Mining Met.*, 40, 229, 1937.
33. **Whitby, K. T.,** A rapid, general purpose, centrifuge sedimentation method for measurement of size distribution of small particles, *Heat Piping Air Cond.*, Part 1, 231, January 1955.
34. **Leschonski, K. et al.,** A new on-line particle size analyzer, Proc. 3rd Salford Conf. Particle Size Analysis, Salford, U.K., September 1977.
35. **Olivier, J. P., Hickin, G. K., and Orr, C.,** Rapid automatic particle size analysis in the sub-sieve range, *Powder Technol.*, 4, 257, 1970/1971.
36. **Roller, P. S.,** Separation and Size Distribution of Microscopic Particles: An Air Analyzer for Fine Powders, U.S. Bureau of Mines, Tech. Paper 490, U.S. Government Printing Office, Washington, D.C., 1931.
37. **Leschonski, K. and Rumpf, H.,** Principle and construction of two new air classifiers for particle size analysis, *Powder Technol.*, 2, 175, 1969.
38. **Rumpf, H. and Leschonski, K.,** Principles and new methods of air separation, *Chem. Ing. Tech.*, 39, 1231, 1967.
39. **Graham, A. L. and Hannot, T. H.,** The micro-particle classifier, *Ceram. Age.*, September 1962.
40. **Wright, B. M.,** A size-selecting sampler for airborne dust, *Br. J. Ind. Med.*, 11, 284, 1954.
41. **Kaye, B. H.,** An investigation into the relative efficiency of different sampling devices, *Powder Met.*, 9, 213, 1962.
42. **Hering, S. V. and Friedlander, S. K.,** Design and evaluation of a new low-pressure impactor. II, *Environ. Sci. Tech.*, 13, 184, 1979.
43. **Barnes, E. C. and Penney, G. W.,** An electrostatic dust weight sampler, *J. Ind. Hyg. Toxicol.*, 20, 259, 1938.
44. **Brock, J. R.,** Experimental work and theory for the thermal force in the transition region, *J. Coll. Int. Sci.*, 25, 392, 1968.
45. **Drehmel, D. C. and Gooding, C. H.,** High gradient magnetic particulate collection, Proc. AIChE, 82nd Natl. Meet., Atlantic City, August 29 to September 1, 1976.
46. **Brown, E. M., McCrone, W. C., Kuhn, R. A., and Forlini, A. L.,** Dispersion staining, *Microscope*, 13, 311, 1963.
47. B.S.I., *Methods for the Determination of Particle Size of Powders*, No. 3406, part 4, British Standards Institute, London, 1963.
48. **Morton, R. R. A.,** Techniques for Particle Measuring using Image Analyzers, NBS Spec. Pub. 431, National Bureau of Standards, Washington, D.C., 1975.
49. **Denee, P. B.,** Detection of respirable dust by SEM back-scattered electron imaging in *Electron Microscopy and X-ray Applications to Environmental & Occupational Health Analysis*, Russell, P. A. and Hutchings, A. E., Eds., Ann Arbor Science Publishers, Ann Arbor, Mich., 1978, 237.
50. **West, P. W.,** The identification and determination of airborne particulates by means of the ring oven technique, *Air Pollut. Control Assoc. J.*, 16, 601, 1966.
51. **Appel, B. R., Colodny, P., and Wesoloski, J. J.,** Analysis of carbonaceous materials in southern California atmospheric aerosols, *Environ. Sci. Tech.*, 10, 359, 1976.
52. **Bigg, E. K., Ono, A., and Williams, J. A.,** Chemical tests for individual submicron aerosol particles, *Atmos. Environ.*, 8, 1, 1974.
53. **Mamane, Y. and Pueschel, R. F.,** A method for the detection of individual nitrate particles, *Atmos. Environ.*, 14, 269, 1980.
54. **De Maio, L. and Corn, M.,** Gas chromatographic analysis of polynuclear aromatic hydrocarbons with packed columns, *Anal. Chem.*, 38, 131, 1966.
55. **Dulka, J. J. and Risby, T. H.,** Ultra-trace metals in some environmental and biological systems, *Anal. Chem.*, 48, 640A, 1976.
56. **De Luca, P. P. and Boddaputi, S.,** Procedures for the identification of particulates in parenteral solutions, Proc. INTERPHEX Meet., New York, September 15 to 17, 1980, 34.
57. **Lee, R. E. and Hein, J.,** Method for the determination of carbon, hydrogen and nitrogen in size fractionated atmospheric particulate matter, *Anal. Chem.*, 46, 931, 1974.
58. **McCrone, W. C.,** Particle Analysis, *Res. Dev.*, 30, September 1967.
59. **Adams, F. C. and Van Grieken, R. E.,** Absorption correction for X-ray fluorescence analysis of aerosol loaded filters, *Anal. Chem.*, 47, 1167, 1975.
60. **Birks, L. S.,** Pinpointing airborne pollutants, *Environ. Sci. Tech.*, 12, 150, 1978.

61. **Altpeter, L. L.,** New analytical prospects for an old detector, Proc. Int. Conf. Environ. Sensing and Anal., Las Vegas, September 14, 1975.
62. **Begnoche, B. C. and Risby, T. H.,** Determination of metals in atmospheric particulates using low volume sampling and flameless atomic absorption spectrometry, *Anal. Chem.*, 47, 1041, 1975.
63. **Benson, P. A. and Gleit, C. E.,** Neutron activation and radiochemical determination of Mo, Cr and Fe content of individual microspheres, *Anal. Chem.*, 35, 1029, 1963.
64. **Cunningham, P. T., Johnson, S. A., and Young, R. T.,** Variations in chemistry of airborne particulate material with particle size and time, *Environ. Sci. Tech.*, 8, 131, 1974.
65. **Davis, W. D.,** Continuous mass spectrometric analysis of particulates by use of surface ionization, *Environ. Sci. Tech.*, 11, 587, 1977.
66. **Voorhees, K. J., Kunen, S. M., Durfee, S. L., Currie, L. A., and Klouda, G. A.,** Characterization of airborne particulates by pyrolysis/mass spectrometry and carbon-14 analysis, *Anal. Chem.*, 53, 1463, 1981.
67. **Desaedeleer, G. G., Winchester, J. W., and Akselsson, K. R.,** Monitoring aerosol elemental composition in particle size fractions for predicting human respiratory uptake, *Nucl. Instrum. Methods,* 142, 97, 1977.
68. **Cahill, T. A.,** Innovative aerosol sampling devices based upon PIXE capabilities, *Nucl. Instrum. Methods,* 181, 473, 1981.
69. **Macias, E. S., Radcliffe, C. D., Lewis, C. W., and Sawicki, C. R.,** Proton-induced gamma ray analysis of atmospheric aerosols for carbon, nitrogen, and sulfur compounds, *Anal. Chem.*, 50, 1120, 1978.
70. **Cronn, D. R., Charlson, R. J., Knight, R. L., Crittenden, A. L., and Appel, B. R.,** A survey of the molecular nature of primary and secondary components of particles in urban air by high resolution mass spectrometry, *Atmos. Environ.*, 11, 929, 1977.
71. **Gavrilovic, J. and Majewski, E.,** Use of ion and electron microprobes for full characterization of particulate matter, *Am. Lab.*, 9, 19, 1977.
72. **Allen, J. and Gould, R. K.,** Mass spectrometric analyzer for individual aerosol particles, *Rev. Sci. Instrum.*, 52, 804, 1981.
73. **Rosen, H. and Novakov, P.,** Raman scattering and the characterization of atmospheric aerosol particles, *Nature (London),* 266, 708, 1977.
74. **Wieser, P., Wurster, R., and Seiler, H.,** Identification of airborne particles by laser induced mass spectroscopy, *Atmos. Environ.*, 14, 485, 1980.
75. **Keyser, P. R., Natusch, D. F. S., Evans, C. A., and Linton, R. W.,** Characterizing the surfaces of environmental particles, *Environ. Sci. Tech.*, 12, 768, 1978.
76. **Abts, L. R., Beyer, R. T., Galletti, P. M., Richardson, P. D., Karon, D., Massimino, R., and Karlson, K. E.,** Computerized discrimination of microemboli in extracorporeal circuits, *Am. J. Surg.*, 135, 535, 1978.
77. **Gatlin, L. A. and Nail, S. L.,** Initial evaluation of focused ultrasound for in situ measurements of small particles in flowing water streams, *J. Parenteral Sci. Technol.*, 35, 227, 1981.
78. **Kinsman, S.,** Instrumentation for filtration tests, *Chem. Eng. Prog.*, 7, 48, 1974.

# *Index*

# INDEX

## A



## B

## C

## D



## E

## M



## N

## O



## P

## T



## U

**V**



**W**



**X**



**Y**



**Z**